Food Microbiology and Food Safety

Practical Approaches

Series Editor:

Michael P. Doyle
Center for Food Safety
University of Georgia
Griffin, GA, USA

Tanya Roberts

Editor

Food Safety Economics

Incentives for a Safer Food Supply

 Springer

Editor
Tanya Roberts
Economic Research Service, USDA (retired)
Center for Foodborne Illness Research and Prevention
Vashon, WA, USA

Food Microbiology and Food Safety
ISBN 978-3-319-92137-2   ISBN 978-3-319-92138-9   (eBook)
https://doi.org/10.1007/978-3-319-92138-9

Library of Congress Control Number: 2018953731

This Springer imprint is published by the registered company Springer Nature Switzerland AG
The registered company address is: Gewerbestrasse 11, 6330 Cham, Switzerland

To my hubby, Paul Kolenbrander, who cooked dinner while I worked on the book and to our grandchildren who are the future: Emily, Lena, Zachary, Eric, Abigail, and Garrett.

Preface and Acknowledgments

The International Association for Food Protection (IAFP) requested that I write this book as part of its scientific series published by Springer. Because of my respect for IAFP, I enlisted the help of many co-authors to discuss economic incentives and how they relate to food safety. Foodborne pathogens are the focus of the book, but the economic principles apply to chemicals and other contaminants.

I started working on food safety economics in the Economic Research Service (ERS) of the United States Department of Agriculture (USDA). The new USDA Undersecretary for Food Safety and Quality Services (FSQS) in the Carter Administration, Carol Tucker Foreman, asked ERS to do a Benefit/Cost Analysis (BCA) for meat and poultry inspection. The inspectors were telling her that infectious diseases were a thing of the past and that they were only inspecting for broken wings and bruises, things that consumers could see for themselves. So why not do away with inspection altogether and save the taxpayer's money?. As a new employee with some experience in health economics, I was tasked with the BCA.

I was fortunate to work with Jack C. Leighty, DVM, Director of the Pathology and Epidemiology Division, who explained veterinary science and directed me to key references. My job was to estimate the public health protection benefits of meat and poultry inspection, compare these to the costs of inspection, and determine whether there was a net benefit. While researching what FSQS meat and poultry inspectors did, I was given a long list of animal diseases that were cause for inspection actions. However, no reference was made to human illnesses associated with these animal conditions. Then there was no Office of Public Health in FSQS, only an Office of Science. Today FSIS has an Office of Public Health and Science.

The BCA project introduced me to the public health issues of meat and poultry inspection which were of great importance to the export industry and had a huge impact on the health of US consumers. I decided to work in this area of public policy. ERS' *FoodReview* published my annual estimates of the US costs of foodborne disease that expanded to include new pathogens and new long-term health outcomes (discussed in Chaps. 6–9 of this book). In 1994, Peggy Foegeding and I were co-chairs of the Council for Agricultural Science and Technology's (CAST) report, *Foodborne Pathogens: Risks and Consequences,* which gave us the opportunity to work with a dedicated team of microbiologists, epidemiologists, and others. We distilled the science about foodborne pathogens, economics, and public policy into the report.

The 1993 outbreak of *E. coli* O157:H7 in the Jack in the Box hamburgers led to many public policy lessons. Parents' advocacy for their children was one outcome (Chap. 16). The new preventive regulations of HACCP were another (Chap. 4). Industry response to innovate or not was on display (Chap. 10). And legal liability litigation flourished (Chap. 17).

The other chapters of this book apply economic concepts to real world issues in food safety. For example, Chap. 2 examines the issue of limited information linking the foodborne illness/food/company/pathogen and the negative impact on economic incentives to provide the socially optimal level of food safety. Chapter 3 uses principal-agency theory to explore food safety in contracts along the food safety supply chain. Chapter 5 analyzes the impact of restaurant ratings prominently displayed in storefront windows. Chapters 11 and 12 explore international regulations and their attempt to control *Campylobacter* and *Salmonella* in chicken. Chapter 13 concludes that the foodborne surveillance system pays off by reducing the size of outbreaks. Chapters 14 and 15 discuss aspects of new regulatory challenges for the Food and Drug Administration, imported food and antibiotics in animal feed, respectively. Chapter 1 introduces the book and the overarching theme of economic incentives and Chap. 18 offers concluding thoughts and future challenges for food safety.

This book has been a pleasure and a challenge to write. I give hearty thanks to my many co-authors! Special thanks go to Arie Havelaar, Brecht Devleesschauwer, Barbara Kowalcyk, and Robert Scharff for tackling the core issues of estimating the societal costs of foodborne illness in the United States and the world. Diogo M. Souza Monteiro deserves special appreciation for starting the writing and coordinating of Chap. 1 as well as taking on the challenge of explaining principal-agency theory to noneconomists. Walter Armbruster earned my appreciation for willingly tackling a new and important topic, antibiotic resistance of foodborne pathogens, as well as coordinating co-author responses on the concluding chapter. Denis Sterns is a new collaborator who was personally involved as a lawyer in the Jack in the Box outbreak. Patricia Buck is a co-founder of the Center for Foodborne Illness Research and Prevention and drew on her personal experience as an advocate for safer food. Johan Lindblad was the chief veterinarian for the Swedish Poultry Meat Association and helped their poultry farmers implement the *Salmonella* controls. Clare Narrod, Derrick Jones, and Peter van der Logt drew on their careers working in government to write their chapters on public policy. And finally, thanks to Robert Scharff for contributing three chapters.

I greatly appreciated conversations with all the co-authors throughout this process and with Mary Ahearn and Abigail Kolenbrander. Terra O'Malley provided welcome relief from the task of formatting my chapters and references. I couldn't have written this book without everyone's dedication and professional expertise.

Vashon, WA, USA   Tanya Roberts

Contents

Part I   Food Safety Economic Incentives in Regulations and in the Private Sector

1   Overview of Food Safety Economics   3
Diogo M. Souza Monteiro, Tanya Roberts, Walter J. Armbruster,
and Derrick Jones

**2   Pathogen Information Is the Basic Problem for Economic
Incentives**   13
Tanya Roberts and Robert L. Scharff

**3   Economics of Food Chain Coordination and Food Safety
Standards: Insights from Agency Theory**   29
Diogo M. Souza Monteiro

**4   Benefit/Cost Analysis in Public and Private Decision-Making
in the Meat and Poultry Supply Chain**   49
Tanya Roberts

**5   Economic Impact of Posting Restaurant Ratings:
UK and US Experience**   67
Derrick Jones

Part II   Economics of Foodborne Illness Metrics: When to Use What

6   Burden and Risk Assessment of Foodborne Disease   83
Brecht Devleesschauwer, Robert L. Scharff, Barbara B. Kowalcyk,
and Arie H. Havelaar

7   The Global Burden of Foodborne Disease 107
Brecht Devleesschauwer, Juanita A. Haagsma,
Marie-Josée J. Mangen, Robin J. Lake, and Arie H. Havelaar

8 The Economic Burden of Foodborne Illness in the United States 123
Robert L. Scharff

9 Improving Burden of Disease and Source Attribution Estimates ... 143
Barbara B. Kowalcyk, Sara M. Pires, Elaine Scallan,
Archana Lamichhane, Arie H. Havelaar,
and Brecht Devleesschauwer

Part III Case Studies in Applied Food Safety Economics

10 Economic Incentives for Innovation: *E. coli* O157:H7 in US Beef. 177
Tanya Roberts

11 Benefits and Costs of Reducing Human Campylobacteriosis Attributed to Consumption of Chicken Meat in New Zealand. 209
Peter van der Logt, Sharon Wagener, Gail Duncan, Judi Lee,
Donald Campbell, Roger Cook, and Steve Hathaway

12 Sweden Led *Salmonella* Control in Broilers: Which Countries Are Following? 231
Tanya Roberts and Johan Lindblad

13 The Role of Surveillance in Promoting Food Safety 251
Robert L. Scharff and Craig Hedberg

14 Economic Rationale for US Involvement in Public-Private Partnerships in International Food Safety Capacity Building 267
Clare Narrod, Xiaoya Dou, Cara Wychgram, and Mark Miller

15 The Political Economy of US Antibiotic Use in Animal Feed 293
Walter J. Armbruster and Tanya Roberts

16 The Role of Consumer Advocacy in Strengthening Food Safety Policy 323
Patricia Buck

17 A Critical Appraisal of the Impact of Legal Action on the Creation of Incentives for Improvements in Food Safety in the United States 359
Denis Stearns

Part IV Concluding Thoughts

18 International Food Safety: Economic Incentives, Progress, and Future Challenges 389
Tanya Roberts

Glossary of Economic Terms 399

Index. 403

Part I
Food Safety Economic Incentives in Regulations and in the Private Sector

Chapter 1
Overview of Food Safety Economics

Diogo M. Souza Monteiro, Tanya Roberts, Walter J. Armbruster, and Derrick Jones

Abbreviations

BCA	Benefit-cost analysis
BIFSCo	Beef industry Food Safety Council
CDC	Centers for Disease Control and Prevention
COI	Costs of illness
DALYs	Disability-adjusted life years
FBD	Foodborne disease
FDA	Food and Drug Administration
FERG	Foodborne Disease Burden Epidemiology Reference Group
FMSA	Food Safety Modernization Act
HACCP	Hazard Analysis and Critical Control Points
HIV/AIDS	Human immunodeficiency virus/acquired immune deficiency syndrome
QALYs	Quality-adjusted life years
USA	United States of America
USDA	United States Department of Agriculture
USDA/FSIS	United States Department of Agriculture, Food Safety Inspection Service
WHO	World Health Organization

D. M. Souza Monteiro
School of Natural and Environmental Sciences Newcastle University, Newcastle Upon Tyne, UK
e-mail: Diogo.Souza-Monteiro@ncl.ac.uk

T. Roberts
Economic Research Service, USDA (retired), Center for Foodborne Illness Research and Prevention, Vashon, WA, USA
e-mail: tanyaroberts@centurytel.net

W. J. Armbruster
Farm Foundation (Retired), Darien, IL, USA
e-mail: waltja@live.com

D. Jones
Independent Economic Consultant, and former Chief Economist and Head of the Analysis and Research Division at the UK Food Standards Agency (Retired), London, UK

© Springer International Publishing AG, part of Springer Nature 2018
T. Roberts (ed.), *Food Safety Economics*, Food Microbiology and Food Safety,
https://doi.org/10.1007/978-3-319-92138-9_1

1.1   Introduction to Food Safety Economics

The goal of this book is to introduce food scientists, policy analysts, academics, industry and nongovernmental organization managers, and consumer groups to the principles and main applications of food safety economics. Safety is a food attribute which depends on a range of environmental, technological, and human factors. By the nature of agricultural and food production, processing, and distribution, there is never absolute control of the processes. Consequently, it is impossible to achieve an absolute safe food product. In other words, there is always a probabilistic process affecting the level of food safety. However, there are processes and practices that have a lower probability of causing food safety hazards. The choice between alternative production and food processing systems is influenced by socioeconomic and cultural aspects as well as profits. Economists can play an important role helping make informed choices.

Food safety is a critical attribute of foods purchased by consumers in retail outlets (from supermarkets to independent grocery shops) and food service vendors (from fine dining restaurants to street food stalls). Consumers generally assume that all foods being sold are completely safe; otherwise, they would not be in the market. In many countries, food safety is being used to differentiate certain products, if the standards used to support the claims have emerged from or are sanctioned by governmental agencies. In Denmark, for example, *Salmonella*-free labels are permitted on poultry if certain standards are met. In the United Kingdom, food safety ratings are posted on the front window of restaurants and other places where prepared food is sold (Chap. 5).

The US Congress delegated food safety to federal inspectors over a century ago, starting with the safety of beef and pork. A 2009 survey of US taxpayers found that 51% of respondents thought that food safety and inspection should be the most important expenditure in the US Department of Agriculture (USDA) budget. They wanted 30% of USDA's budget spent on food safety and inspection. However, the current allocation is but 3% for food safety (Ellison and Lusk 2011). These survey results confirm the great importance of safe food to consumers and the importance of continued delegation of food safety to the federal government. If taxpayers/consumers felt they could handle food safety by themselves, they might support budget cuts for USDA or other national food safety agencies. Alternatively, they would favor allocating the agricultural and food policy budgets to other programs, such as spending on nutrition or research.

Delivering safe food is not easy or costless. The challenge is to deliver an "acceptable" level of food safety at the least possible cost. The problem is that what may be acceptable from a public health perspective to high-risk groups of the population may not be economical (or even feasible) for some businesses or industries. To complicate matters, for most products (e.g., oysters), consumers are unable to distinguish between safe and unsafe food, as pathogens and most physical and chemical contaminants of foods cannot be detected by visual inspection. Food safety is what economists call a "credence" attribute, defined as one that cannot be verified before purchase or immediately after consumption. This leads to asymmetric information, as the vendor of food may know with more precision the average food safety level than the consumer, who is unlikely to have access to that information. The supplier has control over the sanitation in its facilities and knowledge of the results of the pathogen tests it has performed.

All these issues have economic dimensions, which can be summarized in the following question: how can food businesses and governmental agencies deliver the expected level of food safety at the least possible private and public cost? The first section introduces and explains key economic concepts and frameworks that are fundamental to any private manager or public sector agent responsible for the implementation of food safety strategies to prevent foodborne illnesses. The book has three more sections applying and extending these concepts. The second section focuses on the evaluation of the health costs of food safety, overviewing a set of foodborne illnesses metrics. The question posed is how costly are foodborne hazards in a single country or in regions of the world, or even worldwide? In the third section, case studies of applied food safety economics are presented and analyzed, for example: the economic incentives to control pathogens through the legal system, lessons learned about pathogen control in different countries, and the political economy of antibiotics in animal feed. Finally, the last section reflects on the future of food safety in an increasingly connected and globalized food system, where the public/private responsibility for food safety is evolving as new scientific techniques become available.

1.2 Overview of Economic Theory Applied to Food Safety

Notwithstanding other economic and noneconomic factors affecting decisions and investments in food safety, information is critical for the effective management of food safety levels. Business managers and officials in governmental agencies need to understand and know the expected degree of contamination of a given food, risk associated with alternative production processes, and the location of contaminated lots of food to effectively manage food safety. When there are critical information gaps, the ability to make sound decisions is hindered. Furthermore, economists argue that information is a fundamental condition for the operation of competitive markets. Consequently, when information is not available or is not reliable, not only business but also markets and industries may lose their ability to optimize the allocation of resources. This results in sub-optimal levels of the product, service, or attribute traded, i.e., too little food safety. In short, faced with inadequate levels of information, trading partners do not have sufficient economic incentives to deliver adequate levels of food safety.

In Chap. 2, a key revelation is that only 1/1000 cases of US foodborne illnesses can be attributed to any given pathogen, food, or specific company which leads to weak incentives for companies to produce safer food. However, it is important to remember that companies supplying food have a large financial stake in continuing to supply safe food to maximize profits and protect their integrity in the eyes of the consumer. But the inability to link illnesses to companies highlights the severity of the information problem in food markets and justifies policy interventions through regulatory solutions or incentives to provide safer food. One regulatory solution could be to require collection of data on pathogens in the farm-to-fork food supply chain by the federal government to link companies to foodborne illness. Another would be to mandate uploading results of all quality control tests performed by industry and third-party certifiers into an accessible database. However, much of this information is proprietary and its publication may have legal implications and may involve prohibitive costs for business or the taxpayer. While expanded and coordinated testing could be used to profile farms and develop a database to facilitate identification of foodborne pathogens and improve rapid traceback to the company and product causing the illness, the complexity of supply chains raises issues on the viability of this approach. The need for private/public sector collaboration to protect food safety is obvious.

Economists have developed frameworks to inform decisions when there are information failures. One of the most commonly used frameworks is agency or contract theory which is presented and applied to food safety in Chap. 3. The basic idea of this framework is that when there are information asymmetries, they need to be dealt with through an agreed strategy or a contract. Specifically, this framework explicitly recognizes that in the presence of information failures, a buyer may purchase from an inept seller (in economic language, face adverse selection) or the seller may not exert the level of effort she should to deliver the quantity or quality of the good or service traded (i.e., incurs a moral hazard). To overcome this problem, the buyer needs to set incentives to motivate the seller to reveal her true ability and compensate the buyer such that she exerts the contracted level of effort. The emergence of private food standards and their increased use in procurement to select suppliers is a way to mitigate adverse selection. Then, when contracts have clearly defined output and quality attributes in the delivery expectations, as well as obligations of inspections, they implicitly are reducing chances of noncompliance. Still, there are costs to monitoring and enforcing contracts, and less conscientious agents may have an incentive to shirk and deliver unsafe food.

Along with the private sector, the government also is exposed to information failures when it aims to reduce the incidence of foodborne diseases. Regulators have a variety of options at their disposal to influence the food industry and reduce the frequency and impact of foodborne illness. These range from strict command and control regulations to the promotion of voluntary adoption of good agricultural or manufacturing practices. How do governments assess alternative policy options and select between them? This is the topic of Chap. 4 that introduces the benefit-cost analysis (BCA) framework. In a nutshell, the BCA framework systematically accounts for all the benefits and costs of alternative available interventions to reduce the chances of foodborne illness. These options are then compared, and the one with the largest benefit-cost ratio, or net benefit, is recommended. One of the challenges of this framework is the ability to rigorously and completely estimate both the costs and benefits involved in a BCA. These economic values are not necessarily found in markets (e.g., the value of pain and suffering due to an illness) and depend on methodologies that have well-known limitations. An application of this framework is then examined by looking in detail at the regulatory process leading to the adoption of Hazard Analysis and Critical Control Points (HACCP) for US meat and poultry.

Across the world, there has been a significant increase in the proportion of food consumed away from home. In several EU countries, the United States, and Canada, the food service industry now has almost the same amount of aggregate food expenses as the retail industries. This industry is characterized by a wide variety of businesses, ranging from high-end hotels and restaurants to street food stands. With this diversity comes a wide variety of practices and food handling conditions. Again, information asymmetries are rampant, and consumers or patrons of these establishments are often unaware of the food safety risks they expose themselves to. In some cases, the managers of these businesses may also be oblivious of the food safety levels of the ingredients they use in preparing their food. To mitigate these information asymmetries and help consumers make informed choices of the places where they eat away from home, some city governments have introduced restaurant rating schemes that use a label to be placed at the door or any other visible location of a premises indicating the results of hygiene inspections. The economic rationale and description of the operation of these rating schemes are presented in Chap. 5, which provides an overview of the different forms of restaurant rating schemes in the United Kingdom and the United States and reports on the attempts to measure and evaluate their public health impacts in raising hygiene standards and reducing foodborne disease.

Together these chapters introduce the economics of information and its importance to the private business sector, governmental, and consumer decision-making to prevent food safety incidents. What is not discussed in great depth, but is certainly implicit in all these chapters, are the externalities associated with food safety outbreaks. Externalities are costs that agents impose on others through their activities. Failing to internalize costs imposed on others creates a subsidy to companies causing externalities, as it has lower costs of operation than it would if it accounted for such costs. While we do not present a formal conceptualization of the economics of externalities and how they may be considered, the second section of this book examines the impact of food safety on public health costs and losses of productivity due to foodborne illnesses.

1.3 The Societal Burden of Foodborne Illness

The chapters in this section examine the externalities imposed by an excess of foodborne illnesses in society. Food safety is a critical global public good that has important implications for food security, public health, and economies. Globally, foodborne disease (FBD) is a leading cause of mortality and morbidity, causing an estimated 600 million illnesses and 42,000 deaths annually. Children are particularly impacted, accounting for 40% of the overall burden and a third of the deaths.

Foodborne disease can result in long-term health outcomes, such as irritable bowel syndrome, reactive arthritis, diabetes, hypertension, kidney disease, and neurological dysfunction. Combined, these health impacts lead to reduced quality of life, shorter lifespans, increased medical costs, decreased worker productivity, and lower incomes.

Chapter 6 introduces the concepts used in quantifying the societal burden of FBD in dollars and cents. Starting with disease outcome trees for acute salmonellosis and other foodborne diseases, economists identify significant health outcomes and their duration. Valuation of each health outcome is the next step. Key methodological approaches that have been used to generate estimates of the health and economic burden of foodborne disease are discussed. The cost of illness (COI) method includes estimating the medical costs, lost productivity, and lost life expectancy. Another, more comprehensive measure, adds in the lost quality of life caused by the foodborne illness. This metric uses disability-adjusted life years (DALYs) or quality-adjusted life years (QALYs).

In 2006, the WHO launched an initiative to estimate the global burden of FBD, which was carried forward by the Foodborne Disease Burden Epidemiology Reference Group (FERG). Chapter 7 outlines FERG's quantified global and regional burden of 31 foodborne hazards, including 11 diarrheal disease agents, 7 invasive disease agents, 10 helminths, and 3 chemicals and toxins. Baseline epidemiological data were translated into DALYs following a hazard-based approach and an incidence perspective. In 2010, foodborne diseases were estimated to cause 600 million illnesses, resulting in 420,000 deaths and 33 million DALYs, demonstrating that the global burden of FBD is of the same order of magnitude as the major infectious diseases such as HIV/AIDS, malaria, and tuberculosis. It is also comparable to certain other risk factors such as dietary risk factors, unimproved water and sanitation, and air pollution.

Some foodborne hazards were found to be important causes of FBD in all regions of the world, while others were highly localized resulting in a concentrated burden. Despite the data gaps and limitations of these initial estimates, it is apparent that the global burden of FBD is considerable and affects individuals of all ages, particularly children under the age of 5 and persons living in low-income regions of the world. By using these estimates to support evidence-based priorities, all stakeholders, both at national and international levels, can contribute to improvements in food safety and population health.

Foodborne diseases represent a constant threat to public health and a significant impediment to socioeconomic development worldwide. At the same time, food safety remains a marginalized policy objective, especially in developing countries. A major obstacle to adequately addressing food safety concerns is the lack of accurate data on the full extent and burden of FBD.

Chapter 8 introduces economic models used to estimate the societal burden of foodborne illness in the United States for the 30 pathogens identified by the Centers of Disease Control and Prevention (CDC) as a top priority. The COI is estimated at $61 billion annually for medical, productivity, and mortality costs. A higher estimate of $90 billion annually includes an additional QALY cost for reactive arthritis that occurs after infection with *Campylobacter*, *Salmonella*, *Shigella*, and *Yersinia*. *Salmonella* is responsible for 34% of the costs and *Campylobacter* 24% of the costs in the enhanced model. The high costs for these two pathogens suggest new regulatory efforts are needed to control them. USDA/FSIS has implemented new regulations for *Salmonella* and *Campylobacter* in poultry.

Disease burden estimates provide the foundation for evidence-informed policymaking and are critical to public health priority setting around food safety. While significant efforts have been undertaken to quantify the burden of foodborne illness (see Chaps. 7 and 8), there are still significant gaps in our knowledge. Chapter 9 outlines how to improve burden of disease estimates, for example, (1) enhancing the foodborne hazards surveillance infrastructure, (2) improving our understanding of the chronic sequelae associated with foodborne illness, and (3) linking these health outcomes back to specific hazards and foods. These latter linkages are critical as they enable decision-makers to identify and prioritize food safety interventions to prevent and reduce the burden of disease.

1.4 Case Studies in Applied Food Safety Economics

This section of the book explores economic incentives to mitigate food safety hazards through a sample of case studies focusing on regulatory programs to control pathogens, private sector response to regulations, legal responses, public advocacy, and public-private partnerships. Chapter 10 discusses how a 1993 US outbreak of *Escherichia coli* O157:H7 in hamburgers led to new regulations focusing on the prevention of pathogens in the food supply. The economic incentive for industry innovation is emphasized as Jack in the Box struggled to reclaim its customer base and market share. An equipment manufacturer was incentivized to invent a beef carcass steam pasteurization system. The Beef Industry Food Safety Council (BIFSCo) was created in 1997 and sponsors an annual Beef Industry Safety Summit as a platform to share the latest knowledge about how to control pathogens in the farm to fork food chain. An economic model illustrates the risk-cost trade-off for slaughterhouse and fabrication measures to control *E. coli* O157 in ground beef: irradiation of sides of beef, careful hide removal, steam pasteurization of sides of beef, and during fabrication, testing of combo bins of trim and testing of hamburger production.

In Sweden, an outbreak of salmonellosis led to testing meat and poultry to determine its prevalence and a public-private partnership to eliminate *Salmonella* in broilers, which is the object of Chap. 11. The Nordic countries and Denmark followed, but US broiler companies have resisted additional food safety efforts. In both Sweden and Denmark, economists have estimated the costs of *Salmonella* control at US 1 cent/pound of poultry meat at retail. Surveys of US consumers in supermarkets show willingness to pay for *Salmonella* control that is much greater than 1 cent/pound. But US broiler companies are much more powerful than consumers.

In Chap. 12, New Zealand's regulatory programs have had success in reducing foodborne illness caused by *Campylobacter* in poultry. The annual rate of human campylobacteriosis cases identified in New Zealand increased consistently after notification became compulsory in 1980. A large proportion of the cases were attributed to the consumption of chicken meat. In 2006, a *Campylobacter* risk management strategy was developed that concentrated on optimizing broiler chicken processing while reducing carcass contamination, and hence the number of human cases of campylobacteriosis. While it has been very successful with more than a 50% decrease in the number of notified foodborne human cases being achieved by 2008, a greater reduction in the burden of human illness is still sought. By 2015, there has been a net benefit of at least $NZ 67.3 million annually attributable to implementation of the *Campylobacter* risk management strategy.

In Chap. 13, the economic value of foodborne illness surveillance is analyzed. Foodborne illness surveillance systems are designed to collect, analyze, and disseminate information about foodborne illnesses. Consequently, they help solve critical information problems faced by consumers, firms, and government agencies. By providing improved accountability to the market, these surveillance systems create incentives that lead to safer foods and better consumer awareness. For public health officials, surveillance provides a means to identify and mitigate current outbreaks, prioritize resources, and craft better preventative interventions. To illustrate the economic value of surveillance, we provide an analysis of PulseNet, a network of public health and food safety regulatory agencies in the United States run by CDC to track results of tests for major foodborne pathogens. This analysis was performed using updated economic data and models. PulseNet-related activities lead to substantial social benefits due to reductions in illnesses caused by *Salmonella* spp. and *E. coli* O157:H7. Adjusting for underreporting and under diagnosis, as many as 330,840 *Salmonella*- and 17,475 *E. coli*-related illnesses are averted each year due to PulseNet. This leads to economic benefits of up to $5.4 billion annually.

Chapter 14 discusses the market failures associated with food safety provision and why both the public and private sector have been involved in international food safety capacity building. The Food Safety Modernization Act (FSMA) places more responsibility on the private sector to ensure the safety of imported food. It also requires the US Food and Drug Administration (FDA) to provide international food safety capacity building for complying with the new regulations and to monitor and evaluate those efforts. However, public resources to support programs abroad are limited. International capacity building supported by the private sector, which sometimes has its own standards that suppliers need to meet, can complement public efforts. This chapter reviews public and private sector efforts and some of the partnerships that have formed and discusses the limitations of monitoring and evaluation efforts when relying solely on publicly available data. It argues that mobilizing public-private partnerships in monitoring and evaluation can make it easier to capture the collective impact of international food safety capacity building efforts.

Chapter 15 examines the political economy of antibiotics in animal feed in the United States. The contribution of food animals to the rising antibiotic resistance of human pathogens is recognized. The FDA has recently pushed for voluntary withdrawal of antibiotics in feed used for growth promotion and received full industry indications of compliance effective at the beginning of 2017. A major poultry company had earlier dropped growth-promoting use of antibiotics and found no difference in costs between raising poultry with antibiotics in feed and without, once appropriate changes to animal husbandry practices were implemented and natural products with antimicrobial effects were introduced into the poultry diet. An ERS benefit-cost analysis found insignificant costs to the pork and broiler industries in ceasing the use of antibiotics in animal feed as growth promoters.

Chapter 16 explores the role of consumer advocacy in food safety regulations. In the United States, consumer advocates made important contributions to the passage and implementation of both the 1996 Pathogen Reduction/Hazard Analysis and Critical Control Points (HACCP) final rule at USDA and the 2011 Food Safety Modernization Act at FDA. As scientific knowledge improves our understanding of food safety hazards, there is much more to be done if we hope to meet the ongoing and future food challenges, such as the emergence of antibiotic-resistant strains of foodborne bacteria. Arguably, public concern about the health, safety, and sanitation practices of the meatpacking industry can be traced to *The Jungle* written by Upton Sinclair, leading to enacting the Meat Inspection Act and the Pure Food and Drug Act by Congress in 1906. Since then, new challenges have emerged as food production practices and longer supply chains evolved for a population demanding fresher food, more food eaten away from home, and more prepared food items for consumption at home. Large food safety outbreaks associated with *E. coli* O157:H7 spurred the federal government and its state partners to shift food oversight away from a prescribed reactive approach to one that is more proactive and preventive and based on a science- and risk-based approach.

In Chap. 17, the limitations of the legal system in creating economic incentives for improved pathogen control are explored. It takes as a starting point the rise of litigation seeking compensation for persons injured by unsafe food that began in the wake of the 1993 Jack in the Box *E. coli* O157:H7 outbreak. Such litigation is deemed to have a positive impact on the relative safety of food commercially sold in the United States. This was mostly the result of filing of lawsuits which tended to amplify public attention and concern. This in turn drew increased media scrutiny of food companies wanting to protect brand values and avoid stricter regulations. Despite the positive impact of litigation on food safety, the impact has remained limited by several issues, including the high information costs (problem of identifying the company, pathogen, and food that caused the illness) and high transaction costs of suing (time, money, and the emotional trauma of reliving the illness or death of a loved one). As a result, litigation can create incentives for companies to invest in food safety, but effective regulation remains the key.

Finally, Chap. 18 concludes and provides insights on the ways economics can contribute to understanding food safety decision-making in the public and private sectors. It also discusses future challenges to achieving a safer global food supply.

1.5   Summary

In this book, we use economic theory and empirical examples to examine many issues in food safety. The fact that most food safety outbreaks are undetected or impossible to track to their source along with the existence of widespread information asymmetries reveals that private markets may fail and provide a sub-optimal level of food safety. This in turn leads to excessive societal costs of foodborne illness. The economic concepts and frameworks presented in this book help to understand and suggest solutions to address these challenges. Remedies to improve hazard control are the use of contracts in the private sector, adoption of technology such as used in HACCP programs, the development of public/private partnerships, and command and control regulations. Each method has plusses and minuses. In several chapters, we use benefit-cost analysis to examine the costs and benefits of options to control foodborne pathogens. Key to the success of any approach to reduce antimicrobial resistance is the economic incentives for pathogen control.

We trust that this book helps take some of the mystery and complexity out of economic analysis and provides useful examples of how to use economic incentives to increase the safety of the global food supply.

Reference

Ellison BD, Lusk JL. Taxpayer preferences for USDA expenditures. Choices. Quarter 2. 2011. Available online: http://choicesmagazine.org/choices-magazine/submitted-articles/taxpayer-preferences-for-usda-expenditures

Chapter 2
Pathogen Information Is the Basic Problem for Economic Incentives

Tanya Roberts and Robert L. Scharff

Abbreviations

CDC	Centers for Disease Control and Prevention
CIDT	Culture-independent diagnostic tests
DNA	Deoxyribonucleic acid
ERS	Economic Research Service/USDA
FDA	Food and Drug Administration
FSIS	Food Safety and Inspection Service/USDA
FWW	Food and Water Watch
HACCP	Hazard Analysis Critical Control Points
IOM	Institute of Medicine/NAS
LTHO	Long term health outcome
MRC	Mechanically separated chicken
MRSA	Methicillin resistant *Staphylococcus aureus*
NAS	National Academy of Sciences
NRC	National Research Council/NAS
NSLP	National School Lunch Program
USDA	United States Department of Agriculture
UTI	Urinary tract infection
WGS	Whole-genome sequencing

T. Roberts (✉)
Economic Research Service, USDA (retired), Center for Foodborne Illness
Research and Prevention, Vashon, WA, USA
e-mail: tanyaroberts@centurytel.net

R. L. Scharff
Department of Human Sciences, The Ohio State University, Columbus, OH, USA
e-mail: scharff.8@osu.edu

© Springer International Publishing AG, part of Springer Nature 2018
T. Roberts (ed.), *Food Safety Economics*, Food Microbiology and Food Safety,
https://doi.org/10.1007/978-3-319-92138-9_2

2.1 Economic Theory of Information

Economists have long been interested in how a lack of information in the market affects market outcomes (e.g. Stigler 1961). How the market is affected is determined by what information is missing, who is lacking that information, and what incentives are generated due to the lack of information. For food safety, missing information essentially relates to the risk and consequences consumers have from becoming ill due to food consumption. Risk information may be lacking for both consumers and producers. Consumers lack information because they are unable to observe the riskiness of the foods they consume (Wessells 2002). Even when made ill by a food, consumers may misattribute their illnesses to the wrong foods because of pathogen incubation periods. Food firms may also suffer from information problems (Chap. 2). Downstream retailers are unable to fully observe upstream supplier practices in the absence of an effective traceability system (Hobbs 2004; Stranieri et al. 2016). Also, producers are unable to observe characteristics of consumers, including whether they take mitigating actions (e.g., fully cooking their foods) or whether they are immunocompromised. This lack of information on both sides lead to what Elbasha and Riggs (2003) call a double moral hazard problem, where both consumers and producers engage in suboptimal protective efforts due to the fact that they do not bear the full costs of the risks they create.

The most typical case is one where producers and/or sellers of foods have more information about the riskiness of the specific foods sold than consumers. For example, a farmer knows whether or not she is following best practices and a restaurateur or retailer knows whether foods have been purchased from reputable suppliers and whether foods are being held at safe temperatures. The consumer generally is not able to observe industry practices, leading to uninformed, potentially risky choices being made.

The consequences of this type of asymmetric information were the basis of a seminal paper in the economics of information by George Akerlof (1970). Akerlof's study of the used car market, "A Market for Lemons," demonstrated that when the buyer had no easy way to find out information about a particular used car, the market would fail to operate optimally and low-quality used cars would dominate the market. Essentially, dealers of low-quality used cars are free riding on the value created by sellers of high-quality used cars. Because high-quality used cars are more expensive to acquire and consumers cannot assess quality differences, the low-quality dealers will eventually drive the high-quality dealers out of the market. Fortunately, in the current market for foods, the dire predictions of Akerlof's model are not fully realized. As Akerlof himself noted, institutions often arise to counter information problems. For example, government HACCP regulations or private sector third-party audits can be used to ensure that upstream suppliers are engaging in best practices (McCluskey 2000). Industry trade groups may promote this practice through the use of certification programs that downstream brands can use to signal "safety" to their customers (Henson and Caswell 1999). In cases where consumers can accurately link illnesses to foods, consumer experiences can produce information and impose costs on culpable firms. Finally, media coverage and investigations may reveal particularly large outbreaks (Chunara et al. 2012). A formal model of optimal deterrence is provided in Appendix A. The signals created by these market-based remedies are only effective, however, if illnesses can be linked to the foods and firms responsible. As the next section shows, there are great challenges in making those linkages.

2.2   Linking Illnesses to Contaminated Foods: The Principal Information Problem

To fully appreciate why it is difficult to link illnesses to the foods causing them, it is vital to understand the nature of the foodborne illness detection information. If a consumer becomes ill immediately after eating, it is relatively easy to determine what food causes the illness for two reasons: (1) this short reaction time identifies which meal is linked to the illness, and (2) the ingredients are likely to be available on the consumer's plate or in the kitchen for testing. Then the food items can be tested to identify the foodborne pathogen and identify the food company producing the contaminated food. Only a few foodborne illnesses cause a quick reaction time, most notably the toxin produced by the bacterium *Clostridium botulinum*. The problem is that foodborne pathogens have a lag time of a day to a week to a month before causing of acute foodborne illness (Table 2.1). And there is huge variability in how

Table 2.1 Selected foodborne pathogens: symptom delay and infectious dose

Pathogen	Symptom delay	Infectious dose
Bacteria		
Campylobacter	2–5 days	500+ cells
Enterotoxigenic *E. coli* (ETEC)	8–44 h	~10,000,000 cells
Enteropathogenic *E. coli* (EPEC)	4+ h	Low for infants Ten million cells for adults
Enterohemorrhagic *E. coli* (EHEC)	1–9 days	10–100+ cells
Enteroinvasive *E. coli* (EIEC)	12–72 h	200–5,000 cells
Listeria monocytogenes	Hours-months	<1,000 cells
Salmonella	6–72 h	As low as 1 cell
Shigella	8–50 h	10–200 cells
Vibrio vulnificus	12 h to 21 days	1,000 cells
Yersinia	1–11 days to months	~100,000 cells
Parasites		
Toxoplasma gondii	5–23 days	1 cyst[a]
Trichinella	1–4 weeks	2 viable larvae

[a]Guo et al. (2015). Source: data from FDA's Bad Bug Book (2013), available at http://www.fda.gov/

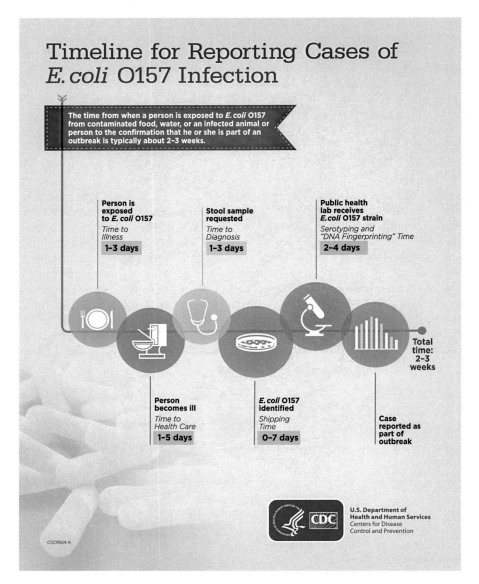

Fig. 2.1 Timeline for Reporting Cases of *E. coli* O157 infection

many bacterial or parasitic pathogens are required to cause illness. According to the US Food and Drug Administration's (FDA) *Bad Bug Book*, just 1 cell of *Salmonella* (perhaps in a chocolate bar where the fat protects the *Salmonella* from the acidity of the human stomach) or 2 viable *Trichinella* larvae can cause illness.

CDC's timeline for reporting cases of *E. coli* O157 takes 2–3 weeks (Fig. 2.1). To find cases in an outbreak of *E. coli* O157 infections, public health laboratories perform a kind of "DNA fingerprinting" on *E. coli* O157 laboratory samples and see if there are matches with other people, contaminated food, water, or infected animal. Public health officials conduct intensive investigations, including interviews with ill people, to determine if people whose infecting bacteria match by "DNA fingerprinting" are part of a common-source outbreak.

About 15 years ago, the child of one of the authors' best friends was in the emergency room with bloody diarrhea occurring every 15 min. Their pediatrician wanted to send the child home. No testing for *E. coli* O157 had been done. Fortunately, she survived. But only after her mother called Dr. Phil Tarr, Seattle Children's Hospital, and he called their pediatrician to talk about her treatment. The lesson to be learned here is that patients, physicians, and reporting systems all play an important role in detecting foodborne illnesses. In this case, neither the parents nor the local pediatrician knew to test for *E. coli* O157.

In other situations, where the symptoms are ambiguous or it is not convenient or affordable to visit a medical clinic, the first steps may not be taken and the case will be unreported. Even more difficult to detect are long-term health outcomes (LTHOs) caused by foodborne pathogens. As explored in Chaps. 6 through 9, the variety of symptoms and the length of time until symptoms occur makes it ever more challenging to detect LTHOs. These LTHOs range from arthritis, diabetes, neurological problems, mental health problems, to kidney failure.

The vast majority of US acute foodborne illnesses—999/1,000—are not traced back to a particular product. This estimate is based on data from CDC. Painter et al. (2013) state that from 1998 to 2008, a total of 13,352 foodborne disease outbreaks were reported. But only 4,887 outbreaks causing 128,269 illnesses had an implicated food vehicle and a single pathogen identified. The 999/1,000 estimate of an inability to link illnesses to a food is derived in two steps: (1) divide these 128,269 identifiable illnesses by the 11 years of data to get the annual identifiable illnesses = 11,661 illnesses; (2) divide the identifiable illnesses by CDC's estimate that 47.8 million US foodborne illnesses occur each year to get the result that 999/1,000 of US foodborne illnesses cannot be identified with a specific pathogen/food combination (47.8 million/11,661 illnesses ≥ 99.9%). This low identification rate of only 1/1,000 cases results in very weak incentives for companies to produce safe food, since companies are so seldom linked to the illnesses they cause.

If the illnesses are not detected and linked to the food company, then there is no mechanism for ill individuals to seek compensation. Economic theory relies on compensation as a remedy when people are injured by their purchases in the marketplace. See Chap. 17 for an examination of legal liability, compensation for foodborne illness, and economic incentives.

Based on the low probability of getting caught for causing a foodborne illness, food sellers have very little incentive to acquire information to control pathogens in the private marketplace. The information required for pathogen control includes not only knowledge of pathogen tests and resources to pay for the tests but also the management knowledge, skill, and the will to implement pathogen control measures. It is a large commitment by the company to use pathogen test results to verify that the pathogen control system is working "as planned" (Buchanan and Schaffner 2015). See Chap. 10 on *E. coli* O157 for an example of a successful pathogen control system, Texas American Foodservice Inc.'s Bacterial Pathogen Sampling and Testing Program.

Buyers, especially consumers, are challenged to understand the scientific literature behind food safety and find the time to read it. And company lobbyists far outnumber the few food safety activists representing consumers in the halls of the Congress and the offices of regulators. In addition, animal producers are used to dealing with animal diseases, but many human pathogens that contaminate animal protein products can live harmlessly in the gut of food animals, for example, *Salmonella*. The shift to controlling pathogens that do not cause animal disease, but cause human illness, is a big change in the farmer's perspective.

In the United States, both consumers and industry have asked for Federal inspectors to establish and police a minimum level of safety. In 1890, US sellers of hog bellies in the European market requested certification that US hog bellies were trichinae-free. After Upton Sinclair raised awareness with *The Jungle*'s description of commercial sausage-making, consumers and the US President asked the Congress to pass Federal food safety regulation and got it in 1906.

2.3   Pathogen Testing and Economic Incentives

Continual advances in pathogen testing technology are improving the ability of food producers to measure food safety. Unnevehr et al. (2004) investigated how new pathogen testing methods influence private incentives to generate and use food safety information. Using an economic framework, they discuss the impact of shifts in supply and demand for food safety information in food markets. Food safety information can be viewed as an input into the production of food. As such, it can be analyzed according to a model of input supply and demand (Fig. 2.2). The supply

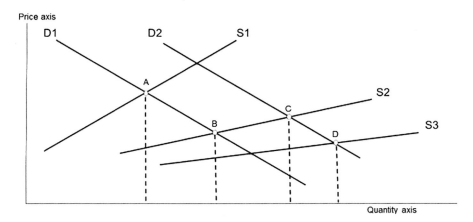

Fig. 2.2 Demand and supply curves for pathogen testing

curve for pathogen information (S1) reflects its marginal cost, which rises as more resources are poured into research and development to develop tests that generate more pathogen information (tests for new pathogens, more sensitive tests, faster tests, etc.).

The demand for information is a derived demand, based on the market demand for food safety. Incentives for safety provisions will be determined by market forces and regulatory initiatives. This derived demand curve (D1) slopes downward, because there is diminishing marginal benefit to additional information. Initial accurate information has very high value, but this value declines as more information is obtained.

In an unregulated private marketplace, there is minimal demand for food safety information. Thus, companies will generate the amount of information determined by the intersection of the supply and demand curves (point A in Fig. 2.2).

Over time, scientific advances are applied to pathogen testing. Costs for production of existing pathogen tests fall, and the supply curve shifts down to S2. As test prices fall, companies move along demand curve and order more tests. The market equilibrium now becomes point B.

When a foodborne illness outbreak occurs and causes a loss of sales and reputation damage, this increases the private demand for food safety information. Companies use pathogen testing to help determine how, when, and where their product became contaminated as well as how to control their production processes better. In addition, public outcry over foodborne illness outbreaks contributes to regulatory initiatives to improve food safety, which also increases the demand for food safety information and pathogen testing. In the United States, both the private marketplace and regulatory initiatives have shifted the demand for food safety information outward, as shown by D2 in Fig. 2.2. The new equilibrium becomes point C with higher prices as the quantity of testing increases and shortages of testing kits occurs.

The higher prices and profits from the shift in demand for food safety information (D2) stimulate the development of new testing and pathogen control technologies. As the investment in new test technologies yields results, new tests enter the marketplace. These new pathogen tests offer companies a broader array of pathogen tests, more sensitive tests, cheaper tests, and/or alternative types of tests. These new tests shift the supply curve down to S3. The new equilibrium becomes point D with lower prices and a greater quantity of pathogen testing.

When either the demand or supply for information shifts, the marketplace moves to new equilibrium (points A to B to C to D) where more food safety information is bought and sold in the marketplace. Over time, prices trend downward as scientific advances create cheaper, faster, and more sensitive pathogen tests.

Some recent improvements in testing for foodborne illness include whole-genome sequencing (WGS) tests and culture-independent diagnostic tests (CIDT). CDC's PulseNet uses both for foodborne pathogens.

- *Whole-Genome Sequencing.* In 2013, PulseNet started using WGS for *Listeria monocytogenes* and found a threefold increase in outbreaks and detected contamination in new foods. WGS provides more detailed and precise data for identifying outbreaks, identifies outbreaks faster, and is an affordable way to obtain high-level information using just one test (CDC 2017).

- *Culture-Independent Diagnostic Tests.* The number of CIDT positive-only infections reported to FoodNet has been increasing markedly since 2013, as clinical laboratories test for more CIDTs (in prepackaged testing kits for multiple pathogens). Initially, increases were primarily limited to *Campylobacter* and STEC, followed by substantial increases in *Salmonella* and *Shigella* beginning in 2015. The pattern continued in 2016, with large increases in the number of CIDT positive-only *Vibrio* and *Yersinia* infections. When including both confirmed and CIDT positive-only infections, incidence rates in 2016 were higher for each of these six pathogens (Marder et al. 2017).

Food safety information has different kinds of economic value to food producers, including avoidance of loss of reputation during an outbreak, capturing price premiums in contracts for supplying safer inputs, increased sales, and/or reduced production costs. Identifying a bad product before it is sold can prevent a costly recall or other types of costs associated with product rejection, including liability for foodborne illness outbreaks. In addition to avoiding losses from allowing a bad product to enter the market, food safety information can also help firms to demonstrate compliance with regulation or to avoid regulatory actions that result in lost production or sales. Profits from product sales can be enhanced by certifying safe (and consistently safe) products, which can ensure market access to a particular buyer and may result in higher prices or less variation in prices obtained over time. Finally, more specific food safety information can help firms to alter their production processes so as to more cost-effectively supply food safety. As food safety in the final product is the result of many different actions in food processing, better understanding of how food safety results from these interlinked actions can lead to better management of food production processes.

2.4   Empirical Examples of Uses of Pathogen Testing Data

This section of the chapter gives some applications of how pathogen testing can be used to provide more information about the source of pathogens and of how pathogen performance standards can improve the food safety performance of private companies. (In Chap. 5, the economic impact of posting restaurant hygiene ratings is examined.)

The US National School Lunch Program (NSLP) provides subsidized or free meals to more than 31 million children each school day. Each year, the NSLP buys about $150 million worth of raw and cooked ground beef. The NSLP standards are strict and include refusing any shipment that tests positive for *Salmonella*. USDA's Economic Research Service (ERS) examined the performance of four groups of ground beef suppliers: commercial suppliers that must be under the FSIS limit (blue

Percent of samples testing positive for *Salmonella spp*

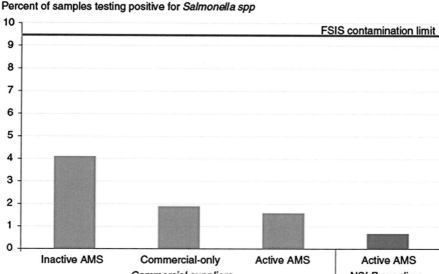

Fig. 2.3 Strict standards nearly eliminate *Salmonella* from ground beef supplied to schools. *AMS* USDA's Agricultural Marketing Service, *NSLP* National School Lunch Program. Source: USDA, Economic Research Service estimates based on 2006–2012 *Salmonella* spp data from USDA, Food Safety and Inspection Service (Bovay and Ollinger 2015)

line in Fig. 2.3), active NSLP suppliers, and two other groups that are approved by USDA's Agricultural Marketing Service, active AMS suppliers and inactive AMS suppliers. The NSLP suppliers had the lowest levels of *Salmonella* in ground beef (0.7%) (Bovay and Ollinger 2015). By setting strict standards, the NSLP lowered the *Salmonella*-contamination rate in its ground beef.

Getting a handle on Salmonella contamination in US poultry. While USDA/FSIS HACCP regulations require *Salmonella* testing (see Chap. 4), FSIS reported the data only in the aggregate (not by individual establishment). Consumer groups acted to make *Salmonella* test data available:

- Food & Water Watch (FWW) began filing Freedom of Information Act requests with FSIS to obtain data that named the individual plants that had failed *Salmonella* testing for broiler carcasses. In 2006 (Food and Water Watch 2006), FWW published a report, Foul Fowl, which is an analysis of *Salmonella* contamination in broiler chickens, featuring data from 1998 through 2005. In 2008 (Food and Water Watch 2008), FWW published More Foul Fowl with FSIS test data for 2006 through January 2008.
- Since 1998, (Consumer Reports 1998) has purchased chicken in US supermarkets and tested them for the pathogens *Salmonella* and *Campylobacter* and published the results by brand name. Over the years 1998–2013, the results ranged from an average contamination rate of 11% to 16% for *Salmonella* and 63% to 43% for *Campylobacter* (Consumer Reports 1998, 2014).

Table 2.2 FSIS tests of *Salmonella* in Chicken, July 1, 2015, through September 30, 2015

Chicken product	Percent positive (%)
Chicken carcasses	1.4
Chicken parts	22.1
Ground chicken	29.3
Mechanically separated chicken	72.7

Source: USDA/FSIS, Quarterly Progress Report on Salmonella and Campylobacter Testing of Selected Raw Meat and Poultry Products: Preliminary Results, July 2015 to September 2015. http://www.fsis.usda.gov

FSIS responded by publishing its *Salmonella* test data in March 2008, after FWW's 2nd Foul Fowl report (Roos 2008). In 2015 (USDA/FSIS 2015), USDA/FSIS implemented a new pathogen testing program for poultry that includes *Campylobacter* as well as *Salmonella*. These new FSIS test data will be published online.

The *Salmonella* tests expand the data beyond carcasses to chicken parts, mechanically separated chicken, and ground and other comminuted chicken (Table 2.2). What sticks out is the low percent positives for whole carcasses at 1.4% vs. the 73% positive rate for mechanically separated chicken (MSC). MSC is essentially a ground chicken skeleton. "The bones are ground up, and the resulting mass is forced through a sieve" p. 15–15 (USDA/FSIS 2017). This industrial process creates a temperature rise that can facilitate pathogen growth in the product and may explain part of the higher pathogen levels for MSC vs. the whole carcass.

MSC is further cooked by the processor or the consumer—which raises the issue of microwave cooking by consumers who are not informed that their meal contains such a high-risk ingredient. The frozen Chicken Kiev outbreaks may be due to this high-risk ingredient of MSC (News Desk 2015a, b).

Another possibility for the lower chicken carcass *Salmonella* level is the controversy of whether carcass sampling has false-negative results much of the time because chicken plants have deliberately used chemicals to alter the test results. Gamble et al. (2016) conclude: "Carryover of active sanitizer to a carcass rinse solution intended for recovery of viable pathogenic bacteria by regulatory agencies **may cause false-negative results**" (emphasis added). In this scenario, the owners of the chicken plants are gaming the system to get false-negative results for the carcass tests.

FSIS posted company identification for *Salmonella* test results in poultry from 2008 to 2013 for establishments that failed to meet the standard for *Salmonella* control (category 3). However, when *Campylobacter* testing was initiated for poultry, FSIS "temporarily" suspended posting the *Salmonella* data on the USDA website. FSIS stated that it would resume posting the *Salmonella* company test data along with the new *Campylobacter* data. As of July 2017, neither the *Salmonella* nor the *Campylobacter* data have been posted on the USDA website.

Pathogen testing throughout the food chain can increase the probability of finding human disease linkages. For example, Canadian researchers at McGill University found a linkage between human urinary tract infections (UTI) and chicken by testing chicken packages in the local supermarkets for the human illness strain of *E. coli*. They found the same strain and concluded that chicken was the likely cause of the human UTI illnesses and the likely reservoir for this pathogen (Racicot Bergeron et al. 2012). Their research illustrates how pathogen testing from farm to fork could be used to make linkages to human illnesses.

Danes test pigs and find match for human MRSA illnesses. Denmark has relatively low levels of livestock-associated Methicillin-resistant *Staphylococcus aureus* (LA-MRSA) compared to the rest of Europe. Still, the LSA-MRSA strain that causes blood infection in humans was found in the environment on 60% of Danish farms. Pig farmers were generally healthy, usually only showing skin or soft tissue infections. However, immunocompromised persons in the rural areas of Denmark had MRSA blood infections that required hospitalization and were sometimes lethal. This study illustrates how whole-genome sequencing tests of the food supply chain can identify reservoirs of human pathogens. LA-MRSA can spread from farm animals to people through direct contact with the animals, through contaminated meat that's produced from the animals, and possibly through air, dust, and water near industrial pig operations (Larsen et al. 2017).

2.5   Economic Incentives of Public Pathogen Information

The US National Academy of Sciences' National Research Council (NRC) was commissioned to evaluate the impact of posting pathogen data by USDA's Food Safety and Inspection Service (FSIS). The NRC's 2011 report, The Potential Consequences of Public Release of FSIS Establishment-Specific Data, identified these economic incentives for better pathogen control via public release of data:

1. Protect brand reputation in food safety.
2. Enhance customer base and profitability.
3. Allow downstream users to identify companies with performance records below/ above industry average.
4. Create economic pressure to improve food safety.
5. Provide insights into strengths/weaknesses of different processing practices.
6. Enhance performance benchmarking.
7. Improve consistency of inspector performance.

"The committee concluded that public release of FSIS establishment-specific data, by themselves or in combination with other privately or publicly available data, could yield valuable insights that go beyond the regulatory uses for which the data were collected" (National Research Council 2011, p. 65).

US scientists and consumer advocates have promoted the creation of a national pathogen database. The National Academy of Science's Institute of Medicine (IOM) recommended "…establishing a centralized, risk-based analysis and data management center in order to improve efficiency and work toward a safer food supply" (Institute of Medicine and National Research Council 2010). This national database and management center recommendation is modeled after those that exist in many European countries, including the Netherlands.

In April 2016 (Safe Food Coalition 2016), the Safe Food Coalition representing consumer advocacy organizations, in a letter to USDA's Secretary Vilsack, recommended that USDA: "Build a platform to allow food producers, processors and retailers to share data as public health partners in controlling antibiotic resistant (AR) bacteria in meat and poultry products." The rationale for this database is multifaceted. First, CDC, the European Union, and the World Health Organization have all recognized that AR zoonotic bacteria pose a substantial threat to public health (see Chap. 15). AR increasingly occurs in *Salmonella* and *Campylobacter*, the two foodborne pathogens causing the greatest number of confirmed foodborne illnesses (Huang et al. 2016). These AR human illnesses are characterized by greater severity, increased risk of hospitalization, bloodstream infection, and treatment failure.

Second, the primary cause of AR foodborne pathogens is the use of antibiotics for growth promotion in food animals. Current data are only on sales of antibiotics. Yet, data on which animals receive treatment, the actual usage of animal antibiotic drugs, and the incidence of antibiotic-resistant pathogens in farming communities are critical to assessing how animal antibiotics have contributed to the growing AR threat. The Safe Food Coalition recommends that FSIS work with the US Food and Drug Administration, CDC, and the other components of USDA to improve the data for on-farm antibiotic use by specific classes of food animals and to link them to AR pathogens in the food supply chain.

2.6 Discussion and Policy Implications

Given the advances in testing methods and technologies, faster and cheaper pathogen tests are available. Expanded and coordinated testing could support a farm to table pathogen database to facilitate more rapid trace back and more targeted food safety interventions. From an economist's perspective, it makes sense to require public pathogen testing data as a condition for selling food in the US marketplace. Currently, food producers are getting a "free ride" by not bearing all the costs of producing this food (Unnevehr 2006). Instead, the ill food consumers are paying the medical bills and losing time from work and other activities along with the inconvenience and suffering of enduring the foodborne illness (see Chap. 8). US public health and regulatory agencies should compare the public health benefits to the costs of pathogen testing and administering such a database, examine the barriers to creating such a database, and develop a plan for overcoming those barriers if the benefits are greater than the costs.

But food producers want to minimize costs, and have an incentive to meet the bare minimum standard, use sleight-of-hand to appear to meet the regulatory or contract requirements, or even play catch-me-if-you-can and disregard regulatory or contractual requirements. A major factor in determining which of these strategies is adopted by a food company depends in large part on the probability of getting caught by the regulator or contractor.

Policymakers that seek to use the power of information in the markets have a couple of options. First, they can attempt to raise the penalties associated with outbreaks that are discovered. This can be achieved through simply publicizing outbreaks when they occur. Though recall efforts often focus on those who are thought to have purchased tainted foods, expanding publicity to reach others may have an effect on the reputational penalty. Second, attempts can be made to increase the probability that illnesses associated with tainted product will be discovered. Discovery of the link between illness and product is the first step of the chain between risk and public awareness. Improved surveillance and epidemiology has had a significant impact in this area, as discussed in Chap. 13. Both of these options are based on the power of information in the market, a tool that is too often over looked or undervalued as a potential solution to the problem.

Appendix: A Model of Optimal Deterrence

The effect of information problems on industry incentives can be illustrated using a simple deterrence model. In the case of foodborne illness, optimal deterrence for a risk neutral business is achieved when the expected penalty for exposing people to an unknown/unnegotiated risk equals the cost of illnesses incurred as a result of the risk, as demonstrated in Eq. (1):

$$P_i \times \text{Penalty}_i = U_i \times \text{ConsumerCost}_i \tag{1}$$

where P_i is the probability a firm producing illness-causing food is caught, Penalty_i is the penalty the firm faces if caught, U_i is the unnegotiated level of risk that consumers face (number of excess illnesses), and ConsumerCost_i is the average economic loss experienced by consumers (including quality of life losses) for each illness.

The penalty for exposing consumers to unnegotiated risk is composed primarily of two parts, litigation costs and reputation costs. In many cases, reputation cost, being primarily an information-based sanction, is the greater of the two. Many Chipotle customers who read that people were made sick from the chain's food, for example, reassessed their risk perceptions associated with Chipotle and, as a result, chose to dine elsewhere. The loss of profits associated with this behavior is a significant penalty.

It is instructive to look at a number of hypotheticals to illustrate the issues involved in deterrence of risky industry food safety behavior. First, if all risks are anticipated and consumers can choose risk levels in a market ($U_i = 0$), no penalties will be necessary since the market will incorporate the cost of risk into prices paid by consumers. As Akerlof's lemons model demonstrates, however, this is not the case. Consumers typically do not have the ability to choose risk levels due to information problems. This is not to say that consumers are completely powerless in the market. Consumers can buy products such as pasteurized eggs for a premium and avoid risky foods (oysters out of season, pink burgers, etc.). The actual value of U_i is unclear. Next, if the penalty (Penalty$_i$) to firm i is set to be equal to the cost to consumers (ConsumerCost$_i$) and the probability a firm providing unanticipated risky food is penalized is 100%, optimal deterrence will occur. This is consistent with litigation that fully compensates for consumer losses. Given that 99.9% of illnesses are not linked to a specific product, this is not a realistic scenario. It is also unclear what the actual value of P_i is because some portion of these unlinked illnesses are caused by cross contamination at home. A third possibility for optimal deterrence is that the penalties that are imposed in a given instance exceed consumer costs by an amount sufficient to deter unnegotiated risk, as shown in Eq. (2):

$$\text{Penalty}_i \geq \left(U_i \times \text{ConsumerCost}_i \right) / P_i. \tag{2}$$

Hypothetically, if 90% of risk is unnegotiated ($U_i = 0.9$) and 99% of illnesses due to the food firm are not discovered ($P_i = 0.01$), the penalty needed to assure optimal deterrence would be 90 × ConsumerCost$_i$. This is not as implausible as it may seem. If company A can allow a contaminated lot to slip through and 100 people are made ill as a result (at an average cost of $5000 each), the company would only have to expect to be penalized $45 million to induce optimal food safety behavior. Though litigation alone is unlikely to reach this level of penalty, reputation costs (especially for large companies) may easily reach this number in some cases. Though high-profile cases such as the Chipotle and PCA may have reached these thresholds, it is unlikely that they have been met in the many lower-profile outbreaks that garner less media attention. Furthermore, many outbreaks occur as a result of bad practices by smaller firms, which have limited assets, exposing them to maximum penalties that are suboptimal to deter risky behavior (the firms are judgment proof). Research is needed to determine the level of deterrence in industry, as a whole.

Abbreviations used in this chapter.

References

Akerlof GA. The market for "lemons": quality uncertainty and the market mechanism. Q J Econ. 1970;84(3):488–500. http://links.jstor.org/sici?sici=0033-33%28197008%2984%3A3%3C48 8%3ATMF%22QU%3E2.0.CO%3B2-6

Bovay J, Ollinger M. Strict standards nearly eliminate Salmonella from ground beef supplied to schools. Amber Waves. 2015 January/February. www.ers.usda.gov.

Buchanan RL, Schaffner D. FSMA: Testing as a tool for verifying preventive controls. Food Prot Trends. 2015;35(3):228–37.

CDC, PulseNet's Whole Genome Sequencing. https://www.cdc.gov/pulsenet/pathogens/wgs.html. Accessed June 2, 2017.

Chunara R, Andrews JR, Brownstein JS. Social and news media enable estimation of epidemiological patterns early in the 2010 Haitian cholera outbreak. Am J Trop Med Hyg. 2012;86(1):39–45.

Consumer Reports. Consumer Reports finds 71 percent of store-bought chicken contains harmful bacteria. February 1998. www.consumersunion.org.

Consumer Reports. The high cost of cheap chicken. March 2014. www.consumersunion.org.

Elbasha EH, Riggs TL. The effects of information on producer and consumer incentives to undertake food safety efforts: A theoretical model and policy implications. Agribusiness. 2003;19(1):29–42. https://doi.org/10.1002/agr.10043.

Food and Drug Administration (FDA), The Bad Bug Book. 2013. Available at http://www.fda.gov/

Food and Water Watch. Foul fowl: an analysis of *Salmonella* contamination in broiler chickens, 2006.

Food and Water Watch. More foul fowl: an updated analysis of Salmonella contamination in broiler chickens, March 2008.

Gamble GR, Berrang ME, Buhr RJ, Hinton AJR, Bourassa DV, Johnston JJ, et al. Effect of simulated sanitizer carryover on recovery of *Salmonella* from broiler carcass rinsates. J Food Prot. 2016;79(5):710–4. https://doi.org/10.4315/0362-028X.JFP-15-461.

Guo M, Buchanan RL, Dubey JP, Hill DE, Lambertini EA, Ying Y, et al. Qualitative assessment for *Toxoplasma gondii* exposure risk associated with meat products in the United States. J Food Prot. 2015;78(12):2207–19.

Henson S, Caswell J. Food safety regulation: an overview of contemporary issues. Food Policy. 1999;24(6):589–603.

Hobbs JE. Information asymmetry and the role of traceability systems. Agribusiness. 2004;20(4):397–415.

Huang JY, Henao OL, Griffin PM, Vugia DJ, Cronquist AB, Hurd S, et al. Infection with pathogens transmitted commonly through food and the effect of increasing use of culture-independent diagnostic tests on surveillance — Foodborne Diseases Active Surveillance Network, 10 U.S. Sites, 2012–2015. MMWR Morb Mortal Wkly Rep. 2016;65(14):368–71. https://doi.org/10.15585/mmwr.mm6514a2.

Institute of Medicine and National Research Council. Enhancing food safety: the role of the Food and Drug Administration. Washington, DC: The National Academies Press; 2010. https://doi.org/10.17226/12892.

Larsen J, Petersen A, Larsen AR, Sieber RN, Stegger M, Koch S, et al. Emergence of livestock-associated methicillin-resistant Staphylococcus aureus bloodstream infections in Denmark. Clin Infect Dis. 2017; https://doi.org/10.1093/cid/cix504.

Marder EP, Cieslak PR, Cronquist AB, et al. Incidence and Trends of Infections with Pathogens Transmitted Commonly Through Food and the Effect of Increasing Use of Culture-Independent Diagnostic Tests on Surveillance — Foodborne Diseases Active Surveillance Network, 10 U.S. Sites, 2013–2016. MMWR Morb Mortal Wkly Rep 2017;66:397–403. https://doi.org/10.15585/mmwr.mm6615a1.

McCluskey J. A game theoretic approach to organic foods: an analysis of asymmetric information and policy. Agric Resour Econ Rev. 2000;29(1):1–9.

National Research Council. The potential consequences of public release of food safety and inspection service establishment-specific data. Washington, DC: The National Academies Press; 2011. https://doi.org/10.17226/13304.

News Desk. CDC: antibiotic resistance increasing in certain Salmonella serotypes, food safety news. 2015a June 9. Available at http://www.foodsafetynews.com/2015/06/cdc-antibiotic-resistance-increasing-in-certain-salmonella-serotypes/#.VyDC0NQrKUk

News desk, CDC final update: 15 *Salmonella* illnesses linked to Barber Foods chicken products. Food Safety News. 2015b Oct 15. Available at www.foodsafetynews.com.

Painter JA, Hoekstra RM, Ayers T, Tauxe RV, Braden CR, Angulo FJ, et al. Attribution of foodborne illnesses, hospitalizations, and deaths to food commodities by using outbreak data, United States, 1998–2008. Emerg Infect Dis [Internet]. 2013; https://doi.org/10.3201/eid1903.111866.

Racicot Bergeron C, Prussing C, Boerlin P, Daignault D, Dutil L, Reid-Smith RJ. Chicken as reservoir for extraintestinal pathogenic *Escherichia coli* in humans, Canada. Emerg Infect Dis. 2012;18(3):415–21. https://doi.org/10.3201/eid1803.111099

Roos R. USDA names chicken plants with *Salmonella* problems. The Center for Infectious Disease Research and Policy (CIDRAP) News, Apr 1, 2008. http://www.cidrap.umn.edu/news-perspective/2008/04/usda-names-chicken-plants-salmonella-problems. Accessed June 3, 2017.

Safe Food Coalition. Letter to USDA's Secretary Vilsack, April 2016.

Stigler GJ. The economics of information. J Polit Econ. 1961;69(3):213–25.

Stranieri S, Cavaliere A, Banterle A. Voluntary traceability standards and the role of economic incentives. Br Food J. 2016;118(5):1025–40.

Unnevehr LJ. "Food Safety as a Global Public Good: Is There Underinvestment?" Plenary Paper, International Association of Agricultural Economists Conference, Australia, August 12–18, 2006. http://ageconsearch.umn.edu/bitstream/25733/1/pl06un01.pdf

Unnevehr L, Roberts T, Custer C. New pathogen testing technologies and the market for food safety information. AgBioForum. 2004;7(4):212–8. Available at http://www.agbioforum.org

USDA/FSIS. Quarterly progress report on *Salmonella* and *Campylobacter* testing of selected raw meat and poultry products: preliminary results, July 2015 to September 2015. http://www.fsis.usda.gov

USDA/FSIS. Process category introduction, May 26, 2017. https://www.fsis.usda.gov/wps/wcm/connect/71a3018b-5f77-46fe-bf78-b3f9cc4eb138/15_IM_Process_Category.pdf?MOD=AJPERES

Wessells CR. The economics of information: markets for seafood attributes. Mar Resour Econ. 2002;17(2):153–62.

Chapter 3
Economics of Food Chain Coordination and Food Safety Standards: Insights from Agency Theory

Diogo M. Souza Monteiro

Abbreviations

BRC   British Retail Consortium
FAO   Food and Agriculture Organization/UN
FSSC   Food Safety System Certification
GAP   Good Agricultural Practices
GFSI   Global Food Safety Initiative
HACCP   Hazard Analysis Critical Control Point
IFS   International Features Standard
SQF   Safe Quality Food
UN   United Nations

Over the past three decades, food markets became increasingly integrated, and contracts between upstream suppliers and downstream manufacturers, retailers, and food service business are increasingly the norm. Economic theory suggests that integrated companies will have fewer foodborne illness outbreaks, since the integrated companies have more control over their supply chain from farm to table. Yet, despite this change in global market structure, there have been many food safety outbreaks in the last decade. The prevention and mitigation of these outbreaks were significantly undermined by the existence of information failures, even within a single company. This chapter introduces agency theory, an economic framework that helps understand the role of information on the vertical contractual relations in the food supply chain. The change in economic incentives under different contract situations is explored. We further discuss how this framework can be used to examine alternative public policies and private strategies to improve supply chain coordination and reduce food safety risks, against some standard established either by the private sector or by government agencies.

D. M. Souza Monteiro (✉)
School of Natural and Environmental Sciences Newcastle University,
Newcastle Upon Tyne, UK
e-mail: Diogo.Souza-Monteiro@ncl.ac.uk

© Springer International Publishing AG, part of Springer Nature 2018   29
T. Roberts (ed.), *Food Safety Economics*, Food Microbiology and Food Safety,
https://doi.org/10.1007/978-3-319-92138-9_3

3.1 Introduction

Following the Jack in a Box food safety outbreak in 1993, a reform of American food safety legislation led to the introduction of mandatory Hazard Analysis and Critical Control Points (HACCP) in meat slaughter and processing plants. Similarly, food safety incidents in the European Union led to an upgrade of European food safety regulation. However, despite these improvements, the United States faced a series of high-profile foodborne disease outbreaks across the fresh produce and food processing industries between 2005 and 2008. These incidents seriously compromised the reputation and trust in the American food safety system. During each of these crises, it became apparent how poor the level of information in the system was, as it took days to uncover the source of the problem. Moreover, as was the case of the tomato outbreak in 2006 (that was actually caused by a jalapeno), the system failed to identify the real culprits in a timely manner.

Notwithstanding the epidemiologic, forensic, and legal elements required to contain and minimize the impact of food safety incidents, it is critical to understand the impact of information on the prevention of foodborne disease outbreaks. If authorities and food businesses could quickly identify the origin and extent of food safety outbreaks, their consequences could be substantially minimized. In the age of the Internet and advanced information systems, where we can easily and almost instantly access news of what is happening around the world and access different types of information at our fingertips, this may come as a surprise and a paradox. However, if one takes a closer look at the complexity of modern food chains, one realizes there are both strong economic and legal incentives to conceal information. The legal element is particularly evident in the case of food safety due to liability laws, particularly the negligence and liability laws in the United States.[1] If an agent in the supply chain can be solely and entirely liable for a food poisoning incident, she might be tempted to conceal any information that may lead to a prosecution or an accusation.

Information is thus a key element in a safe food chain. However, it is imperative to distinguish and understand different types of information. First consider the information about the presence (or the absence) of attributes in a product. If we could observe with certainty at the point of sale or consumption whether a food is contaminated, we would avoid consumption of the product, and businesses would only offer safe food. This is not however the case; in most instances food safety levels cannot be inferred from simple observation of a product. Second consider the information about the production or processing method originating the product. The presence of certain attributes of a product is determined by the production process or method. In other words, some processes are more effective in assuring a given level of quality (and safety) than others. Finally, we have to consider the person or business producing the product. Clearly, not all producers have the same competence and commitment to produce and deliver a product. The same production method or process can have very different attributes depending on the agent that is

---

[1] See Chap. 18 in this book for a detailed treatment of the legal aspects of food safety.

using it. These different aspects of information have different economic values and implications. Therefore, they cannot be dealt with in the same way.

In an ideal world information would be freely and readily available to all private and public agents in the food system such that they could make effective and sound decisions. In reality information is *costly*. Businesses and consumers need to incur expenses to determine the presence or absence of attributes in food, to record and store information and, finally, to transmit and share it. Then, as we will see below, private information plays a key role in bargaining and contracting which also needs to be taken into account. Consequently, unless agents are properly compensated, they have a strong incentive to withhold information. Likewise, if the probability of being liable for a foodborne illness is low, there are limited incentives to adopt preventive costly measures (see Chap. 18). For example, if a highly qualified and competent operator in a slaughter house is not compensated (or given enough time) to comply with food safety protocols (for instance, HACCP) and register any events occurring in a given day, food safety may be compromised and the information he may have acquired will be lost forever. Similarly, if a farmer cannot get a higher price for his product after a significant investment in biosecurity on his feedlot operation, he may not have an incentive to maintain a higher level of prevention. Finally, consider a ready-meal manufacturer testing a product for a pathogen and then reporting the results. This information is highly valuable to the buyer, but the costs of testing are often incurred by the seller who further risks losing revenue should the tests be positive. Clearly, buyers and consumers have the right to expect a high level of food safety and to be informed on any potential risks of contamination; however they should also recognize that assuring such high safety levels and getting access to information are not free.

Economists have long realized the importance of information in market transactions. One of the conditions for the existence of competitive markets is that there is free and accurate information available to all the agents in the marketplace. This is because having full information about the product or service transacted is essential for a complete valuation of what is transacted and for a rational decision. However, the reality is that in most markets, there is imperfect information on the attributes of a given product, on the most effective process to produce a good or a service, or on the ability of an agent to do a job. Moreover, this imperfect information is also prevalent within organizations transacting in a marketplace. Thus, in real markets information is often unobservable and unverifiable which challenges the ability of agents to make accurate valuations of the product or service they aim to purchase. Recognizing this reality and the fact that in a lot of situations we delegate on others the execution of tasks from which we benefit economists has developed agency theory.[2]

This theory, also known as theory of incentives, helps us understand and take into account imperfect information when we want to understand how organizations out-

---

[2]This chapter introduces agency theory for noneconomists. The next section presents this theory in a nontechnical fashion. Readers that have an economics or management background or want to have a more technical introduction to this theory are encouraged to read intermediary level business economics textbooks. An excellent text covering this theory is Laffont and Martimort (2002).

source the production to another firm or agent. In essence, this theory explains how a buyer and a seller may negotiate a contract when they don't have complete information on the other's ability and diligence to deliver the product with the agreed specification. More specifically, agency theory helps us think about the incentives a buyer needs to put in place to motivate a seller to assure he delivers the product with the agreed levels of quality and price level. Note that this theory does not only apply to market transactions, rather we can also use it to think about how a governmental authority might motivate an industry or consumers to adopt preventive measures to reduce food safety outbreaks.

This chapter introduces agency theory and how it can help us understand the challenges of coordinating food safety in food systems. In the next section, we (1) discuss the reasons why the principal agent framework is appropriate to examine food safety in supply chains, (2) describe the basics of an agency model and its key features, and (3) introduce the limitations of the framework. Then Section 3 provides a couple of applications of this framework, discussing the difference between a business-to-business and a government-to-business case and the challenges of contracting with multiple agents. Section 3.4 describes some of the international challenges of food safety control discussing some of the main private standards, and Sect. 3.5 concludes.

3.2   Agency Theory and the Economics of Information

Food systems are instrumental for the provision of food security, defined as: "when all people, at all times, have physical and economic access to sufficient *safe* and nutritious food that meets their dietary needs and food preferences for an active and healthy life" FAO (1996). Food safety is a key aspect of food chains; however as suggested above, buyers and consumers cannot infer the true quality or safety level of a product by direct observation. This is because safety is a credence attribute of food. Derby and Kirby (1973) define credence goods or attributes as those for which the seller knows more about the quality of the product or service it is selling than the buyer. In credence attributes, there is asymmetric information about the true level of food safety, because one of the contracting agents knows more about the characteristics of the product being transacted than its trading counterpart.

Writing on the economics of food safety, Antle (2001) claims there are actually two important food safety information issues in food supply chains. First there is *symmetric imperfect* information, as both suppliers and buyers may ignore the actual level of say *Salmonella* contamination of a given batch of burgers. Second there is asymmetric information as each agent has private information on their ability and efforts to mitigate food safety hazards which they withhold from their counterparts. Hirschauer (2004) links these two information issues as he identifies and describes two main hazards associated with food safety outbreaks: (1) technological hazards, linked to uncertainty about the process of contamination of food, and (2) moral hazards, which are related to opportunism of suppliers and buyers who use their private information and/or shirk on efforts to prevent food contamination.

Technological hazards may be thought of as the source of symmetric imperfect information, while moral hazards are linked to asymmetric information.

Technological innovation and increased awareness of suppliers of their process may reduce the degree of technological hazards. Even if there is uncertainty about how food safety outbreaks emerge, there is also considerable knowledge on how incidents can be prevented or mitigated. In fact, there are now a host of manufacturing practices (sanitation of workers and tools), control processes (such as the Hazard Analysis and Critical Control Points) and technologies (e.g., irradiation) that can effectively increase the level of food safety.

However, some of these techniques require significant investments, which increase the cost of operation and reduce business profitability. In other words, businesses need to have clear incentives to justify investments in food safety. This is where it is important to understand the impact of moral hazard, which requires a deeper understanding of the economics of information and the theory of incentives. The goal of this theory is to explain how we may organize the transaction between a buyer (the principal) who pays a price for a product whose characteristics depend on the effort of a seller or supplier (the agent).The challenge is to define the set of incentives that need to be written in a contract to ensure that an agent accepts to deliver a product with the expected level of quantity (and quality or food safety level) required by the principal. The buyer or principal has thus two main challenges: (1) to assure participation of the agent in the contract and (2) to motivate the buyer to exert the level of effort that ensures the volume of production and/or the level of quality required is delivered. Thus, implicit to the definition of the price proposed by the principal when making an offer is the ability to contract of the level of effort that the supplier (agent) needs to exert to deliver the product with the attributes required by the buyer (principal). Following we present a standard model to further explain the mechanics of an agency theory model.

3.2.1 Principal Agent Model and Contracting

The food supply chain can be seen as a sequence of supplier[3]-buyer pairings, where a buyer wants to obtain a product from an upstream supplier to sell to a downstream buyer or to the final consumer. Both the supplier and buyer maximize their profits, which translates into the buyer trying to obtain the product at the least possible price and the supplier trying to get the highest net benefit[4] from the sale of its output. In modern food chains, these transactions are increasingly governed by contracts between businesses operating at different points in the chain. Essentially, these con-

---

[3] In this section the words "supplier" and "seller" will be used interchangeably to designate the agent that produces and sells an output to a party downstream.

[4] In other words, to maximize the difference between the price paid by the buyer per unit and the costs of production. These costs include the additional efforts required to produce higher quality, the costs of implementing and managing a quality system, as well as costs of recording and sharing information.

tracts specify a level of output to be delivered by the supplier to the buyer and a price to be paid for such delivery to the former. For example, retailers' often contract with farmers or cooperatives to supply fresh produce to their shops. While initially these contracts mainly specified quantities and prices, increasingly they also specify the production and processing methods as well as the quality attributes of the output to be delivered, namely, its level of food safety.

Recall the three types of information introduced above. In a world of perfect information, the buyer would know with certainty the level of quality of a product, the reliability of a production and processing method, as well as the ability of the supplier to deliver the agreed levels of output and food safety. Let's focus on the later aspect of information and how it affects transactions.[5] Should there be perfect information on the ability and the level of effort exerted by the supplier, when producing the product to be delivered, the level of food safety would be completely observed or inferred. Consequently, a buyer would be able to design a complete and efficient contract to deliver a good with the required specifications at the least possible cost. However, as suggested above, the reality is that buyers do not have perfect and complete information on the true quality of the product they purchase upstream. They can determine the quality but only at a cost. In other words, when designing a supply contract, the buyer faces uncertainty on whether the supplier is actually capable of delivering the product with the required specification. Furthermore, the buyer cannot observe the actions of the seller; therefore, he also faced uncertainty on the seller's commitment to exert the level of effort required to deliver a product with higher level of quality.[6] There is imperfect information because the supplier has private information that is critical to the buyer, but that it is not in his best interest to disclose.

Agency theory helps us understand and model these transactions under imperfect information. This theory determines the incentives that a buyer needs to put in place in order to (1) attract the suppliers that have the ability to deliver the output with the quality attributes required and (2) motivate the seller to exert the level of effort that minimizes the cost of the output and the risks associated with quality failures.[7] More formally, when designing a contract under imperfect information, buyers need to take into account two key risks (Barros and Martinez-Giralt 2012):

- *Adverse selection* is defined as the risk associated with the failure of contracting a supplier that is truly able to deliver the required product or service with the level of quality required. The risk faced by the principal is to select an agent that is unfit for the job or service contracted.
- *Moral Hazard* is the risk associated with shirking in the contracted levels of effort required to produce the volume and quality of output. Given that the buyer

---

[5] The remaining elements unfold from this.

[6] The problem in here is that for an agent or seller effort is costly, so the least amount of effort exerted the larger the return. Since more quality (or food safety levels) requires more effort, unless the buyer creates the right incentive, the seller will not necessarily exert the level of effort required to deliver a safer food.

[7] See application in Chap. 10 on *E. coli* O157 and Jack in the Box's required testing.

cannot fully observe the effort of the supplier, she has to provide an incentive or a punishment to discourage the seller from shirking on the agreed upon effort. In other words, the problem is to assure the agent exerts the level of effort required to deliver the level of quality expected by the buyer and written in the contract.

These well-known issues have two consequences: (1) prevent buyers from making complete rational decisions and (2) affect the ability of markets to perform efficiently. In the absence of complete information on the attributes of a product, it is impossible to assess their true value. Consequently, the buyer may be paying more than the actual value of the product. This is because the seller is getting an information rent, due to inability of the buyer to observe both the capacity and commitment of the seller. From the presentation so far, it should be clear there is interdependency between buyer and seller, as both want to get something the other has. However, each agent also has a private interest, aiming to maximize her utility or profits. This results in a conflict of interest, because the value and joint utility they get from the transaction depend upon unobservable attributes (of the agents but also of the product they trade). Should the buyer be able to observe with certainty the seller's competency and effort, then she would be able to design a contract that would maximize the utility from the transaction for both parties. The agent would not be able to hide the true value of the good as it would become apparent. The reality however, particularly on food safety, is quite different, as it is virtually impossible to both ascertain the true ability of an agent to perform and to observe his actions.

So, what can be done? Basically, the idea of agency theory is to factor these asymmetries in information when designing a contract. This translates into forcing the supplier to reveal his true ability and to give him a clear disincentive to shirk on the level of effort required to deliver a higher level of quantity or quality. A key and implicit assumption of agency theory is that the contracts can be resolved in a court of law. In other words, should the buyer or the seller fail to comply with the terms of the contract, they can bring the case to a legal authority that will be able to resolve the dispute.

In economics, an agency theory problem is typically analyzed in mathematical terms as a constrained maximization problem.[8] Weiss (1995) pioneered the adoption of this framework to the economics of food safety, and since then there have been a variety of applications. Table 3.1 describes in words the main features of an agency model.

Table 3.1 draws on Starbird (2005) and Elbasha and Riggs (2003), and while it offers a simplified version of the problem, it enables us to draw some important lessons. The first thing to take into account is that in an open economy, there is always an alternative market where the seller can sell the product. This is important, because if a buyer cannot afford to pay at least U, the seller will not participate in the contract. Moreover, there is a distribution of ability (or competency) in the seller's market. It is reasonable to assume the buyer knows the nature of this distribution,

---

[8] Agency theory is an application of noncooperative games. Thus1 an alternative way of modeling this problem is using game theory and finding the optimal strategies for each party.

Table 3.1 Framing the relation between buyers and sellers

Partners	Problem/objective function	Choice variables	Comments
Buyer	Wants to design a contract to buy a quantity X of a product valued at p from the seller. She offers a price w to the seller ($p > w$). However, the product is unsafe with probability π. Assume the buyer will bear the total costs L^a of a food safety incident. The buyer knows that not all sellers are alike, as some are safer than others. Moreover, the probability of a food safety incident can be reduced with more effort from the seller	The buyer chooses the price w to write in a contract to get the X units of product. However, given that the quantity and safety level of the product are linked to the effort exerted by the seller, the level of effort also needs to be considered in a contract	The level of food safety adds considerable complexity to the buyer-seller transaction. The buyer not only has to consider the price to pay to maximize his profits but also how to avoid a possible loss caused by an outbreak
Seller	The seller has an alternative market on which to sell the product that gives him a total profit of U. He has private information on his ability and amount of effort to deliver the product with the required level of food safety. He will accept the contract if his net benefit is at least as high as the net gain from the alternative market	The seller choses the level of effort such that: 1. he get at least the value U 2. higher effort levels will lead to more compensation	The first condition is known as the participation constraint. The second is named incentive compatibility constraint. What this setup clearly shows is that imperfect information imposes additional costs to the buyer, as the supplier will only exert higher effort if he is compensated

[a]The literature associates this loss to the costs of recalling and disposing products, legal fees, regulatory fines, and loss of reputation (see Starbird 2005; Elbasha and Riggs 2003)

but she cannot observe the actual competency of a given seller. Thus, she needs to design the contract such that only the high-quality sellers participate. Clearly, this makes these transactions more costly than buying in a spot market. A further complication (and additional cost) arises from the incentive compatibility constraint, which compensates the supplier for exerting a high level of effort.

Now consider the probability of a food safety outbreak π is a function of the level of food safety θ. The level of food safety can be defined and measured in a number of different ways. For example, it may reflect the level of pathogen contamination of a product, handling practices, exposure to possible contaminants, and contamination by chemical or physical agents. For convenience, assume that it relates to the degree to contamination by pathogens. Thus to a low level of contamination corresponds a high level of food safety, and conversely high levels of contamination have low safety levels. Further assume that the level of food safety is a function of the supplier's level of effort e and a stochastic term ε. Consequently, the supplier does not have absolute control of the process, and regardless of his efforts, there is always a probability that an incident will occur.

In agency models, the distribution of types of agents in the market is simplified, and it is assumed there are only two types of suppliers: high effort (e^H) and low effort (e^L). A high- effort seller is more efficient and therefore will produce a safe product at a lower cost.[9] That is, this agent is more effective and efficient in delivering food safety. However, unless the buyer compensates the high-effort supplier, the supplier will exert the lower effort. This is in essence the incentive compatibility constraint. Formally we have $\theta(e^H,\varepsilon) > \theta(e^L,\varepsilon)$ and $p[\theta(e^H,\varepsilon)] < p[\theta(e^L,\varepsilon)]$, that is, a high-effort supplier will deliver, on average, a safer product and therefore have a lower probability of originating an outbreak. The stochastic element is a key issue as it adds uncertainty to the problem of the buyer and leads to further opportunities for the high-effort seller to disguise as low effort.

Another feature of the model deserving attention is the loss and how it is distributed. When there is strict liability, the loss needs to be entirely borne by the food chain. In fact, strict liability implies that if a product had a defect and caused an injury, then the agent producing such product will be held accountable even if the event was accidental (see Chap. 18 for a more detailed legal account of the issue). What is clear though is that the way outbreak losses are distributed matters. In our simplified formulation, we suggest that the cost of a food safety hazard is entirely borne by the buyer, but the loss might just as well be entirely passed on to the seller or be shared between the buyer and the seller. What is clear is that if the loss is partially or totally assumed by an external party (say the government or the consumer), there won't be enough incentives to invest in preventive actions. Also, if the legal system fails to punish the culprits of food safety hazards (as suggested in (Starbird 2005; Mahdu et al. 2015) then neither buyers nor suppliers will feel compelled to make the necessary investments to exert higher levels of effort to increase the food safety levels of their foods.

There are many ways to solve agency problems. The most common approach is to first determine the price to be paid to the seller and then to assure the desired level of effort is indeed exerted. Importantly, the existence of asymmetric information leads to an opportunity for the party with information to get a rent[10] which makes the transaction inefficient when compared to a case of full information.

In short, agency theory provides us with a framework to think about transactions when there is asymmetric information on the ability and diligence of a seller to deliver a good demanded by a buyer. The aim of this framework is to provide guid-

---

[9] Here there is an implicit assumption that quantity and food safety are both increasing in levels of effort. In reality, this may not be the case, and efficiency in producing higher volume may be decoupled from food safety.

[10] In economic terms, a rent is a value that needs to be paid in excess of market price to obtain a service or a product. In order to assure the supplier exerts a high-effort level in food safety precaution, the buyer needs to pay more than she would in the spot market. In this case, the rent is due to information asymmetries.

ance on how to design a contract that is acceptable to the seller and makes him deliver the level of effort required by the buyer in terms of food safety levels. So it helps us think about the appropriate incentive structure to assure supply chains deliver the expected level of food safety.

3.2.2   Limitations of Agency Theory

Agency theory provides very clear and important lessons to our understanding of transactions under imperfect information. However it does have some limitations. First because there is a wider heterogeneity in suppliers than the simplified dichotomy we typically use. Moreover, in practice the contracts will have other conditions beyond just defining a price and a quantity or quality to be delivered. So the framework should be used as a guide to help us think about conditions of the contract and what should be its specifications. Second, the model we presented is static, meaning it considers transactions in only one time period. There are models in the literature that consider the possibility of contracting in multiple periods; however even in those models, it is assumed that the terms of the contract remain constant across periods.[11] Moreover, most of these dynamic models don't allow for agents to change their type or the principal to learn about the ability and diligence of the supplier, which in practice would mean that an inefficient agent cannot improve or the principal to use information in past periods to improve the contract. In practice, however, governments and buyers can offer incentives for suppliers to improve their ability to perform. This limits the possibility of gathering information on the level of performance of the supplier, which would allow the contract to be revised and made more efficient. A third problem regards monitoring and enforcing the contract which relies on the assumption that it is possible to observe and verify compliance. If a court of law cannot verify or validate the evidence on the cause of a food safety outbreak, it will not be possible to find a culprit guilty as charged, which means that the contract may not truly bind the parties.

3.3   Applications of the Theory of Incentives

One of the main features of agency theory is its flexibility. Just as the problem can be defined in terms of a buyer and seller, it can also be specified in terms of a government facing an industry. In that case the principal is the government or a regulator and the agent is the industry. What is the main difference between a government and a private entity as principals? Section 3.3.1 below gives a tentative answer to this question.

---

[11] Readers' interested in these types of models are referred to Laffont and Martimort (2002), Chap. 8.

Also, in modern supply chains, retailers and food service businesses often contract with several different suppliers. For example, it is unlikely that the supply of tomatoes to a supermarket chain will be sourced from a single supplier throughout the year. Also, when we say the government faces an industry, we implicitly assume there are several different businesses involved. So how can this framework help think about providing incentives to a group of suppliers or to a set of organizations in a given industry? This will be discussed in the section 3.3.2.

3.3.1 Government vs. Industry as Principals

The main difference in having the government as the principal is that governments consider the welfare of both the industry and consumers. In other words, the government needs to assure that society maximizes its welfare, which in economic terms is the sum of industry profits and consumer utility. Clearly, there is a potential conflict between consumers' utility and industry profitability. To see this, consider the additional costs (and corresponding decrease in profitability) the industry incurs to deliver safer food. Then compare these with gains in the utility of consumers and society from additional food safety. For example, consumers will be happier if they avoid pain associated with food outbreaks, but also this means they will have a lower number of sick days (which increases productivity) and also lower medical and hospitalization costs. If the loss of profits to the food industry due to higher costs of food safety prevention is higher than the gains in consumer's utility and societal welfare, then, from a pure economic perspective, society as a whole would be worse-off. So when considering interventions to improve food safety levels, governments need to carefully assess how to weigh the costs and benefits to industry and consumers of different policy interventions to mitigate food safety.

At the heart of the matter, once again, is information on the actual cost (to industry and consumers) of prevention. If consumers knew that the industry is not investing heavily in food safety, then they will infer that food is not as safe and either take their own precautionary measures, avoid those products they perceive as unsafe, or pay less for them. This creates an incentive for the industry to invest on food safety. Elbasha and Riggs (2003) suggest that insofar as the degree of safety is unobservable and non-verifiable, the industry has little incentive to invest in prevention. So, the rule of the government may need to be more subtle. Rather than imposing a level of food safety, the government might make public information about the industry food safety levels. This can be done by investing in monitoring and inspection policies and then reporting results to consumers. Consequently, when it comes to food safety, the government might be better off focusing on finding and disclosing information rather than on direct intervention.[12] For example, the government might con-

---

[12] Note, however, that the government might be considering a subsidy to promote food safety or specifying the optimal level of a fee or penalty to minimize food safety outbreaks.

sider designing an inspection policy aiming at revealing the true level of food safety of an industry.

Starbird (2005) uses a principal-agent framework to determine the optimal inspection policy that makes a seller exert a higher effort to deliver safe food. The inspection policy is characterized by the sampling method, the acceptance rate, and the penalty from failing to pass inspection. While in his model Starbird does not directly analyze the relation between a regulator and the industry, it does shed light onto how the government may influence the parties in a food chain to invest in food safety measures. It is implicit there that if the government mandates a stricter sampling policy, then the probability of detection of unsafe samples increases which leads the suppliers to increase their effort levels and decrease the odds of a food safety hazard. By the same token, if the government increases the penalties or the amount of public contributions to food safety outbreak costs, then it also incentives the adoption of more precautionary measures.

The government and the legal system can create incentives for the industry to adopt safer food production and processing practices through the negligence and liability laws. As Stearns (2017) suggests in Chap. 18 of this book, avoiding litigation is another instrument liberal societies have to incentivize a safer food marketplace. Liability laws affect the way risks are allocated throughout the supply chain. Notwithstanding potential inequities emerging from opportunism of businesses with more resources to shift their risk upstream, making someone in the supply chain accountable for managing the food safety risks is an alternative to direct intervention by governments. In the United Kingdom, the Food Safety Act of 1990 is an example of how a government can incentivize the industry to invest in food safety by holding businesses selling directly to consumers accountable for the safety of the food they sell (Food Standards Agency 2009).

In short, as one of the problems the government faces is uncertainty over the cost structure and the industry and firms' ability to exert higher effort, agency theory is a very suitable framework to understand how to incentivize the industry to deliver safer food. This theory helps us understand the role of government and the impact of different policy options aiming at increasing food safety levels. Governments do not necessary need to intervene directly with regulation to attain a given level of food safety, rather they may be more effective if they create an incentive structure that forces industry to take due diligence and protect against possible outbreak costs.

3.3.2 Contracting with Multiple Agents

The problem with contracting with different businesses is that information asymmetries not only persist but multiply. Economists have developed agency models to deal with the case where a principal proposes a contract to a set of suppliers. There are a number of possible approaches; the first is to treat the group as an individual, which means that a similar structure to the one above is considered. A second approach is to offer individual contracts to each supplier in the group. In this second case, the model may consider whether agents are homogenous or heterogeneous. Consider the case of a producers' organization supplying fresh produce to a retailer, producers will have different abilities and willingness to exert effort. Thus, the food safety level will not be constant for every supplier. In fact, unless the identity of each producer is preserved, a low-effort producer may free-ride on the high-effort members of the organization. By the same token, when the government is considering interventions to boost food safety with an industry, it needs to take into account differences across businesses and how less efficient firms may jeopardize the industry efforts to respond to incentives.

One of the ways buyers and governments can mitigate free-riding when dealing with multiple agents is by making compensation to suppliers a function of the level of food safety in the market. The rationale is that if the buyers or consumers know that the risk of getting sick from foodborne pathogens is higher, they will not be willing to pay as much for food. Linking the payment of each individual to the overall performance of the group creates a strong disincentive to free-ride. Hamilton and Zilberman (2006) analyzed this issue in the context of collective reputation associated with environmental labels. While they don't use an agency framework, their insights are quite important and applicable. In their model, they propose that the price paid per unit of output has a fixed component and a variable premium which depends on the purity (the proportion of output that is of high quality). In the context of food safety, this degree of purity can be thought as the proportion of product that passes inspection. Of course, if a given industry or producer organization has a poorer food safety record, then buyers won't pay as much for their products. This creates an incentive for members of the industry or organization to invest in food safety and monitor the effort levels of their members.

To summarize, often a buyer needs to contract with a group of suppliers. These often are heterogeneous in both ability and effort levels, which complicates the design of the contract, as along with adverse selection and moral hazard, the buyer has a potential free-riding problem. This possibility needs to be taken into account, and one way to address it is to decouple the payment and include a component associated with a measure of the suppliers' group performance.

3.4 International Private Standards

As food chains become increasingly global, buyers face additional information gaps when contracting with overseas suppliers. For example, the European Union countries import a significant amount of fresh produce and vegetables from Latin American and African countries. A significant amount of US food imports is governed by contracts with grocery and food retail service.

Dealing with international suppliers increases the complexity of assuring a safe food supply for buyers will not only have more uncertainty about selecting and motivating suppliers to deliver required food safety standards but also have to take into account the ability of local private or public agents to monitor and enforce contracts. There are often striking differences in the minimum food safety requirements across countries. Thus, the regulatory food safety standard levels in exporting countries may not provide the assurances required by buyers, particularly when exporters are in developing countries and buyers are food retailers or food service businesses from the European Union, Japan, or United States which have higher food safety requirements. One of the ways food manufacturers and retailers have addressed this issue is by setting private standards that include food safety provisions and impose them as a condition to offer a contract. These standards are typically designed by individual buyers (for instance, retailers) or, more commonly, by industry associations. An example is the family of manufacturing standards designed by the British Retail Consortium (BRC), an association of British grocers (representing both large multiples and small independent grocers). These standards define requirements that suppliers need to comply with if they accept a supply contract. Among these conditions are a clear demonstration of the commitment of the managerial team of the supplier to implement a food safety program, a food safety management plan, and a registry on how procedures have been implemented. Moreover these private standards require an audit or a certification process through which the supplier agrees to be monitored by an independent third party agent that will visit the potential supplier and verify whether requirements are being followed.

While these private food safety standards are being used on contracts with both domestic and international suppliers, they are particularly useful when dealing with the latter. Recognizing there was a multiplication of food safety standards designed by different grocers and food manufacturers or their associations and there was an increasing cost of compliance imposed on suppliers, in 2000 a group of 650 international food retail, manufacturing, and food service companies came together to form the Global Food Safety Initiative (GFSI). The goal of this initiative is to assure consumers across the globe access safe food (Sansawat and Muliyil 2012). Following an agreement between seven global food retail companies, any supplier to a manufacturer, retailer, or food service company complying with one of the benchmarked schemes approved by GFSI will no longer be required to comply with other

Table 3.2 Food production and manufacturing schemes recognized by GFSI

Manufacturing schemes	Primary production schemes	Primary and manufacturing schemes
1. British Retail Consortium (BRC) Global Standard 2. Food Safety System Certification (FSSC 22000) 3. International Features Standard 4. Safe Quality Food Code (SQF) 5. Best Aquaculture Practices Standards 6. Global Red Meat Standard	1. CanadaGAP 2. GlobalG.A.P. 3. Safe Quality Food code: (a) Module 2: System elements (b) Module 5: Food safety fundamentals (c) Module 7: Food safety fundamentals	1. PrimusGFS

Adapted from Sansawat and Muliyil (2012)

schemes. The expectation is that this will open options to suppliers of different supply chains and reward those that have invested in improvements of their quality and food safety levels. Thus, the GFSI is becoming an umbrella organization that integrates competing private schemes and contributes to a standardized program to improve food safety levels of supply chains associated with major international retailers, food service, and food manufacturing companies.

The schemes that are recognized and benchmarked under GFSI are the British Retail Consortium Global Standards, the Food Safety System Certification (FSSC 2000), the International Features Standard (IFS), and the Safe Quality Food (SQF) Code. Along with these broader internationally accepted schemes that comprise both primary production and food manufacturing quality and safety assurances, there are also other sectorial, primary, and company-specific schemes that have been recognized under GFSI. These are summarized in Table 3.2 below.

It is beyond the scope of this chapter to provide a detailed comparison between each of the schemes recognized and benchmarked by GFSI. Interested readers are referred to Sansawat and Muliyil (2012), to the GFSI website, or to each of these schemes webpages that have detailed information on the requirements of each of these quality and safety standards. It is also outside the expertise of the author to compare the ability of each of these schemes to deliver an adequate level of food safety. It is nevertheless worth highlighting what the schemes have in common and how they relate to the economic framework discussed above. All aspiring suppliers to companies subscribing with the GFSI benchmarked standard are required to adopt the following procedures (with slight differences across schemes):

1. *Agree a contract*: In other words accept a proposal by the buyer to supply a given agricultural or food product.
2. *Optional preaudit*: Most schemes offer the supplier an opportunity to be visited by an auditor to assess their quality system and how it conforms with the standard requirements.
3. *Certification audit*: In this step, an independent auditor or certifier visits the supplier and performs an extensive and detailed examination of the production or processes as well as the quality scheme. The goal of this certification is to determine whether the supplier is in full compliance with the quality standard or scheme requirements and whether there are nonconformities. If these exist, the supplier will be informed of what needs to be rectified and given a timeline to perform such changes.
4. *Audit*: this is to confirm that (if any) the nonconformities were rectified and any recommended changes were implemented.
5. *Certification document*: Once the independent certification body has verified that the supplier is fully complying with the requirements of the standard and that all required adjustments were implemented, the supplier is issued a certificate of compliance and can start supplying the buyer.
6. *Recertification*: The certification is issued on an annual basis, so each year this process is repeated.

Clearly the process required by the food quality and safety assurance benchmarked and recognized by GFSI relates to the economic framework described in this chapter. As we just saw, an agro-food producer aiming to supply a major retailer, manufacturer, or food service business will need to accept a contract proposal to the potential supplier. As a buyer imposes compliance with one of the standards benchmarked by GSFI as a condition to get a contract, she is (at least implicitly) taking into account uncertainty associated with adverse selection. This is the reason why most schemes offer a preaudit, which is really a clever way to prescreen a supplier and gather information about his type. Then the certification and audits are actually a mechanism to mitigate moral hazard. The annual recertification is also a mechanism designed to create an incentive for the supplier to keep or improve his effort levels. So, by designing and imposing compliance with a standard as a condition to offer a contract, buyers address the adverse selection problem, forcing the supplier to reveal their true capacity. Moreover, imposing an audit on supplier allows the collection of information that will enable the buyer to learn the risks of each supplier. Also, through certification (that is verification of the extent to which the supplier is complying with the standard), the buyer has a mechanism to detect shirking on the expected level of effort.

While these international food quality and safety standards have obvious merits, they also carry a number of caveats. First it is not clear how growers and consumers are represented and have a say on the development of these standards. So the standards may not be taking fully into account practical knowledge producers may have of their operation and may be creating unrealistic expectations as well as disenfranchising a critical element of food systems.

Second, and possibly more importantly, is the role played by the certifiers which is worth further examination. From the description of the schemes, it seems that the certifiers act as gatekeepers to the buyer, excluding or limiting access to contracts all suppliers that do not conform with the standard requirements. But the preaudit, possibly carried out by the same entity that issues the certification documents, has elements of a consulting or advisory service that provides information to the supplier on what needs to be done to pass the audit. This may lead to a conflict of interest. So do certifiers act as inspectors or as consultants? How do they relate to the supplier and the buyers? If all the certifier is doing is examining conformance with the standard, the supplier may be tempted to adopt the type of misbehavior indicated by Stearns in Chap. 18 of this book, where he says that a supplier may be complying with a standard and still not doing enough to prevent food safety incidents. Furthermore, less scrupulous certifiers may be tempted by bribery from suppliers or to extort additional payments to provide the certification document. This is particularly relevant for contracts with suppliers from countries with limited institutional or governance infrastructures or with sophisticated criminal organizations as it became apparent in the food horse scandal in the United Kingdom (Levitt 2016) or been recently reported on the New Zealand (Roy 2016) and Mexican Avocado industries (The Economist Explains 2016).

When there is a potential conflict of interest and collusion between the certifier and supplier or between the certifier and the buyer, the benefits of food quality schemes and their impact on the level food safety risks being undermined. Tirole (1986) studied this possibility and shows that a monitoring organization (such as a third party certifier) may misreport the actual level of effort and food safety of a supplier by colluding with one of the contracting parties at the expense of the other. This is not a trivial issue, and the opportunism of a third party certifier needs to be taken into account, when they provide signals to the buyer on compliance with the terms of the contract. Note that the third party certifier can be thought of as a supplier of a service to the buyer, the supplier, or both and that it has both hidden information on his ability to monitor a quality standard and on the effort required to do it effectively. Interestingly, Stearns in Chap. 18 of this book provides an example of a lawsuit following an outbreak in the peanut industry, where a buyer relied on a certification document obtained from a private third party certifier hired by the seller to inspect the food safety conditions of the operation.

Third, by designing and enforcing compliance under a given set of rules, international buyers are selecting only those suppliers who can adopt and meet the specified quality standard. However this is not costless, as limiting the pool of potential suppliers increases the costs of contracting. Also, regarding the societal level of food safety, while potentially higher-risk suppliers are excluded from high-value supply chains, they will not necessarily leave the market, rather they may move to lower-value supply networks. In this case the probability of an outbreak will not necessarily diminish. Moreover, these alternative channels may be supplying lower income segments of the population, raising concerns on equity as people with less protection may be facing a higher risk of getting foodborne illnesses. Finally, the increasing interest of media on any food outbreak and its ability to amplify a message without much concern about the actual dimension, limitation, or even source of the problem can lead to spillover effects across all the industry.

A fourth issue related to the use of international private standards contracts with international developing country suppliers regards the jurisdiction on which disputes can be settled. There are really two issues: one is whether noncompliance with a private standard can be considered by a legal court and the second is which legal court would be selected. This is a challenging issue because the authority to penalize a supplier that failed to deliver safe food is not clear-cut. As we saw above, one of the key assumptions of agency theory is that contracts can be resolved in a court of law; on its absence, it may be impossible to enforce a contract.

In short, private food standards have emerged in the last couple of decades in response to regulations that hold retailers, food service, and other businesses selling food directly to consumers accountable for any incidents that may occur. Increasingly, these standards are comparable and have a global scope, often being more demanding than national legislations. As retailer and food service firms impose these standards to their suppliers, they are really addressing the information asymmetries raised and explained by agency theory. In fact, by imposing a standard as a condition for supply, a buyer is forcing their suppliers to reveal their types. The standard becomes a screening device. Furthermore, by having a third party verifying whether the standard is being followed, the buyer is addressing the compliance issue.

Nevertheless, international standards have limitations and may not effectively prevent food safety outbreaks. First, it is important to recognize that the standards are designed to mitigate the risks of liability and reputation of the firms or industry associations that own them and not necessarily to incentive an optimal level of food safety. Second, only the consumers and customers of the businesses that are imposing the standards on their suppliers are safer. While the number of retail and food service business adhering to these international standards is growing and commands a larger market share, there is still a significant amount of food that is sold outside these supply networks. Third, the monitoring of these standards is not continuous, but rather discrete and at most once every quarter, which may not create a strong enough incentive for high levels of compliance.

3.5 Final Remarks

As the food system becomes increasingly complex, careful management of food safety risks is ever more important. A key element to prevention and containment of food safety outbreaks is information. This chapter introduces the economics of information and agency theory. This theory provides key insights on how information about an agents' ability and diligence in performing a given task needs to be considered when contracting food safety levels between suppliers and buyers in a food chain. Unless the agent has an incentive to reveal his type and exert the level of effort required to deliver an acceptable level of food safety, there is a chance he may not deliver what is expected. It is important to understand that this theory is not addressing the problem of getting information on a given product or process, but rather on how to take into account private information when contracting with an agent (person or business) to produce a product with the desired specification.

Understanding this problem and its implications for modern food system is vital as otherwise private and public systems designed to improve food safety levels won't be as effective. In fact, failing to recognize the value of private information may lead producers and processors of food to under invest in prevention and ignore the true level of contamination of the products they sell. The flexibility of this framework helps us understand the characteristics of explicit contracts between buyers and sellers or between regulators and industry but also the implicit contract between a consumer and a food supplier. For instance, using this framework, one may realize that the consumer may also be opportunistic and shirk on her/his own precaution efforts. In fact, consumers may assume all the food available for consumption is safe to eat and trust that a system is in place to assure that is the case. Unsafe food does not have a market, and therefore consumers will not necessarily realize the trade-off between the risks of buying a contaminated food and the costs incurred by the industry to assure a low risk. Governments rightly expect and require that all segments of the population, regardless of their socioeconomic status, have access to safe food. However, just as the consumer, regulators will need to recognize business motivations and challenges and that without the right set of incentives, it may be impossible to get further improvements in the delivery in food safety.

Acknowledgments I am very grateful to Derrick Jones and Tanya Roberts for giving me the opportunity to write this chapter as well as for their helpful comments and encouragement. Derrick also gave me great suggestions on how to simplify the technical language and very helpful editorial suggestions. I am particularly in debt to Tanya for challenging me to reflect how agency theory related to private food safety standards. Sterling Andrew Starbird, with whom I learned about the application of agency theory to the economics of food safety, generously offered to revise an early draft and provided inestimable insights and comments. I am particularly thankful for his suggestions and corrections to the second section of this chapter. Any remaining errors, mistakes, or omissions are of course my own.

References

Antle JM. Economic analysis of food safety. In: Gardner B, Rausser G, editors. Handbook of agricultural economics, vol. 1. Amsterdam: Elsevier Science; 2001. p. 1083–136.

Barros PP, Martinez-Giralt X. Health economics: an industrial economics perspective. London: Routledge; 2012.

Darby MR, Karni E. Free competition and the optimal amount of fraud. J Law Econ. 1973;16(1):67–88.

Elbasha EH, Riggs TL. The effects of information on producer and consumer incentives to undertake food safety efforts: a theoretical model and policy implications. Agribusiness Int J. 2003;19:29–42.

FAO [Food and Agriculture Organization] Rome Declaration on World Food Security and World Food Summit Plan of Action. World food summit 13–17 November 1996, Rome.

Food Standards Agency. The food safety act 1990—a guide for food businesses 2009 edition. London: Food Standards Agency; 2009.

Hamilton SF, Zilberman D. Green markets, eco-certification and the equilibrium fraud. J Environ Econ Manage. 2006;48(2):978–96.

Hirschauer N. A model-based approach to moral hazard in food chains. Agrarwirtschaft. 2004;53(5):192–205.

Laffont JJ, Martimort D. The theory of incentives: the principal agent model. Princeton, NJ: Princeton University Press; 2002.

Levitt T. Three Years on the Horsemeat scandal: 3 lessons we have learned. The Guardian, 2016. Available at: https://www.theguardian.com/sustainable-business/2016/jan/07/horsemeat-scandal-food-safety-uk-criminal-networks-supermarkets. Assessed 1 Jan 2017.

Mahdu, O, Boys, KA, Geyer, L, Ollinger, M. Penalties for foodborne illness: jury decisions and awards in foodborne illness lawsuits. 2015. http://purl.umn.edu/205810. Assessed 1 Feb 2017.

Roy EA. Avocado shortage fuels crime wave in New Zealand. The Guardian 2016. Available at: https://www.theguardian.com/world/2016/jun/15/avocado-thieves-shortage-crime-fruit-black-market-new-zealand. Assessed 1 Feb 2017.

Sansawat S, Muliyil V. Comparing global food safety initiative (GFSI) recognised standards. SGS Group Management, 2012. https://www.foodprocessing.com/assets/wp_downloads/pdf/white-paper-comparing-gfsi-standards.pdf. Assessed 3 July 2018.

Starbird SA. Moral hazard, inspection policy and food safety. Am J Agric Econ. 2005;87(1):15–27.

Stearns D. A critical appraisal of the impact of legal action on the creation of incentives for improvements in food safety in the United States. In: Roberts T, editor. Food safety economics: incentives for a safer food supply. New York: Springer; 2017.

The Economist Explains. The link between avocados and crime. 2016. Available at: http://www.economist.com/blogs/economist-explains/2016/10/economist-explains-9. Assessed 1 Feb 2017.

Tirole J. Hierarchies and bureaucracies: on the role of collusion in organizations. J Law Econ Organ. 1986;2:181–214.

Weiss MD. Information issues for principal and agents in the market for food safety and nutrition. In: Caswell JA, editor. Valuing food safety and nutrition. Boulder: Westview Press; 1995. p. 69–79.

Chapter 4
Benefit/Cost Analysis in Public and Private Decision-Making in the Meat and Poultry Supply Chain

Tanya Roberts

Abbreviations

AI	Avian Influenza
BCA	Benefit/Cost Analysis
CCP	Critical Control Point
CDC	Center for Disease Control and Prevention/USA
EO	Executive Order from the President of the USA
ERS	Economic Research Service/USDA
FSIS	Food Safety and Inspection Service/USDA
HACCP	Hazard Analysis and Critical Control Points
ICMSF	International Commission on Microbiological Specifications for Food
LTHO	Long-term health outcomes
OIRA	Office of Information and Regulatory Affairs/OMB
OMB	Office of Management and Budget/USA
PPIA	Poultry Products Inspection Act/USDA
USDA	United States Department of Agriculture

Benefit/cost analysis (BCA) is a tool that can be used to examine either public or private decision-making. What differs is what variables are included in each BCA. Hazard Analysis and Critical Control Point/Pathogen Reduction (HACCP/PR) regulation was established in 1996 for US meat and poultry. From the US government's perspective, the estimated benefits of HACCP were a reduction in foodborne illnesses of the American public citizens and their expenses (medical costs, productivity losses, and pain and suffering) vs. the costs to industry of implementing the HACCP regulations. The public health protection benefits were estimated to be in the billions of dollars, while the industry costs were in the millions of dollars.

T. Roberts (✉)
Economic Research Service, USDA (retired), Center for Foodborne Illness
Research and Prevention, Vashon, WA, USA
e-mail: tanyaroberts@centurytel.net

© Springer International Publishing AG, part of Springer Nature 2018
T. Roberts (ed.), *Food Safety Economics*, Food Microbiology and Food Safety,
https://doi.org/10.1007/978-3-319-92138-9_4

A private firm's decision whether to invest in poultry production in China, however, examines benefits and costs using variables to determine profitability. Some of the factors to be considered in this private BCA include (1) the increasing demand for poultry meat in China and other Asian countries with rising incomes; (2) the demand for the wide range of chicken parts in China; (3) the cost savings of raising chickens in China, including lax environmental regulations and enforcement; (4) the possibility of selling chicken breasts (and frozen chicken products) in the US market at a premium price that more than covers all costs, including transportation; and (5) competition with increasingly global food companies in Asian markets.

4.1 Different Applications of Benefit/Cost Analysis in the Meat and Poultry Industry

BCA is a useful decision-making tool in a variety of contexts, not the least of which is the meat and poultry industry (see Box 4.1). BCA for public decision-making compares the benefits vs. costs for impacts in three general categories: households, industry, and government sectors of the economy. These benefits and costs may be easily quantifiable in the marketplace or not easily quantifiable for nonmarket benefits, such as public health benefits of reducing foodborne illnesses in the population. In contrast, a private decision will be based on the market benefits and costs of relevance to the private firm.

Box 4.1 Benefit/cost analysis and use in decision-making

Benefit/cost analysis (BCA) was invented by the French civil engineer and economist, Jules Dupuit, in the mid-nineteenth century (Ekelund et al. 2000). His goal was to develop a way to compare the scale of public works projects, such as bridges, flood control, and sewers. Dupuit was the first to develop the idea of the maximization of utility, and he developed the mathematics of marginal costs and marginal benefits. The word "utility" is an economic term meaning satisfaction or benefit. He used BCA to compare bridge options, A, B, and C, of different sizes and scale to determine which provided the most satisfaction or benefit compared to the costs of building bridge A, B, or C. This comparison included the returns on investment of different scales of projects. Dupuit generalized BCA to cover any kind of market transaction, or even nonmarket transactions or decision-making, such as marriage.

A central feature of Dupuit's BCA is choice among alternative activities for a person deciding how to spend his/her time and money or a business or a government deciding how to use its resources. In each case, the goal is to maximize satisfaction (utility), or profits, or the public good. BCA is a tool for considering alternative ways of spending time and money. The implicit question is "What course of action yields the most satisfaction, or profits, or public

---

(continued)

good in the long run, or the short run?" The time horizon in both examples discussed in this chapter is a long-run consideration. Decision-makers can use BCA prospectively or retrospectively, i.e., either before or after the decision is made.

The US Army Corps of Engineers was an early user of BCA in the development of large rivers, such as the Mississippi. In 1981, however, President Reagan signed Executive Order No. 1991 that created the Office of Information and Regulatory Affairs (OIRA) in the Office of Management and Budget (OMB). All regulatory agencies were required to prepare impact analyses "for any regulations that are likely to result in annual effects on the economy of $100 million or more." All proposed rules were to be analyzed through a BCA, with the requirement that the BCA contain the following information:

- Description of potential benefits of the regulation, including benefits that cannot be quantified in monetary terms and identification of those likely to receive the benefits
- Description of potential costs of the regulation, including adverse effects that cannot be quantified in monetary terms and identification of those likely to bear the costs

Presidents Clinton and Obama expanded upon these basic concepts for BCA in other Executive Orders in 1993 and 2011, EO 12866 and EO 13563, respectively.

In 1996, a prestigious consensus report by economists on the usefulness of BCA stated:

"Benefit-cost analysis can play a very important role in legislative and policy debates on improving the environment, health, and safety. It can help illustrate the tradeoffs that are inherent in public policymaking as well as make those tradeoffs more transparent. It can also help agencies set regulatory priorities.....

Benefits and costs of proposed major regulations should be quantified wherever possible. Best estimates should be presented along with a description of the uncertainties. Not all benefits or costs can be easily quantified, much less translated into dollar terms. Nevertheless, even qualitative descriptions of the pros and cons associated with a contemplated action can be helpful. Care should be taken to assure that quantitative factors do not dominate important qualitative factors in decision-making." (Hahn et al. 1996, pp. 1–2)

In this chapter, I describe and contrast the use of BCA in regard to (1) a public decision to promote the reduction of foodborne illness and (2) a private decision to evaluate the investment decision associated with entering a new market. The comparison of the two examples in the meat and poultry industry highlights the differences in BCA associated with a private decision relative to a public decision-making process.

4.2   Public Decision-Making for Pathogen Control under HACCP

The 1993 Jack in the Box outbreak of *E. coli* O157:H7 contamination of burgers was the catalyst for Federal government action proposing the HACCP regulations. Parents exerted political pressure on companies via legal action and on members of congress and regulators with their protests about the deaths and illnesses of their children (see Chaps. 10, 16, and 17). In addition, a new president inhabited the White House and President Clinton sent his new Secretary of Agriculture, Mike Espy, to speak to the Washington state legislature where the outbreak was discovered. This political pressure and public outrage resulted in USDA's Food Safety and Inspection Service (FSIS) proposing a Federal program, HACCP, be applied to control pathogens in meat and poultry in 1995.

Why was a HACCP system chosen as a regulatory tool? Ever since the space program developed a high level of pathogen control in food for the astronauts, food scientists and microbiologists have studied and extended the science. Traditional methods of pathogen control in foods include drying, salting, curing, sugaring, heating, and cooling. Canning was an innovation during the Napoleonic wars. HACCP, however, more rigorously controlled the pathogen "kill step" during cooking, irradiation, or freeze-drying. In addition, prevention steps to reduce pathogen contamination of raw ingredients were included in HACCP.

The international community was exploring applications of HACCP to other parts of the food supply. In 1988, the International Commission on Microbiological Specifications for Food (ICMSF) endorsed the use of HACCP systems in food production, processing, and handling in its report, "HACCP in Microbiological Safety and Quality". In 1993, the Food and Agriculture Organization/World Health Organization Codex Alimentarius Commission adopted a HACCP document that still serves as a guide for countries to incorporate HACCP principles into their food industries.

The HACCP system is science-based and uses a systematic approach to identify of specific hazards and measures for their control or prevention to ensure the safety of food. The preventive measures must be described in detail, and people who execute them must be trained. HACCP involves careful recording of all details and actions to provide documentation that the system is in operation and in full control of all hazards in food processing. The Codex Alimentarius Commission defined the 7 principles and 12 steps that must be applied during the development of the HACCP plan and in the implementation of the HACCP system (Codex Alimentarius Commission 1997). The seven principles of HACCP for food production and processing are:

1. Conduct a hazard analysis.
2. Determine the Critical Control Points (CCPs).
3. Establish critical limit(s).
4. Establish a system to monitor control of each CCP.
5. Establish the corrective action to be taken when monitoring indicates that a particular CCP is not under control.

6. Establish verification procedures to confirm that the HACCP system is working effectively.
7. Establish documentation concerning all procedures and records appropriate to these principles and their application.

The 7 basic principles are implemented into the system through the 12 steps:

1. Assemble HACCP team.
2. Describe product.
3. Identify intended use.
4. Construct a flow diagram.
5. On-site confirmation of flow diagram.
6. List all potential hazards associated with each step, conduct a hazard analysis, and consider any measures to control identified hazards (Principle 1).
7. Determine Critical Control Points (Principle 2).
8. Establish critical limits for each CCP (Principle 3).
9. Establish a monitoring system for each CCP (Principle 4).
10. Establish corrective actions (Principle 5).
11. Establish verification procedures (Principle 6).
12. Establish documentation and record keeping (Principle 7).

The industry costs of compliance with the HACCP regulation will vary with each company, depending on the company's comparative advantage in achieving pathogen control and prevention (Roberts 2005). For example, does the company already have a vice president for pathogen control with a college degree in quality assurance? Does the company maintain a database on pathogen testing in its supply chain and know which pathogens occur with the company's food products? Does the company have knowledge of feasible control options? Does the company have an edge on technological innovations and could invent better solutions for pathogen control in slaughter and/or processing plants (see Chap. 10)? Does the company already have a HACCP system in place?

HACCP is the scientific and management/regulatory response to the problem of meat and poultry companies' needing an economic incentive to reduce pathogens in the products they sell. From an economist's perspective, Federal regulations must establish that the private marketplace fails to provide the significant level of protection, in this case the protection against foodborne illness (OMB 1996). The specific kind of market failure in this case is an externality. "An externality occurs when one party's actions impose uncompensated benefits or costs on another" (p. 6, (OMB 1996)). Foodborne illness victims are very rarely compensated for the costs of their foodborne illnesses, deaths, and/or long-term health outcomes (LTHOs).

As discussed in Chap. 2, the probability of linking an acute illness to a food and company in the United States is limited to an outbreak investigated by Centers for Disease Control and Prevention (CDC) and is estimated at 1 case out of every 1,000 cases of foodborne illnesses that occur, meaning that 999/1,000 cases cannot be linked in the United States. This linkage is needed before compensation will be paid by the company (see Chap. 17).

When uncompensated damages are imposed upon private citizens by the marketplace, this is called a "market failure" or "externality" in economic jargon. In the United States, Federal regulators must establish that citizens are harmed by unregulated private marketplace before regulations can be promulgated (OMB 1996). The method for estimating the benefits and costs of a regulatory proposal is the benefit/cost analysis (BCA).

4.3   Use of BCA in HACCP

Approved for meat and poultry in 1996, USDA's HACCP program was accompanied by a BCA as required for significant regulations estimated to cost over $100 million (Food Safety and Inspection Service/USDA 1996). In the prospective HACCP/Pathogen Reduction (PR) case, the question asked in the BCA is how do the public health protection benefits stack up against the costs to the beef, poultry, and pork industries of complying with the pathogen control regulations? The information requirements are extensive and must include information (or assumptions) on the incidence of foodborne illness in the United States, in which foodborne pathogens are associated with specific health outcomes in the short run (acute) and long run, as well as the economic estimates of the costs of each illness.

USDA's Economic Research Service (ERS) has a long history of estimating the societal costs of foodborne illness. In 1994, Roberts and Unnevehr published medical costs and productivity losses for four bacterial pathogens (*Salmonella*, *Campylobacter jejuni* or *Campylobacter coli*, *Escherichia coli* O157:H7, and *Listeria monocytogenes*) and four parasites (*Toxoplasma gondii*, *Trichinella spiralis*, *Taenia saginata*, and *Taenia solium*) (Roberts and Unnevehr 1994). ERS updated these estimates and published a detailed report to provide analytical support for USDA's HACPP regulation for meat and poultry (Buzby et al. 1996).

Three principles guide economic analysis of regulations aimed at improving health and safety (Roberts and Unnevehr 1994). The first is that benefits from the regulation need to be measured and compared with costs, because regulatory costs are opportunity costs. That is, the resources used could have been applied elsewhere, with potentially greater health benefits. For example, an expenditure of $100 million that is expected to prevent four deaths may not be very sensible if that $100 million could have prevented 50 deaths by being spent in another application. The second principle asserts that health and safety regulations typically do not aim to save the lives of specific people who would otherwise die, but rather aim at reducing the level of risk of illness and death faced by large populations. That view intertwines with the third principle: the benefits of a regulation do not represent the value of keeping a specific person alive, but rather the value of reducing those risks. The most theoretically appropriate way to value a risk reduction is to ask what affected individuals are willing to pay for it. Analysis of risk in the labor market by Viscusi (Viscusi 1993) has successfully estimated a value of a statistical life for the risk of death. Another method, the cost of illness, sums the estimated medical costs and lost earnings due to the illness.

Taken together, the three principles recognize that regulators aim to act on behalf of society to reduce societal risk by spending taxpayers' money for setting and enforcing regulations. Once it is recognized that regulation delivers an outcome (small risk reductions) that people may also purchase in other public and private venues, one can ask whether publicly delivered risk reductions appear to be worth it to the relevant populations. For example, people spend their own money to reduce health risks when they have regular physician visits, when choosing among different cars with different safety records, when choosing among risky jobs, or when engaging in activities to reduce their risk of a foodborne illness.

In considering the FSIS BCA, ERS identified three types of societal costs often evaluated in BCA: costs incurred by ill individuals/households, industry, and the regulatory and public health sector (Table 4.1). Often, traditional cost-of-illness analyses are low estimates that include only individual/household's medical costs and cost of lost productivity. The costs of food safety regulation include industry and government expenditures for the design and implementation of, and compliance with, such programs. In fiscal year 1994, the Federal government budgeted $1.2 billion on food safety regulatory activities, such as inspection and laboratory testing. The food industry also incurs millions of dollars of expense to comply with food

Table 4.1 Societal costs of foodborne illness for public decisions

Costs to individuals/households
Human illness costs
Medical costs
Physician visits
Laboratory costs
Hospitalization or nursing home
Drugs and other medications
Ambulance or other travel costs
Income or productivity loss for
Ill person or person dying
Caregiver for ill person
Other illness costs
Travel costs to visit ill person
Home modifications
Vocational/physical rehabilitation
Child care costs
Special educational programs
Institutional care
Lost leisure time
Psychological (psychic) costs
Pain and other psychological suffering
Risk aversion
Averting behavior costs
Extra cleaning/cooking time costs
Extra cost of refrigerator, freezer, etc.
Flavor changes from traditional recipes (especially meat, milk, egg dishes)
Increased food cost when more expensive but safer foods are purchased
Altruism (willingness to pay for others to avoid illness)
Industry costs

(continued)

Table 4.1 (continued)

Costs of animal production
  Morbidity and mortality of animals on farms
  Reduced growth rate/feed efficiency and increased time to market
  Costs of disposal of contaminated animals on farm and at slaughterhouse
  Increased trimming or reworking at slaughterhouse and processing plant
  Illness among workers because of handling contaminated animals or products
  Increased meat product spoilage due to pathogen contamination
Control costs for pathogens at all links in the food chain
  New farm practices (age-segregated housing, sterilized feed, etc.)
  Altered animal transport and marketing patterns (animal identification, feeding/watering)
  New slaughterhouse procedures (hide wash, knife sterilization, carcass sterilizing)
  New processing procedures (pathogen tests, contract purchasing requirements)
  Altered product transport (increased use of time/temperature indicators)
  New wholesale/retail practices (pathogen tests, employee training, procedures)
  Risk assessment modeling by industry for all links in the food chain
  Price incentives for pathogen-reduced product at each link in the food chain
Outbreak costs
  Herd slaughter/product recall
  Plant closings and cleanup
  Regulatory fines
  Product liability suits from consumers and other firms
  Reduced product demand because of outbreak
      Generic animal product—all firms affected
      Reduction for specific firm at wholesale or retail level
  Increased advertising or consumer assurances following outbreak

Regulatory and public health sector costs for foodborne pathogens

Disease surveillance costs to
  Monitor incidence/severity of human disease by foodborne pathogens
  Monitor pathogen incidence in the food chain
  Develop integrated database from farm to table for foodborne pathogens
Research to
  Identify new foodborne pathogens for acute and chronic human illnesses
  Establish high-risk products and production and consumption practices
  Identify which consumers are at high-risk for which pathogens
  Develop cheaper and faster pathogen tests
  Risk assessment modeling for all links in the food chain
Outbreak costs
  Costs of investigating outbreak
  Testing to contain an outbreak (e.g., serum testing and administration of immunoglobulin in persons exposed to hepatitis A)
  Costs of cleanup
  Legal suits to enforce regulations that may have been violated
Other considerations
  Distributional effects in different regions, industries, etc.
  Equity considerations, such as special concern for children

Source: Economic Research Service/USDA Bacterial Foodborne Disease: Medical Costs and Productivity Losses/AER-741 9

Table 4.2 Annual societal medical costs and productivity losses estimated for four foodborne pathogens in meat and poultry, 1993

Bacterial pathogen	Number of illnesses	Number of deaths	Societal costs for meat and poultry (billion $)	90% attributable to manufacturing (billion $)
Campylobacter coli or jejuni	1,031,250–1,312,500	83–383	0.5–0.8	0.45–0.72
Escherichia coli O157:H7	6000–12,000	120–130	0.2–0.5	0.18–0.45
Listeria monocytogenes	763–884	189–243	0.1–0.2	0.09–0.18
Salmonella	348,000–2,880,000	348–2,880	0.3–2.6	0.27–2.3
Total	1.4–4.2 million		1.1–4.1	0.99–3.69[a]

Adapted from Ref: FR, p. 38964 Federal Register/Vol. 61, No. 144/Thursday, July 25, 1996/Rules and Regulations
https://www.fsis.usda.gov/wps/wcm/connect/e113b15a-837c-46af-8303-73f7c11fb666/93-016F.pdf?MOD=AJPERES
[a]Note: Total does not add due to rounding

safety rules and regulations. If new regulations are added to the current system, industry compliance costs will be higher. Societal benefits of food safety regulation arise from improvement of individuals' health status. From an economic perspective, these benefits include, at least, savings in disease prevention and mitigation expenditures, increases in worker productivity, and reduction in pain and suffering.

ERS cost estimates included in the HACCP final regulation represent the maximum benefits that could be obtained if the microbial infections or intoxications were eliminated for these pathogens (Table 4.2). To estimate medical costs and productivity losses, ERS relied on the US Centers for Disease Control and Prevention (CDC) for estimates of acute illness, deaths, and the percentage foodborne for each pathogen. ERS used four severity categories for estimating the costs of acute illnesses: those who did not visit a physician, visited a physician, were hospitalized, or died prematurely (see (Buzby et al. 1996) for the details).

The long-term health outcomes (LTHOs) are included in the cost estimates for *E. coli* O157:H7 (hemolytic uremic syndrome (HUS) and kidney failure), fetal listeriosis (brain damage), and congenital toxoplasmosis (mental retardation and loss of vision). ERS relied on collaborations with medical doctors and academicians as well as the medical literature for an understanding of the lifetime consequences of these foodborne illnesses. (Note that Part II of this book is devoted to more recent estimates of the costs of foodborne illness, including an increased understanding of the many LTHOs.)

ERS estimated the societal costs of foodborne illness at $4.5–7.5 billion annually (1993) in the United States (Food Safety and Inspection Service/USDA 1996, p. 38964); however FSIS considered that only four pathogens would be reduced by the HACCP rule. The public health protection benefits of HACCP are based on reducing the risk of foodborne illness due to these bacterial pathogens: *Campylobacter jejuni/coli*, *Escherichia coli* O157:H7, *Listeria monocytogenes*, and

Table 4.3 Summary of estimated annual industry compliance costs for HACCP [$ Thousands]

Cost category	Year 1	Year 2	Year 3	Year 4	Year 5+
I. Sanitation SOP's					
Plans and training	2,992				
Observation and recording	8,345	16,691	16,691	16,691	16,691
II. E. coli sampling					
Plans and training	2,627				
Collection and analysis	8,716	16,122	16,122	16,122	16,122
Record review	406	752	752	752	752
III. Compliance with standards					
Salmonella and generic *E. coli*		5,472–16,899	5,353–25,753	5,811–26,079	5,811–26,079
IV. HACCP					
Plan development		3,769	27,755	35,464	
Annual plan reassessment			69	448	1,179
Initial training		1,270	8,284	18,435	
Recurring training		64	542	1,877	2,799
Recordkeeping (recording, reviewing, and storing)		3050	18,479	42,478	54,097
V. Additional overtime		189	837	1,711	2,125
Total	23,086	47,379–58,806	94,884–115,284	139,789–159,934	99,576–119,844

Ref: FR, p. 38956 Federal Register/Vol. 61, No. 144/Thursday, July 25, 1996/Rules and Regulations

Salmonella. These four pathogens are the cause of 1.4–4.2 million cases of food-borne illness each year (Table 4.2). FSIS estimated that 90% of these cases were caused by contamination occurring at the manufacturing stage that can be addressed by improved process control in the slaughterhouse and processing plants. These four pathogens cost the US society from $0.99 to $3.69 billion each year. The high and low range occurs because of the uncertainty in the estimates of the number of cases of foodborne illness and death attributable to the four pathogens.

FSIS estimated that implementation of HACCP by industry would take 5 years before the full public health protection would take place. In year 5, the annual industry costs are estimated at $100–120 million annually (Table 4.3). Recording keeping is over half of the estimated costs, followed by actions to meet the *Salmonella* performance standard, sanitation, and training. In Table 4.4, FSIS calculated the present value of these industry costs using a discount rate of 7%. Note that these industry costs did not vary with the effectiveness of reducing pathogens, since FSIS did not know with certainty the effectiveness of the proposed HACCP requirements in reducing foodborne illness.

Hence, FSIS calculated the projected health benefits for a range of effectiveness levels, where effectiveness refers to the percentage of pathogens eliminated at the manufacturing stage. The link between effectiveness in reducing pathogen levels in manufacturing and health benefits is assumed to be proportionate. Because of the wide range in estimates for the number of foodborne illness cases, each effectiveness level had a low and high estimate for the societal health benefits. These esti-

Table 4.4 Present value of 20-year costs and benefits for HACCP ($ billions)

Effectiveness in reducing pathogens (%)	Public health benefits (billion $)		Industry costs (billion $)
	Low	High	
10	0.71	2.66	0.97–1.16
20	1.43	5.32	0.97–1.16
30	2.14	7.98	0.97–1.16
40	2.85	10.64	0.97–1.16
50	3.57	13.30	0.97–1.16
60	4.28	15.96	0.97–1.16
70	4.99	18.61	0.97–1.16
80	5.71	21.27	0.97–1.16
90	6.42	23.93	0.97–1.16
100	7.13	26.59	0.97–1.16

Note: Analysis assumes zero benefits until year 5. All elements of the HACCP-based program will be in place 42 months after publication of the final rule
Ref: FR, p. 38956 Federal Register/Vol. 61, No. 144/Thursday, July 25, 1996/Rules and Regulations

mates of societal health benefits are shown in Table 4.4, as the present value of a 20-year benefit stream. FSIS concluded that even at a low rate of 20% effectiveness in reducing these four pathogens, the public health protection benefits are greater than the costs. (For more detail on both costs and benefits plus a sensitivity analysis of the estimated public health protection benefits, see (Crutchfield et al. 1997).

This HACCP proposal's goal was to reduce the actual risk of pathogens associated with meat and poultry and thereby improve the health of the US population. What has been the impact of HACCP for meat and poultry on the foodborne illnesses tracked by FoodNet in CDC (CDC 2017)? When the 1996–1998 CDC data is compared with the latest CDC data of 2015, here are the results:

- *Campylobacter* cases declined by 26%.
- *Listeria* cases declined by 45%.
- *E. coli* O157 cases declined by 44%.
- *Salmonella* cases declined only 4%.

This is a simplistic comparison, but the poor performance in controlling *Salmonella,* a main target of HACCP, suggests that there is more to be done to reduce the incidence of foodborne disease in the United States. And FSIS is currently implementing new pathogen testing rules for both *Salmonella* and *Campylobacter* in poultry (see Chap. 12).

4.4 BCA for Private Investment in Chinese Chicken Production

Private decision-making can be complex or simple, depending on the objectives. If it is a major decision, such as deciding to invest in poultry production in another country, then the differences in the liability laws and even the environmental laws may be important considerations, to say nothing of the complexities of selling chicken products in a different culture. In contrast to the use of BCA in the public context, in the private decision-making situation, benefits and costs are limited to those affecting the firm in the relevant marketplace. To better understand this, we consider the case of the Chinese chicken market. First, it is useful to address three questions to better understand the context of the Chinese chicken industry: What is happening in Chinese households? What is happening in Chinese industry? And what is happening in Chinese government regulations?

What is happening in Chinese households? The population of China is the largest in the world at 1.4 billion. Household income is increasing, and consumption of animal protein products is increasing. In the 1960s meat, poultry, and offal consumption provided only 4% of calories for the Chinese people, growing to 12.1% for meat (and offal) and 1.7% for poultry in the 2000s (Ortega et al. 2015). The Chinese diet continues to include even more animal protein, due to increased incomes, industrialization, and urbanization. Quick serve restaurants, such as KFC (which now has more restaurants located in China than in the United States), McDonalds, and Pizza Hut, are popular in China (Pant 2017).

China has close ties to the rest of Asia, and many of these countries are also experiencing increased prosperity and income growth. For example, India is expected to pass China's population in the next couple decades. In summary, Asia is a dynamic region whose countries have the largest populations in the world, and they are moving into the modern industrialized world. All these households present a growing market.

Another consideration is that chicken by-products (feet, combs, intestines, etc.) are used in traditional Chinese dishes and herbal medicine. It is easier to provide these by-products to Chinese customers in a fresh condition, if the birds are grown in China. Since US consumers eat more chicken breast than other parts, it may be cheaper to send the frozen chicken breasts back to the United States and use the rest of the bird in products sold to the Chinese market. Finally, US poultry sold in China enjoys "a reputation of being safe and of high quality" (Ortega et al. 2015) and is possibly sold at a higher price.

What is happening in Chinese industry? China is a mix of modern and traditional industry with significantly different cost structures. For example, the costs of grow-out for white-feather broilers in an industrialized confinement system is cheaper than traditionally raised yellow-feather chicken in China, due largely to their different genetics. Two production variables in Table 4.5 dramatize these lower costs: (1) average harvest time and weight and (2) average placement density. The time to harvest for industrialized white-feather chicken is 40 days vs. 90 days for yellow-feather Chinese chicken raised outside. This means that twice as many industrialized flocks can be produced each year. In addition, each industrialized flock produces twice as much meat in a year since these industrial chickens are more densely housed and are harvested at a heavier weight. Together this means that a given sized industrial chicken house can produce *four* times as much chicken meat as the same sized traditional Chinese outdoor facility for chicken. The table also shows that the feed costs are lower per pound of chicken in the industrialized facility, another important factor since feed costs are roughly half of the costs of production for chicken meat. While the cost of the industrialized confinement facility needs to be

Table 4.5 Production variables for white- vs. yellow-feathered chicken in China

Production variable	White-feathered chicken	Yellow-feathered chicken
Origin	90% of grandparent stock imported	Local Chinese species
Ave. harvest time and weight	40 days at 2.2–2.7 kg	90 days at 1.5–1.9 kg
Ave. placement density	10–16 birds/m^2	8–10 birds/m^2
Feed to meat ratio (kg)	1.7–1.8:1	2.5:1
Immune system	Weak	Robust
Consumption outlets	Fast food restaurants, factory cafeterias,s chools, supermarkets, and food processing	Purchased as live chickens in traditional wet markets for consumption at home
Meat quality, flavor, and avian influenza (AI) risk	Mild flavor, may have white ribbons in breast meat, low risk for AI because raised indoor	Best flavor, high risk of AI because raised outdoors and sold live in wet markets

Source: Adapted from Anderson and Inouye (2017)

factored in and the cost amortized over the length of time the facility will be used, the industrialized broiler meat will be cheaper to produce and offer cheaper prices to Chinese consumers.

One poultry company, Tyson, has long had an interest in chicken genetics and a more recent interest in China. In 1974 Tyson bought the Vantress breeding lines from Cobb, the oldest poultry breeding company in the world (Table 4.6). This assured Tyson of advanced genetics for its industrialized chicken production. But to keep up with breeding genetics, in 1994 it bought 100% of Cobb's stock from the US pharmaceutical company, Upjohn. In 2007 Tyson-owned Cobb-Vantress announced an alliance with Hendrix Genetics.

On the China front, in 2001 Tyson International Holding Co. collaborated with Shandong Zhucheng Waimao Co., Ltd. to build the first modern food processing plants in China. One of the plants was developing partially cooked chicken products, presumably for sale as frozen chicken nuggets, strips, and patties or in frozen chicken meals. The chicken nuggets are manufactured to the specifications (% chicken white or dark meat, % fat, % chicken skin, % mechanically separated chicken, etc.) of the buyer such as KFC or McDonalds. Recently, the partially cooked technology has been used in frozen meals sold in the United States, such as chicken Kiev and chicken cordon bleu, and resulted in foodborne illness outbreaks because the breading is partially cooked, yet appears fully cooked, thereby confusing consumers (Fig. 4.1, Barber Foods). In 2014, the alliance of Tyson's Cobb-Vantress with Hendrix Genetics broke ground in China for a new grandparent operation to produce chicks for grow-out operations in China.

Table 4.6 Tysons/Cobb timeline: investments in broiler genetics and the Chinese market

1916 – Cobb founded in Massachusetts as the world's oldest poultry breeding company
1960 – First shipment of grandparent stock sent overseas to Cobb Breeding Company in the United Kingdom (followed by Argentina, Africa, Spain, Korea, Germany and Eastern Europe, Brazil, South Africa, Russia, Bangladesh, Indonesia, the Philippines, and Sri Lanka)
1974 – Cobb is purchased by the Upjohn Company. Tyson Foods, Inc. acquires the Vantress breeding lines
1983 – Cobb-Vantress is formed as a joint venture between Tyson Foods, Inc. and the Upjohn Company and positions Cobb as an international leader in poultry breeding. New research complex is developed in Jane, Missouri
1994 – Tyson Foods, Inc. acquires 100% of Cobb's stock from the Upjohn Company
2001 – Collaboration between Tyson International Holdings Co., Ltd. and Shandong Zhucheng Waimao Co., Ltd. is the 1st modern food processing company in China. The operation has two cutting edge processing plants and a separate marinating plant. The technology is at the forefront of implementing advanced technology and methods to develop and produce par-fried chicken products. Example is Fig. 4.1
2007 – Alliance announced between Cobb-Vantress and Hendrix Genetics
2008 – Jiangsu Tyson Foods Co. Ltd. founded near Shanghai. The company operates the entire live production, including breeder production, hatchery, broiler, and feed production as well as processing facilities
2011 – Tyson Foods assumes full ownership of the three plants in Shandong, which serve quick-service restaurants and retail outlets
2014 – Cobb breaks ground in China for a new grandparent operation

Sources: About Tyson China, available @ www.tysonfoods.com, accessed May 2017 (Tyson website 2017)

How Cobb has become world leader, available @ www.cobb-vantress.com, accessed Jan 2017 (Cobb-Vantress website 2017)

Fig. 4.1 Example of a par-cooked chicken product that is breaded but *not* fully cooked and that caused a US outbreak

Chinese labor may be cheaper than US labor. That may be true in China's interior, but workers in the coastal areas have had steadily rising wages. Another input available perhaps at lower cost is chicken feed. China scours the seas in large ships and harvests fish to turn into fish meal sold for broiler feed (Jacobs 2017). Perhaps the fish meal will be cheaper in China than US chicken feed which is largely corn and soybean meal.

What is happening in Chinese government regulations? The Chinese government has started a major infrastructure project that will build new highways into the west of the country where labor costs are lower. This will encourage some of the broiler production to move further inland. And given the new roads, processed chicken will be able to reach markets faster in the more populous coastal areas as well as the seaports for shipping to other markets.

China also has fewer food safety and environmental laws. The food safety scandals in China as well as in exports to the United States are notorious, for example, the melamine contamination of milk and the pet food treats that killed the dogs. Even with fewer laws on the books, lax enforcement is an even bigger problem. As discussed in Chap. 15, production of raw ingredients for antibiotics has contaminated much of the water supply in China. Much of the land is contaminated with human waste/animal manure or chemicals (Calvin and Gale 2006). Worker safety laws are also weaker, which means fewer lawsuits filed by farmers against big poultry corporations. Legal liability laws are also lax in China compared to the United States.

The Chinese government has closed many live bird (wet) markets where yellow-feather chickens are sold and where the risk of human illness from avian influenza (AI) that cycles through birds, hogs, and people in these markets is higher. Instead China is encouraging its population to buy packaged poultry sold in grocery stores. Chicken from the Chinese confinement operations that is put into the supermarket packages has a much lower chance of AI. The yellow-feather, traditional Chinese birds, however, are generally considered to have a better flavor than industrial white-feather birds.

The US Federal Government recently opened the market to Chinese broilers as being "equivalent" to the United States. Although raw chicken would need to be labelled "Made in China," processed chicken can be sold in the United States without any label which opens the market to processed chicken nuggets as well as frozen chicken patties and frozen entrees.

To perform a benefit/cost analysis, the private company (1) identifies the project to be evaluated and then (2) identifies the most important variables (discussed above) that are expected to affect either the production costs or the price and quantity of chicken products likely to be sold. The level of detail in estimating the benefits and costs, as well as the time horizon of when the investment is expected to be profitable, also needs to be identified.

In 2016, Tyson Foods produced 23.3% of US ready-to-cook chicken, followed by Pilgrim's Pride at 18.8%, Sanderson Farms at 8.7%, and Purdue Foods at 8.2% (Alonzo 2016). Another indication of market power is industry expenditures on lobbying: in 2010, Tyson Foods spent $2.69 million on lobbying, outspending the other big meat companies (Leonard 2014, p. 286).

The "revolving door" between industry and regulators also continues in China policies: the top regulator in FSIS/USDA, Al Almanza, retired on July 31, 2017;

Table 4.7 Timeline of JBS' purchases of meat and poultry companies around the world

1953 – JBS opens butcher shop in west Central Brazil
2001–5 – JBS expands meat slaughter operations in Brazil and buys Swift-American capital, Argentina's largest beef producer and exporter
2007 – JBS buys Swift & Co. in US and Australian beef, pork, and lamb markets
2008 – JBS buys Smithfield and Five Rivers Cattle Feeding, making JBS the largest cattle feeder in the world
2009 – JBS buys controlling interest in Pilgrim's Pride, Corp., the 2nd largest US poultry company with operations also in Puerto Rico and Mexico
2010 – JBS buys Tatiara Meats in Australia as well as McElhaney Feedlot in Arizona, USA
2013 – JBS buys XL Foods beef processing plant and Lakeside Feeder feed yard in Alberta, Canada
2015 – JBS buys Moy Park poultry producer in the United Kingdom and Cargill's pork business in the USA
In 2017 JBS is the world's largest animal protein producer/exporter serving more than 300,000 customers in more than 150 nations
2017 – Pilgrim's Pride buys GNP in the United States selling premium branded chicken products. From Denmark company Plumrose, JBS buys Plumrose US, selling bacon, ham, and deli meat
2017 – Al Amanza, former head of the Food Safety and Inspection Service, USDA, joins JBS as Global Head of Food Safety and Quality Assurance. He will be tasked with maintaining and expanding access to export markets globally
2017 – JBS is mired in corruption scandals in Brazil, including an allegation that Brazil's Federal bank gave "the company favorable loan terms starting in June 2007 to acquire other meat companies around the world"

References: JBS website: http://jbssa.com/about/history/. Accessed August 2017; Runyon L. JBS, world's largest meat company, mired in multiple corruption Scandals in Brazil. KUNC Fresh Air, Aug 3, 2017, http://kunc.convio.net/form

JBS (the world's largest animal protein producer) immediately hired him and sent out a press release on August 3, 2017 (Table 4.7) stating that Almanza was hired as "Global Head of Food Safety and Quality Assurance." One of Almanza's last actions at USDA was to tentatively determine that China's "… poultry slaughter inspection system is equivalent to the system that the United States has established under the Poultry Products Inspection Act (PPIA)" (FSIS/USDA 2017). The Chinese chicken ruling has another provision that opened exports of beef to China, another benefit to global animal protein producers. In summary, by producing broilers in China, Tysons opens this new consumer market, spreads its price risk across global markets, and helps to "stabilize" world poultry prices, possibly at higher levels.

4.5 Conclusion

Food safety regulations only exist because consumers lack information on whether they will become ill when eating a food item. If the consumer could see pathogens, foodborne illness would not be a problem since no company could stay in business selling contaminated food. In contrast, industry has good information on the level of pathogen control in its food items. If it lacks sufficient information, a company can establish a testing and monitoring system and build this database. But governments have scarce resources for it (Roberts 2013). Industry does not encourage government to allocate resources to building this database, e.g., lack of support for traceability. Companies do not want to increase the probability of being caught causing illness and to be forced to pay damages to consumers who are sickened.

The major restaurant chains, however, do not want to risk their reputation by buying and selling products produced under highly suspect conditions, thereby risking their brand reputation. Chipotle provides a vivid example of the damage that can be caused to a brand through foodborne illness outbreaks (see Chap. 17). Yet, the question remains, how many years did Chipotle get away with causing illness before the company got caught? This question is especially relevant since only 1 out of 1,000 cases of foodborne illnesses are linked to the causative food and company. This low level of 1/1,000 identification of foodborne illnesses means that companies have a low detection rate and low incentive to provide completely safe food. Rather, the companies choose a moderate level of food safety as the goal.

Given the role of the Federal government in food safety regulations, the primary method of weighing the public health benefits of regulation with the costs of industry providing more food safety is discussed. Benefit/cost analysis is the tool used by many governments, one that requires thoughtful use to make sure that the important variables are identified and quantified to the extent possible. Transparency is crucial, so that others can understand the assumptions made in the analysis, the general direction of bias in the estimates (underestimate or overestimate), and technical details such as the choice of discount rate (interest rate used to discount future benefits and costs to a present value). A poorly done BCA can be detected by seeing if important variables are left out of the analysis, by examining the assumptions, by looking at the level of detail and care in the analysis, and by reflecting on the aims of the BCA.

In this chapter, the juxtaposition of a public vs. a private BCA illustrates the different goals and different variables used in each BCA. The HACCP example illustrates how the government can evaluate its protection of the public vs. imposing costs on industry. The Chinese chicken example showcases how an important variable in HACCP, public health protection, may be of minor importance in a private sector BCA. In fact, the lax environmental laws in China vs. the United States can be a positive reason to invest in China, since the probability of being sued is almost nonexistent.

BCA can be prospective as the examples discussed here or retrospective. In Chap. 11, New Zealand examines the benefits and costs of a *Campylobacter* regulatory program retrospectively to determine the success/failure of the regulation.

Acknowledgments Many thanks to review comments from Mary Ahearn, Walter Armbruster, and Diogo M. Souza Monteiro. All omissions and misstatements, of course, remain my errors.

References

Alonzo A. Top 5 broiler producers dominate U.S. production, Poultry International, July 2016.

Anderson L, Inouye A. China – Peoples Republic of poultry and products semiannual 2017 – Chinese year of the rooster, GAIN report number: CH17005, USDA/Foreign Agricultural Service, 2017 Feb 1. Available at https://gain.fas.usda.gov.

Buzby JC, Roberts T, Lin C-T J, MacDonald JM. Bacterial foodborne disease: medical costs and productivity losses, USDA/ERS, AER-741. 1996 Available at www.ers.gov.

Calvin L, Gale F, Food safety improvements underway in China, Amber Waves, ERS/USDA, Nov 2006. Available at www.ers.com.

CDC. Foodborne Diseases Active Surveillance Network (FoodNet): FoodNet 2015 Surveillance Report (Final Data). Atlanta, GA: U.S. Department of Health and Human Services, CDC; 2017.

Cobb-Vantress website, How Cobb has become world leader. Available at www.cobb-vantress.com. Accessed Jan 2017.

Codex Alimentarius Commission. Hazard Analysis and Critical Control Point (HACCP) system and guidelines for its application. 1997. http://www.fao.org/docrep/005/y1579e/y1579e03.htm

Crutchfield S, Buzby JC, Ollinger M, Lin C-TJ, An economic assessment of food safety regulation: the new approach to meat and poultry inspection. ERS/USDA, AER-755, 1997.

Ekelund RB, Herbert RF, Jules Dupuit. Engineer and Economics, Public lecture at the University of Montreal, 2000 Mar 24. Available at http://frenchinfluence.over-blog.fr/article-jules-dupuit-engineer-and-economist-74661116.html

Food Safety and Inspection Service/USDA. Pathogen reduction; hazard analysis and critical control point (HACCP) systems; final rule. Federal Register. Washington, DC: USGPO. 1996;61(144):38805–989. https://www.fsis.usda.gov/wps/wcm/connect/e113b15a-837c-46af-8303-73f7c11fb666/93-016F.pdf?MOD=AJPERES

FSIS/USDA. FSIS announces proposed rule on eligibility of the People's Republic of China to export poultry products to the United States, FSIS constituent update, June 16, 2017. https://www.fsis.usda.gov/wps/wcm/connect/eb673f22-1182-4c02-814b-a11a29495b56/ConstiUpdate061617.pdf?MOD=AJPERES&CONVERT

Hahn RW, Noll R, Stavins R, Schmalensee RL, Lave LB, Portney PR, et al. Benefit-cost analysis in environmental, health, and safety regulation: a statement of principles. Washington, DC: AEI Press; 1996.

Jacobs A. China's appetite pushes world's fish stocks to the brink, New York Times, 2017 April 29. Available at https://www.nytimes.com/2017/04/30/world/asia/chinas-appetite-pushes-fisheries-to-the-brink.html?_r=0

JBS website.: http://jbssa.com/about/history/. Accessed Aug 2017.

Leonard C. The meat racket. New York: Simon & Schuster; 2014.

Office of Management and Budget (OMB). Economic analysis of Federal regulations under Executive Order 12866, January 11, 1996. Available at https://www.whitehouse.gov/omb/inforeg_riaguide. Accessed 8 Nov 2017.

Ortega DL, Wang HH, Chen M. Emerging markets for U.S. meat and poultry in China. Choices. 2015;30(2):1–4. Available at https://www.aaea.org/publications/choices-magazine

Pant M. We are ready for the next thirty years, CEO speech at the annual meeting of Yum China stockholders, 2017. Available at www.yumchina.com

Roberts T. Economics of private strategies to control foodborne pathogens. Choices. 2005;20(2):117–22. www.choicesmagazine.org/2005-2/safety/2005-2-05.htm

Roberts T. Lack of information is the root of U.S. Foodborne illness risk. Choices 2013;28(2). http://www.choicesmagazine.org/UserFiles/file/cmsarticle_300.pdf

Roberts T, Unnevehr L. New approaches to regulating food safety, FoodReview. USDA/ERS. 1994;17(2):2–8.

Runyon L. JBS, world's largest meat company, mired in multiple corruption scandals in Brazil. KUNC Fresh Air, Aug 3, 2017. www.kunc.org.

Tyson website, About Tyson China. Available at www.tysonfoods.com. Accessed May 2017.

Viscusi WK. The value of risks to life and health. J Econ Lit. 1993;31:1912–46.

Chapter 5
Economic Impact of Posting Restaurant Ratings: UK and US Experience

Derrick Jones

Abbreviations

DHS	Department of Health Services
FHIS	Food Hygiene Inspection Scheme
FHRS	Food Hygiene Ratings Scheme
FSA	Food Standards Agency/UK
FSS	Food Safety Scotland
LAs	Local Government Authorities/UK

5.1 Introduction

Improving food hygiene standards is important. In the UK, there are more than 500,000 cases of food poisoning each year traced to known pathogens. This figure would more than double if it included food poisoning cases from unknown pathogens (Food Standards Agency 2014). In the USA, foodborne pathogens cause an estimated 48 million cases of illnesses, 128,000 hospitalizations, and 3000 deaths each year. The Centers for Disease Control estimates that nearly 50% of foodborne disease outbreaks are connected to restaurants or other commercial food outlets (Marler Clark 2017).

Restaurant rating schemes have evolved to offer a potential contribution toward making food safer. They differ in their operational details and design, but share common features and the same general intention—to display a simple hygiene score for a restaurant, based on the results of an official hygiene inspection, in order that consumers have access to the information at the point where they are making decisions about where to eat.

D. Jones (✉)
Independent Economic Consultant, and former Chief Economist and Head of the
Analysis and Research Division at the UK Food Standards Agency (Retired), London, UK
e-mail: derrickjoneseconomics@gmail.com

© Springer International Publishing AG, part of Springer Nature 2018
T. Roberts (ed.), *Food Safety Economics*, Food Microbiology and Food Safety,
https://doi.org/10.1007/978-3-319-92138-9_5

5.2   Outline

The following section briefly outlines the economic theory related to the use of restaurant rating schemes and the potential benefits that flow from them. Subsequent sections of the chapter describe the introduction of rating schemes at a national level across the UK and how they have been evaluated. Further sections then discuss the experiences from the USA and evidence how these have been evaluated. Finally, some general lessons are set out.

5.3   Economic Theory

In terms of economic theory, ratings schemes are intended to address potential market failure (Economics Online 2017) caused by asymmetric information (World Bank 2003). Market failure does not necessarily mean a market fails to work entirely—rather that it is not working as well as it could. In this instance, asymmetric information arises where there is imperfect knowledge—in particular where one party in a potential transaction has different information to another and is able to exploit this advantage. Where the quality of what is being offered is uncertain, this can, over time, lead to only poorer-quality goods being traded, to the detriment both of buyers and to traders who are driven out of the market by less scrupulous competitors. In his much-cited paper, Akerlof (1970) uses used cars as an example of this phenomenon. However asymmetric information and quality uncertainty issues have a wide range of real-world applications and implications for how markets operate. For food businesses, food hygiene is an important attribute of the quality of their products. However, if the costs associated with ensuring high hygiene standards are difficult to signal to customers in the quality of their products, there is a risk that this may lead to pressures to lower hygiene standards—to the detriment both of customers and ultimately to businesses too, as prices will reflect the assumed lower quality of the products and drive out businesses who would otherwise be willing to produce and sell higher-quality products but are unable to do so. Traditionally this problem has been addressed by regulation and inspection regimes to ensure businesses meet legislated hygiene standards. By additionally giving consumers access to the results of such inspections, consumers are able to make informed choices about where they eat, as the hygiene standards achieved by each establishment at the time of inspection can be clearly seen and understood. The information is intended to be displayed at the restaurant but is also often readily available via the Internet. By posting restaurant ratings, the market failure resulting from quality uncertainty is reduced, and businesses are incentivized to maintain and improve hygiene standards. This should result, over time, in rising hygiene scores and, linked to this, to reductions in cases of foodborne disease acquired from food purchased at such food businesses. Changes in food hygiene scores for businesses can readily be tracked and compared over time, while changes in the number of foodborne disease cases are by their nature more difficult to measure.

The different forms of restaurant rating schemes in the UK and the USA and the attempts to measure and evaluate their impacts are discussed in the next section.

5.4   UK Experience with Posting Restaurant Ratings

The UK's Food Standards Agency has in the last few years introduced standardized food hygiene rating schemes, having learned valuable lessons from a wide variety of alternative schemes previously developed and adopted at a local level in the UK, together with the experiences of other countries, most noticeably from Denmark (Yu 2008) and the USA.

In the UK, the Food Hygiene Rating Scheme (FHRS) operates in England, Wales, and Northern Ireland and was formally launched in November 2010. Scotland operates a different scheme, the Food Hygiene Information Scheme (FHIS), which was piloted from November 2006 and subsequently rolled out in January 2009. The FHRS and FHIS were originally both introduced on a voluntary basis, meaning that local government authorities (LAs) were not legally compelled to implement the schemes and similarly display of the stickers by businesses was not a legal requirement. (Local hygiene inspections are carried out by local authorities in the UK. In 2015/2016 there were 419 UK LAs with responsibility for food controls: 354 in England, 22 in Wales, 11 in Northern Ireland, and 32 in Scotland). However, the schemes were very quickly adopted by local authorities, and some administrations have subsequently legislated to make the FHRS compulsory. Display of FHRS information has been a legal requirement in Wales since November 2013 and in Northern Ireland since October 2016. In England, the FSA is building a case toward making the FHRS compulsory. In Scotland, the FHIS continues to be run on a voluntary basis.

The schemes are a partnership between the Food Standards Agency—the Government Department responsible for food safety—and local authorities, who carry out food business inspections. The schemes provide consumers with information about hygiene standards in food premises at the time they are inspected to check compliance with legal requirements. The FHRS rating or FHIS result given to the business reflects the inspection findings. The transparency that the schemes provide enables consumers to make informed choices about where to eat out or shop for food and aims to incentivize businesses to improve hygiene standards.

Under the FHRS, businesses are given one of six ratings on a numerical scale from "5" (very good hygiene standards) at the top to "0" (urgent improvement required) at the bottom. Under the FHIS, businesses are given either a "Pass" result or an "Improvement required" result. Further information about the FHRS and FHIS is available on the FSA and Food Safety Scotland (FSS) websites (Food Standards Agency 2015c; Food Standards Scotland 2015).

Fig. 5.1 Illustrative FHRS
hygiene sticker showing a
"very good" (5) rating

An example of a posted score sticker for the FHRS (operating in England, Wales, and Northern Ireland) is given above in Fig. 5.1.

5.5   UK Evaluation Evidence on Restaurant Ratings

The FSA commissioned a number of independent evaluation and research studies to assess the impact of the introduction of the schemes on hygiene scores and also to seek the views of businesses, local authorities, and consumers (Feeney and Stewart 2015; NOP and Gfk 2011; Young and Gibbens 2012; Gibbens and Spencer 2013). A major evaluation study of FHRS and FHIS was commissioned in 2011 and ran until mid-2014. This explored the impact of FHRS and FHIS on local authorities, consumers, businesses, food hygiene compliance, and the incidence of foodborne disease (Vegeris 2015; Salis et al. 2015). Another study investigated the display of ratings at premises using annual surveys to establish the proportion of businesses voluntarily displaying FHRS ratings or FHIS Pass results at their premises. Telephone interviews with businesses were also conducted to explore the reasons for display/non-display and the perceived impact of display (Food Standards Agency 2015a). Consumer attitudes studies were also undertaken to explore consumer views on extending the scope of the schemes to include those involved in business-to-business trade and looked at what they wanted in terms of food hygiene information and how this should be presented (TNS BMRB 2012). A further study of consumer and small business attitudes in 2013 gathered views from consumers and small businesses on the impact of introducing mandatory display of FHRS ratings in Northern Ireland (TNS BMRB 2013). Consumer awareness, use, and recognition of FHRS have also been tracked through the Food and You surveys and through the Biannual Public Attitudes Tracker (Food Standards Agency 2017).

One of the FSA's aims in introducing the FHRS and FHIS was to provide an incentive for businesses to improve hygiene standards so that they comply with the requirements of food hygiene law. However, it was recognized that consumer awareness of the scheme would also potentially be an important driver to help incentivize hygiene improvements. Such improvements should also feed through to reducing the number of cases of food poisoning in the UK. Thus the major evaluation study looked at both the impact on hygiene scores and also evidence on changes in the number of officially reported cases of food poisoning. The following sections draw from the FHRS study findings and the nontechnical summary report, prepared by the FSA (Food Standards Agency 2015b).

The impact on the hygiene scores was investigated using data on levels of compliance with hygiene legislation in food businesses that is collected annually by the FSA from local authorities across the UK. Three measures were used: The proportion of "poorly compliant" premises (businesses that had compliance levels at the time of the last inspection equivalent to a FHRS rating of either 0 or 1); The proportion of "broadly compliant" premises (businesses that had compliance levels at the time of the last inspection equivalent to a FHRS rating of 3, 4, or 5); and the proportion of "fully compliant" premises (businesses that had compliance levels at the time of the last inspection equivalent to a FHRS rating of 5) (so "fully compliant" premises are a subset of those that are "broadly compliant").

The impact of the schemes on food poisoning was investigated through the number of formally notified food poisoning reports, confirmed Campylobacter laboratory Reports, and confirmed Salmonella laboratory reports. The number per million of population was calculated for each of these three reporting measures.

The impacts were assessed using a statistical technique known as difference-in-difference. This involved comparing data for two groups of local authorities: one group that had introduced the FHRS or FHIS and an equivalent group that had not. For hygiene standards, the evaluation compared changes in the proportion of "poorly compliant," "broadly complaint," and "fully complaint" businesses. For food poisoning, the change in the number of cases of food poisoning per million of population was compared. Comparisons were made using data for 2011/2012 and 2012/2013—1 year and 2 years after the schemes were introduced.

The trends in food hygiene standards and the number of food poisoning cases in each of the two groups of local authorities were first assessed for the period before 2011/2012. This was to check there were no major differences between the groups in terms of general trends, both in direction and rate of change, before the FHRS was introduced or the FHIS was rolled out. Allowance was made for additional factors, including population age and density, business density, and the numbers of local authority staff dealing with food hygiene, each of which might also have had an impact on the hygiene standards and the number of cases of food poisoning during the evaluation period. These factors were included in the statistical analysis and adjustments made to reflect their impact, in order to isolate and estimate the changes arising from the FHRS and FHIS. It was also recognized that the take-up of both the FHRS and the FHIS by local authorities occurred more quickly than originally anticipated, with 95% adopting FHRS and 75% adopting FHIS by the end of 2012/2013. This rapid take-up restricted the number of authorities that could be

Table 5.1 Impact on compliance rates in local authorities that had adopted the FHRS or the FHIS

Time after rollout	Proportion of "poorly compliant" businesses (%)			Proportion of "broadly compliant" businesses (%)			Proportion of "fully compliant" businesses (%)		
	Actual	Est. without FHRS/ FHIS	Impact of FHRS/ FHIS	Actual	Est. without FHRS/ FHIS	Impact of FHRS/ FHIS	Actual	Est. without FHRS/ FHIS	Impact of FHRS/ FHIS
FHRS in England, Wales, and Northern Ireland									
1 year	5.8	7.7	-1.9^{***}	91.0	89.0	2.0^{***}	49.6	47.8	1.8
2 years	4.7	6.4	-1.7^{**}	92.1	90.6	1.5	54.7	51.4	3.3^{***}
FHRS in England only									
1 year	4.6	6.3	-1.7^{***}	92.7	90.9	1.8^{***}	58.3	56.3	2.0
2 years	4.0	5.5	-1.5^{**}	93.4	92.2	1.2	65.8	62.4	3.4^{***}
FHIS in Scotland									
1 year	7.0	8.2	-1.2	86.8	86.0	0.8	34.6	32.7	1.9
2 years	7.1	7.6	-0.5	86.4	86.2	0.2	36.0	32.6	3.4

**Statistical significance at the 95% confidence level
***Statistical significance at the 99% confidence level
Source: Amended from FSA nontechnical summary report (Food Standards Agency 2015d)

included in the analysis and meant that any impact found may be an underestimate of the full impact.

The results for the impact on hygiene standards are given in Table 5.1.

For FHRS as a whole (England, Wales, and Northern Ireland) in the first year, the increase in the proportion of businesses that were "broadly compliant" was statistically significant, increasing to 91.0%. This is 2% higher than it is estimated would have happened without FHRS. (Statistical significance is an expression of the likelihood that a result or relationship is caused by something other than mere random chance. This can be assessed at different levels of likelihood, in this case at 95% and 99% confidence levels. If a result is statistically significant at the 95% level, then there is a 1 in 20 chance of getting such a result randomly. At 99% level this increases to a 1 in 100 chance). Similarly, the increase in the proportion of businesses that were "fully compliant" in the second year was statistically significant moving to 54.7%, which is 3.3% higher than would be expected without FHRS. The findings also show a greater reduction in the proportion of "poorly compliant" businesses for the group of local authorities operating the FHRS. For both years, this was statistically significant. Similar findings were also found in England on its own.

For the FHIS in Scotland, although the general pattern was the same, the differences in compliance levels in local authorities operating the scheme compared with those that were yet to launch it were not found to be statistically significant.

The overall pattern from the evaluation of hygiene scores is therefore promising, showing a statistically significant reduction in the proportion of poorly compliant businesses across England, Wales, and Northern Ireland and corresponding improvements in the proportion of broadly and fully compliant businesses. The results for

Table 5.2 Measures of Impact of the FHRS (England and Wales) on foodborne illnesses

	Estimated impact on the number of formally notified food poisoning reports (per million population)	Estimated impact on the number of confirmed *Salmonella* laboratory reports (per million population)	Estimated impact on the number of confirmed *Campylobacter* laboratory reports (per million population)
1 year after the rollout	−267**	2	−99
(Counterfactual)	(616)	(46)	(515)
2 years after the rollout	89	2	82
(Counterfactual)	(233)	(43)	(349)

**Statistical significance at the 5% level
Sample sizes: The number of local authorities for the impact 1 year after the FHRS rollout is 199, 198, and 204 for the food poisoning, Salmonella and Campylobacter outcomes, respectively
Source: Amended from Table 6.1 from evaluation of the impact of the Food Hygiene Rating Scheme and the Food Hygiene Information Scheme on food hygiene standards and foodborne illnesses final report (Salis et al. 2015)

FHIS in Scotland followed a similar general pattern as that found for FHRS, but the results are not statistically significant.

Due to significant data limitations, the efforts to evaluate impacts on food poisoning focused on FHRS. The results of the estimation of the overall impact of the FHRS on foodborne illness-related outcomes in England and Wales are summarized in Table 5.2.

The findings are not clear-cut and need to be treated with caution. The only impact found to be statistically significant relates to the food poisoning outcome 1 year after the introduction of the FHRS. In this period, the FHRS was found to have reduced the incidence of food poisoning in English and Welsh local authorities. In these areas, the number of formally notified food poisoning reports was estimated to be lower, by 267 units every million people, compared to what it is estimated it would have been, had the scheme not been rolled out. This is known as counterfactual analysis, where a comparison is made between what actually happened and what would have happened, in the absence of an intervention. In this case, the intervention in question is the impact of FHRS. The hypothetical alternative scenario (the counterfactual) against which the impact was evaluated was estimated at 616 reports for every million people. In other words, if the FHRS had not been in place, more cases of food poisoning would have been expected than were actually reported.

The finding indicating that the FHRS reduced the incidence of food poisoning in the population of England and Wales is consistent with the expectations of the theory underpinning the FHRS, which suggests that improvements in businesses' compliance with food hygiene law requirements should result in a reduction in the incidence of foodborne illnesses. However, significant data limitations undermine the validity of the estimates of the impact of the FHRS on foodborne illnesses. The number of reported food poisoning cases is known to be significantly lower than the actual numbers that occur as many people do not visit their doctor when they become ill. There is also an absence in most reported cases of information on the location where the illness was contracted including whether it was acquired in the home or outside the home. Additionally, for *Campylobacter* and *Salmonella* it was not possible to distinguish between cases attributable to food and to those attributable to other sources. Another factor is that were illness is reported is not necessarily where infection occurred. This is particularly an issue for holiday destinations or LAs with large commuter populations. It is therefore difficult to measure current levels and any changes in levels of disease and to attribute them to specific causes.

Thus while a statistically significant result was found 1 year after FHRS rollout, which suggests the scheme reduced the incidence of food poisoning, there was no collaborating evidence from the laboratory reports to suggest that the scheme reduced the recorded incidence of either *Campylobacter* or *Salmonella*. Given the data limitations and that only one of the results was statistically significant, these findings on the impact on reported cases of food poisoning must be treated with caution. A clearer pattern may emerge over time as additional data become available.

The introduction of voluntary restaurant rating schemes in the UK has been widely welcomed, and since its introduction on a voluntary basis, legislation has been introduced in Wales and Northern Ireland to provide for mandatory display of ratings at food premises (Food Standards Agency 2013, 2015c).

The FSA is currently working to strengthen the case for mandatory display in England.

5.6   US Experience with Posting Restaurant Ratings

Unlike the UK, the USA does not operate a national restaurant rating scheme. It does however have a long history of operating restaurant rating schemes at a local level across the country, dating back to at least the 1940s (Ho 2012). Rating schemes became popular in a very short period, and by one estimate, roughly 400 US cities had grading systems in place in 1951 (Ho 2012) (p589). Several states have uniform statewide restaurant grading systems, used to calculate either numerical scores or letter grades, which must be prominently posted by restaurants. The first state to enact such a statewide system was South Carolina in 1995. Tennessee and North Carolina later enacted legislation imposing similar statewide systems. However, many states have no such requirements, and therefore the introduction of any scheme is purely a matter for local government, at city or county level. California follows this approach and has a plethora of cities with grading systems, the most well-known being the letter grading system for restaurants used in Los Angeles. Similar letter grading-based systems have been widely adopted by other cities across the USA (Roberts 2016).

In more recent times, the role of restaurant rating schemes has been widely recognized to both inform consumers and motivate business owners, as a part of the targeted transparency (Weil et al. 2013) and "nudge" agendas (The Economist 2012), using the publication of information, positive reinforcement, and indirect suggestion, rather than direct legislation, to achieve policy goals.

While the USA has a long history of restaurant health inspection and using restaurant rating schemes, surprisingly little detailed evidence exists that has attempted to evaluate the impact of the schemes on hygiene scores or on public health through reducing cases of foodborne disease. The best evidence comes from the study of the impacts of the scheme operated in Los Angeles County (Fung et al. 2015).

The legislation enacting the scheme in Los Angeles County was introduced by the Los Angeles County Board of Supervisors (the governing body of Los Angeles County) in 1997 and came into effect in early 1998. Its introduction followed public outcry from the broadcast of a three-part TV expose on restaurant hygiene in Los Angeles, which used "hidden camera" techniques to reveal a variety of unsanitary practices in restaurant hygiene that were reported to be common in restaurants throughout Los Angeles County, despite the existence of a restaurant hygiene monitoring system by the county.

The new legislation required the public posting of restaurant hygiene grades (A, B, or C) based on Los Angeles County Department of Health Services (DHS) inspections. By making these grades public, the Board of Supervisors sought to reduce the effects of foodborne diseases by putting competitive pressure on public eating establishments with poor hygiene practices to improve their performance or risk losing customers. Although the transparency requirement was adopted at the county level, individual cities within the county were not required to adopt the ordinance (all but ten had chosen to do so by the end of 2005). The running of the scheme has subsequently been refined; in 2013 the LA county Department of Public Health Environmental Health Division implemented an electronic inspection system for restaurants, markets, and other food facilities (County of Los Angeles Department of Public Health 2014). Results for inspections that took place after 2013 can also be viewed online (Los Angeles County Environmental Health 2017).

The system builds directly on the health inspections conducted regularly by the DHS. Health inspections cover a range of very specific practices, including food temperatures, kitchen and serving area handling and preparation practices, equipment cleaning and employee sanitary practices, and surveillance of vermin. Each violation receives one or more points. Cumulative points are then deducted from a starting score of 100. A score from 90 to 100 points receives an A, 80 to 89 a B, and 70 to 79 a C. Cumulative scores below 70 require immediate remedial action by the restaurant owner, which may include suspension of the owner's public health permit and closing of the restaurant (Fig. 5.2).

The transparency system requires restaurants to post the letter grade arising from the most recent inspection on the front window. Restaurants receive two or three unannounced inspections and one reinspection, upon request, per year. Thus, although the posting of grade cards entails relatively small costs, the system relies on a large number of inspections (about 75,000 in 2003) and therefore means a sizable enforcement budget for the DHS.

Fig. 5.2 Illustrative
County of Los Angeles
Public Health inspection
display poster, showing an
"A" rating. Used with
permission from County of
Los Angeles Department
of Public Health

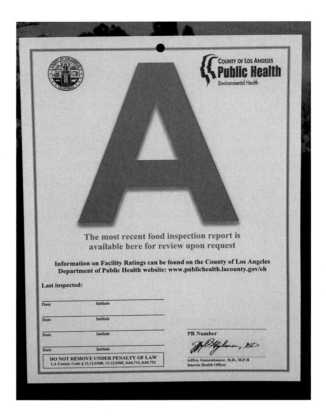

The introduction of the new transparency system led to fairly rapid and significant changes in the overall grade distribution in county restaurants. (The results of inspections had previously been available to the public, but only on request). When the program began, 58% of restaurants received an A grade, a number that grew to 83% by 2003. The incentives to improve are significant. Researchers Ginger Zhe Jin and Phillip Leslie analyzed the impact of restaurant grades and found that after grade posting, restaurants receiving an A grade experienced revenue increases of 5.7% (other factors held constant); B-grade restaurants had increases of 0.7%; and those with a C grade had declines in revenue of 1%. The introduction of grades also improved hygiene at franchised units in chain restaurants (Fung et al. 2015).

More importantly from a public health perspective, studies found significant decreases in foodborne-illness hospitalizations. Hospital discharge data on foodborne-disease hospitalizations were analyzed for Los Angeles County and, as a control, compared with the rest of California during the period 1993–2000. Ordinary least-squares regression analysis was carried out to measure the effect of the grading program on these hospitalizations. After adjusting for underlying time and geographic trends (in order to isolate and remove other effects), the impact of the restaurant hygiene grading program was associated with a 13.1% decrease in the number of foodborne-disease hospitalizations in Los Angeles County in the year following implementation of the program (1998). The result was statistically significant at ($p < 0.01$). (A p-value of "0.01" means that there is a 99% ($1-0.01 = 0.99$) chance of it being true). This decrease was sustained over the next 2 years (1999–2000) (Simon et al. 2005). In another study, the authors estimated the reduction to be 20% (Jin and Leslie 2003).

The results from Los Angeles County therefore strongly suggest that restaurant hygiene grading with public posting of results can be an effective intervention for reducing the burden of foodborne disease. That does not mean rating schemes are without fault however. By their nature, rating schemes are attempting to compress and convey a whole range of data into a single figure. As with any intervention, there is always the need to consider the risk of unintended consequences and how to deal with them. For example, there is a potential trade-off in the use of resources between the effort required to maintain inspection and reinspection rates of all restaurants and the desire to focus efforts on dealing with the worst cases. In an article making the case for restaurant hygiene grade cards, Jin and Leslie note that the focus and importance given to grade boundaries could encourage more lenient marking at the boundaries. There is some evidence from the Los Angeles County data to suggest that while the overall marks have risen over time, the shape of the distribution of hygiene marks has also shifted. Before the introduction of grade cards, hygiene scores followed a smooth bell-shaped distribution. After the introduction of grade cards, there was a dramatic upward spike in the distribution at the score of 90, the cutoff score for obtaining an A grade. There was also a downward spike at 89. A similar pattern also occurred around the cutoff for a B grade. Jin and Leslie note one interpretation of this pattern, which was also consistent with the anecdotal evidence from inspectors, was that inspectors chose to "bump up" a score of 89 to 90 so that the restaurant was not punished because of one point. As long as inspectors do not bump up restaurants which deserve even lower scores, this would be a mild form of grade inflation (Jin and Leslie 2005). This finding suggests ongoing monitoring is needed, to ensure that any grade inflation does not become worse over time.

Other limitations or flaws in the way different hygiene scoring systems operate in the USA have also been identified. Ho (2012) (op cit) carried out detailed investigation and identified a series of potential problems, including grade inflation in San Diego (virtually all restaurants obtained an A rating) and significant inconsistencies among inspectors' scores in New York (due in part to the changes and complexity of the scheme).

As an aside, it is interesting to note that New York is currently also working on using algorithms to study online restaurant reviews, to help identify foodborne disease outbreaks that might otherwise not be officially reported—in other words, using other sources of data, in addition to official inspections, to assist targeting of interventions (Harrison et al. 2013).

5.7   General Lessons and Conclusions

In terms of economic theory, hygiene rating schemes are intended to address potential market failure caused by asymmetric information about the quality of hygiene in food businesses. The schemes are intended to convey a summary of a range of hygiene information in a straightforward and readily understood manner, to assist consumers and motivate businesses to improve. Designing a scheme is by no means straightforward, and there is no perfect system. A number of approaches are currently in operation. In the UK, the FHRS used in England, Wales, and Northern Ireland use a 0–5 scale, while the Scottish FHIS uses a simpler pass/fail system. The USA does not have a national scheme, and US cities and counties have adopted a range of approaches, often displayed either as a simple score or an overall grading letter. Voluntary schemes, such as the FHRS (as originally introduced), are generally considered easier to get up and running, but voluntary approaches run the risk of being ignored by poorer performing businesses. An alternative of having a plethora of local schemes can lead to confusion about consistency between areas. Yet it does also provide the opportunity to attempt to assess the effectiveness and impact of different approaches.

Evaluation studies in the UK and the USA both suggest restaurant ratings schemes can have a real and positive impact on raising hygiene scores over time. Additionally, there is good evidence from the detailed study of data from Los Angeles County supporting the case it has had a statistically significant impact in reducing cases of food poisoning, as measured through hospital admissions. In the UK, the evaluation evidence on reported public health impacts to date has been less clear. Given their nature, measuring the impact of hygiene scoring systems of foodborne disease cases is always going to be difficult to detect reliably and robustly.

Finally, no scheme is perfect. In attempting to address informational asymmetry and quality uncertainty, it is important to consider potential unintended consequences, in terms of issues such as grade drift or not focusing actions on poorest-performing businesses. However, such issues can be addressed through careful monitoring and adjusting how schemes are run. It is clear that food businesses have an increasing role in feeding the population, and food hygiene rating systems can play an important part in ensuring consumers are quickly and simply informed about hygiene standards and similarly that businesses are incentivized to maintain and improve their performance.

Acknowledgments   I am grateful for the helpful comments I received when drafting this chapter from colleagues at the Food Standards Agency. I am also grateful to the FSA and to the County of Los Angeles Department of Public Health for their respective permission to include illustrations of the Food Hygiene Ratings Scheme sticker and inspection display poster.

References

Akerlof GA. The market for "Lemons": quality uncertainty and the market mechanism. Q J Econ [Internet]. 1970 [cited 2017 May 23];84(3):488. Available from: https://academic.oup.com/qje/article-lookup/doi/10.2307/1879431

County of Los Angeles Department of Public Health. County of Los Angeles Department of Public Health Environmental Health Reference Guide For The Food Official Inspection Report [Internet]. 2014 [cited 2017 May 29]. Available from: http://publichealth.lacounty.gov/eh/docs/retailfoodinspectionguide.pdf

Economics Online. Types of market failure [Internet]. Economics Online. 2017 [cited 2017 May 23]. Available from: http://www.economicsonline.co.uk/Market_failures/Types_of_market_failure.html

Feeney J, Stewart C. Food hygiene rating scheme- update and next steps [Internet]. 2015 [cited 2017 May 29e]. Available from: https://www.food.gov.uk/sites/default/files/fsa150306.pdf

Food Standards Agency. Food outlets will be forced to display hygiene ratings - BBC News [Internet]. 2013 [cited 2017 May 29]. Available from: http://www.bbc.co.uk/news/uk-wales-politics-25119724

Food Standards Agency. New UK food poisoning figures published. Food Standards Agency [Internet]. 2014 [cited 2017 May 29]. Available from: https://www.food.gov.uk/news-updates/news/2014/6097/foodpoisoning

Food Standards Agency. Business display of food hygiene ratings in England, Northern Ireland and Scotland. Food Standards Agency [Internet]. 2015a [cited 2017 May 23]. Available from: https://www.food.gov.uk/science/research/ssres/foodsafetyss/fs244011a-0

Food Standards Agency. Food hygiene rating scheme and food hygiene information scheme - impact on hygiene standards of food businesses and on the incidence of food poisoning (non-technical summary) [Internet]. 2015b [cited 2017 May 29]. Available from: https://www.food.gov.uk/sites/default/files/fhrs-fhis-impact-study-non-technical-summary-report-2015.pdf

Food Standards Agency. Frequently asked questions about the food hygiene rating scheme. Food Standards Agency [Internet]. 2015c [cited 2017 May 29]. Available from: https://www.food.gov.uk/multimedia/hygiene-rating-schemes/ratings-find-out-more-en/fhrs

Food Standards Agency. FSA non-technical summary report: food hygiene rating scheme and food hygiene information scheme - impact on hygiene standards of food businesses and on the incidence of food poisoning [Internet]. 2015d. Available from: https://www.food.gov.uk/sites/default/files/fhrs-fhis-impact-study-non-technical-summary-report-2015.pdf

Food Standards Agency. Biannual public attitudes tracker survey Food Standards Agency [Internet]. 2017 [cited 2017 May 29]. Available from: https://www.food.gov.uk/science/research/ssres/publictrackingsurvey

Food Standards Scotland. Food Hygiene Information Scheme. Food Standards Scotland [Internet]. 2015 [cited 2017 May 29]. Available from: http://www.foodstandards.gov.scot/food-safety-standards/food-safety-hygiene/food-hygiene-information-scheme

Fung A, Graham M, Weil D. The transparency policy project: restaurant hygiene [Internet]. 2015 [cited 2017 May 29]. Available from: http://www.transparencypolicy.net/restaurant-hygiene.php

Gibbens S, Spencer S. Business display of fodd hygiene rating in England, Wales & Northern Ireland report prepared for Food Standards Agency [Internet]. 2013. Available from: https://www.food.gov.uk/sites/default/files/758-1-1495_FHRS_Report_2013_-_FINAL_-2-_July_2013.pdf

Harrison C, Jorder M, Stern H, Stavinsky F, Reddy V, Hanson H, et al. Using online reviews by restaurant patrons to identify unreported cases of foodborne illness—New York City, 2012–2013 [Internet], vol. 63. Centres for Disease Control and Prevention. 2013 [cited 2017 May 29]. p. 441–5. Available from: https://www.cdc.gov/mmwr/preview/mmwrhtml/mm6320a1.htm

Ho DE. Fudging the nudge: Information disclosure and restaurant grading. Yale Law J [Internet]. 2012 [cited 2017 May 29]. Available from: http://dho.stanford.edu.

Jin GZ, Leslie P. The effect of information on product quality: evidence from restaurant hygiene grade cards. Q J Econ [Internet]. 2003 [cited 2017 May 29];118(2):409–51. Available from: https://academic.oup.com/qje/article-lookup/doi/10.1162/003355303321675428

Jin GZ, Leslie P. The case in support of restaurant hygiene grade cards [Internet]. vol. 20. Choices. 2005 [cited 2017 May 29]. p. 97–102. Available from: http://www.choicesmagazine.org.

Los Angeles County Environmental Health. Los Angeles county environmental health public portal [Internet]. 2017 [cited 2017 May 29]. Available from: https://ehservices.publichealth.lacounty.gov/

Marler Clark. Foodborne illness: Food poisoning [Internet]. 2017 [cited 2017 May 23]. Available from: http://www.foodborneillness.com/

NOP, Gfk. The display of Food Hygiene Ratings in Wales Client: Food Standards Agency Wales. 2011 [cited 2017 Nov 28]. Available from: https://www.food.gov.uk/sites/default/files/732-1-1244_FHRS_Report-13_December_2011_-_FINAL.pdf

Roberts MT. Food law in the United States. New York: Cambridge University Press; 2016.

Salis S, Jabin N, Morris S. Evaluation of the impact of the food hygiene rating scheme and the food hygiene information scheme on food hygiene standards and food-borne illnesses final report [Internet]. 2015. Available from: https://www.food.gov.uk/sites/default/files/fhrs-fhis-eval2011-14foodborne.pdf

Simon PA, Leslie P, Run G, Jin GZ, Reporter R, Aguirre A, et al. Impact of restaurant hygiene grade cards on foodborne-disease hospitalizations in Los Angeles County. J Environ Health [Internet]. 2005 [cited 2017 May 29];67(7):32–6, 56–60. Available from: http://www.ncbi.nlm.nih.gov/pubmed/15794461

The Economist. Nudge nudge, think think [Internet]. The Economist. 2012 [cited 2017 May 29]. Available from: http://www.economist.com/node/21551032

TNS BMRB. Citizens forum: expanding food hygiene information [Internet]. 2012. Available from: http://webarchive.nationalarchives.gov.uk/20150507232331/http://www.food.gov.uk/sites/default/files/multimedia/pdfs/citizens-forum-report-2012.pdf

TNS BMRB. Citizens forum: mandatory display of FHRS [Internet]. 2013 [cited 2017 May 29]. Available   from: https://www.food.gov.uk/sites/default/files/multimedia/pdfs/citizensforum-fhrs.pdf

Vegeris S. The food hygiene rating scheme and the food hygiene information scheme [Internet]. 2015. Available from: https://www.food.gov.uk/sites/default/files/fhrs-fhis-eval2011-14.pdf

Weil D, Graham M, Fung A. Targeting transparency. Science (80-) [Internet]. 2013 [cited 2017 May 29];340(6139):1410–1. Available from: http://science.sciencemag.org/content/340/6139/1410.full?ijkey=Zeo7suycYcqBg&keytype=ref&siteid=sci

World Bank. Asymmetric information [Internet]. 2003 [cited 2017 May 29]. Available from: http://siteresources.worldbank.org/DEC/Resources/84797-1114437274304/Asymmetric_Info_Sep2003.pdf

Young V, Gibbens S. Food hygiene rating scheme (FHRS) and food hygiene information scheme (FHIS) – display of ratings and inspection results in England, Northern Ireland and Scotland Client: Food Standards Agency [Internet]. 2012. Available from: https://www.food.gov.uk/sites/default/files/758-1-1303_FS244011A_GfK_report_-_FINAL_FOR_PUBLICATION.pdf

Yu M. Information note food hygiene information system in selected places (IN19/07-08) [Internet]. 2008 [cited 2017 May 29]. Available from: http://www.legco.gov.hk/yr07-08/english/sec/library/0708in19-e.pdf

Part II
Economics of Foodborne Illness Metrics: When to Use What

Chapter 6
Burden and Risk Assessment of Foodborne Disease

Brecht Devleesschauwer, Robert L. Scharff, Barbara B. Kowalcyk, and Arie H. Havelaar

Abbreviations

DALY	Disability-adjusted life year
EFSA	European Food Safety Authority
FDA	Food and Drug Administration
FSMA	Food Safety Modernization Act
GBD	Global Burden of Disease
QALY	Quality-adjusted life year
QoL	Quality of life
RASFF	Rapid Alert System for Food and Feed
SMPH	Summary measure of population health
VSL	Value of statistical life
VSLY	Value for a statistical life year
WHO	World Health Organization
WTA	Willingness to accept
WTP	Willingness to pay

B. Devleesschauwer (✉)
Department of Public Health and Surveillance, Scientific Institute of Public Health (WIV-ISP), Brussels, Belgium
e-mail: brecht.devleesschauwer@wiv-isp.be

R. L. Scharff
Department of Human Sciences, The Ohio State University, Columbus, OH, USA
e-mail: scharff.8@osu.edu

B. B. Kowalcyk
Department of Food Science and Technology, The Ohio State University, Columbus, OH, USA

RTI International, Research Triangle Park, NC, USA

Center for Foodborne Illness Research and Prevention, Raleigh, NC, USA

A. H. Havelaar
Department of Animal Sciences, Institute for Sustainable Food Systems, Emerging Pathogens Institute, University of Florida, Gainesville, FL, USA
e-mail: ariehavelaar@ufl.edu

© Springer International Publishing AG, part of Springer Nature 2018
T. Roberts (ed.), *Food Safety Economics*, Food Microbiology and Food Safety,
https://doi.org/10.1007/978-3-319-92138-9_6

YLD   Years lived with disability
YLL   Years of life lost

6.1 Introduction

Food safety is a critical global public good that has important implications for public health, economies, and food security. Globally, foodborne disease is a leading cause of mortality and morbidity, causing an estimated 600 million illnesses and 42,000 deaths annually (Havelaar et al. 2015). Children are particularly impacted, accounting for 40% of the overall burden and a third of all deaths. Foodborne disease can result in long-term health outcomes, such as irritable bowel syndrome, reactive arthritis, diabetes, hypertension, kidney disease, and neurological dysfunction (Batz et al. 2013; Porter et al. 2008; Roberts et al. 2009). Combined, these health impacts lead to reduced quality of life, shorter life spans, increased medical costs, decreased worker productivity, and lower incomes. The impact is substantial, but estimates vary by the number of pathogens, the perspective, and aspects included (e.g., tangible versus intangible costs). Regardless of the approach used, existing estimates are conservative—i.e., the true burden and costs are likely to be higher than presented.

The impact of food system failures is actually much higher than medical costs and lost productivity. Meeting food safety requirements is essential to gaining market access, particularly for developing economies (Grace 2015). The inability to meet food safety requirements has rippling effects, resulting in lower incomes, decreased purchasing power, and reduced access to food which, in turn, can lead to increased medical costs and decreased worker productivity. Recalling contaminated products that do make it to market is very costly, resulting in product losses, loss of markets and consumer confidence, damage to reputation, court cases, and company closures (Hussain and Dawson 2013; Pozo and Schroeder 2016; Ribera et al. 2012). A 2011 survey found that 77% of industry members who had experienced a recall within the past 5 years estimated the related costs to be $30 million, with 23% reporting even higher costs (Grocery Manufacturers Association 2011). In addition, over 81% of respondents described the financial consequences of a recall as either "significant" or "catastrophic" (Grocery Manufacturers Association 2011). In fact, as shown in Fig. 6.1, stock values dropped an average of 1.24% 5 days after the formal announcement of a Class I recall; for a firm with 472 million shares valued at $20/share, this would result in a $109 million loss (Pozo and Schroeder 2016). Similarly, there are significant costs to the government as public health and regulatory agencies respond to failures in the food safety system, including epidemiological investigations, product tracing efforts, enhanced environmental sampling and inspections, and ensuring of the effectiveness of recalls.

In response to the bovine spongiform encephalopathy crisis and other food safety incidents in the 1990s, the Council of the European Union and the European Parliament adopted Regulation (EC) No. 178/2002, known as the

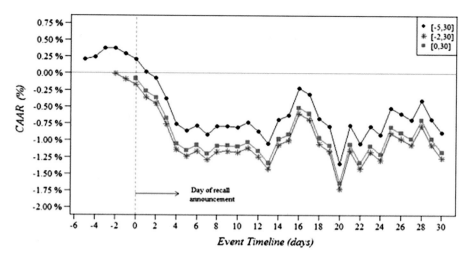

Fig. 6.1 Impact of Class I meat and poultry recalls on stock prices—USA 1993–2013 (Pozo and Schroeder 2016). The average loss in market equity 5 days after recall equaled $109 million

General Food Law of 2002 (http://eur-lex.europa.eu/legal-content/EN/ALL/?uri=CELEX:32002R0178). One of the key principles of the food law is that "measures adopted by the Member States and the Community governing food and feed should generally be based on risk analysis." The Regulation further created the Rapid Alert System for Food and Feed (RASFF), and the European Food Safety Authority (EFSA), an independent agency that provides scientific advice and risk assessments to relevant bodies in the European Commission, the European Parliament and Member States. In 2010, recognizing the importance and infrastructure needs around food safety, the United States Congress passed the Food Safety Modernization Act (FSMA), the first comprehensive reform of the Food and Drug Administration's (FDA) food safety oversight since 1938 (FDA 2017). FSMA mandates FDA to adopt a science-based, risk-informed approach to food safety and holds the food industry more accountable for producing safe products.

Central to the risk-informed framework are risk analysis and burden of disease estimates, which provide the foundation for decision-making and allocation of resources. Information gathered on the burden of foodborne disease provides important data for risk assessment and, subsequently, scientifically grounded risk reduction strategies. For example, burden of disease estimates are currently being used by the FDA to designate high-risk foods that will be prioritized in product tracing measures and design data-driven preventive controls and food safety standards (FDA 2014). In this chapter, we present an overview of risk analysis and disease burden as both are the foundation of a risk-based food safety system. Subsequent chapters present an overview of the research that has been conducted by the World Health Organization (WHO) and in the United States to provide useable estimates of the health impact and economic burden of foodborne disease.

6.2   Risk-Based Food Safety

Due to the complexity and changing nature of the food supply, ensuring its safety has been identified as a *wicked problem*, i.e., a problem that arises in complex and interdependent systems and that is difficult or impossible to solve because of incomplete, contradictory, changing, or incomprehensible requirements (Institute of Medicine 2012). Indeed, the food system is multifaceted, with a large number of stakeholders with diverse interests. The international food production and distribution systems play a major role in the global economy, with significant impacts on income, employment, rural and urban economies, and the environment. Historically, the approach to ensuring food safety has been reactive—responding to crises as they occur—rather than preventive (Koutsoumanis and Aspridou 2016). In the United States, food oversight is distributed across 15 federal and thousands of state and local agencies and regulated by a patchwork of regulations that can be difficult to navigate. Internationally, many countries lack the infrastructure needed to meet international food safety standards which, in turn, impacts trade and local access to safe food. To address the food challenges of the twenty-first century, the paradigm must shift to an integrated, multidisciplinary, systems-based approach that is informed by the best available science, and focus on prevention is needed. At the same time, there is a very real need to utilize limited resources so that they effectively address the most important issues and provide the greatest benefits to the most people. Risk analysis, which consists of risk assessment, risk management, and risk communication (Fig. 6.2), provides an integrated and structured framework for supporting decision-making; it is internationally accepted as the best approach to food safety (FAO 2006).

A risk-based food safety system is one that uses "a systematic means by which to facilitate decision-making to reduce public health risk in light of limited resources and additional factors that may be considered" (Havelaar et al. 2007; National Research Council 2010). Central to the risk-based framework (Fig. 6.3) is an understanding of the risks and burden of disease. Once we understand the burden, we can begin to quantify, attribute, and rank the risks. From there, we can establish public

Fig. 6.2 Components of risk analysis

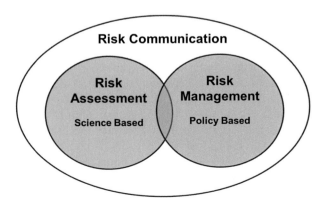

Fig. 6.3 Framework for a risk-based food safety system (National Research Council 2010)

health goals—such as the United States Healthy People 2020 goals or the United Nations Millennium Development Goals—and determine potential prevention and control interventions. We must then evaluate each intervention or policy to determine its ability to positively impact public health at a reasonable cost in a fair manner. After we have identified our prevention and control strategies, we must set priorities and allocate resources to those that will have the biggest public health impact. Finally, we must measure the effectiveness of our efforts in meeting public health goals and objectives.

Risk assessment is used to quantify and characterize risk, which is defined to be a function of the probability of exposure (incidence) and the effect of that exposure (severity) (Codex Alimentarius Commission 2006). The classic risk assessment paradigm assesses exposures and characterizes hazards across the supply chain to predict risk to human health (Fig. 6.4). There are four steps in a risk assessment: hazard identification, exposure assessment, hazard characterization, and risk characterization. *Hazard identification* focuses on identifying the hazards, transmission pathways, associated health effects, and at-risk populations of concern and requires information on the hazard characteristics; exposure routes; epidemiologic link between foods, hazards, and illness; health outcomes (acute and chronic); and sensitive populations. *Exposure assessment* focuses on estimating the probability of exposure and the dose of the pathogen in the food at the moment of consumption. Information needs include data on food consumption trends; the ecology of the hazard, including the prevalence and concentrations of pathogens across the food supply; and processing, packaging, storing, and preparation practices and their impact on hazard growth/die-off. *Hazard characterization* (or dose-response assessment) focuses on estimating the probability, severity, and duration of adverse events due to the presence of the hazard in the food. Typically, data from human and animal models or outbreaks are used to develop a dose-response curve, which estimates the

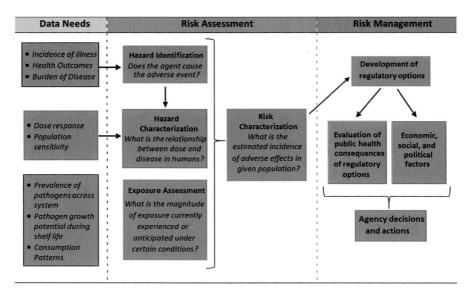

Fig. 6.4 Risk assessment paradigm and areas of focus. Adapted from National Research Council (2009)

relationship between dose, or level of exposure to the hazard, and the incidence and severity of the effect (WHO/FAO 2009). *Risk characterization* combines the information from the hazard identification and characterization and exposure assessment to produce a complete picture of risk, that is, an estimation of the incidence and severity of effects likely to occur in a population due to exposure and the attendant uncertainty.

Two general approaches, based on the data sources used in model construction, are used to assess human health risk (National Research Council 2010). In the *top-down*, surveillance-based approach, information on human disease gathered from epidemiological systems is used directly to estimate risk at the point of consumption (Fig. 6.5). The metrics for likelihood and severity are estimated using population attributable fractions derived from information gathered from epidemiological systems, such as surveillance or cohort studies. Thus, a top-down approach relies on the availability of epidemiological data. In the *bottom-up* approach, estimates are derived using the classic risk assessment paradigm that assesses risk using exposure and dose-response information. In theory, both approaches should result in similar estimates for likelihood and severity; in reality, significant data gaps and biases and uncertainty in the metrics make that unlikely (Bouwknegt et al. 2014). The approach selected will likely depend on the risks under consideration and available data. For example, epidemiologic data are typically less specific to assess risks of exposure to specific food products such as a particular brand of raw milk cheese, making the bottom-up approach more appealing. Alternatively, epidemiological data are typically more reliable to estimate the total incidence of disease by a foodborne

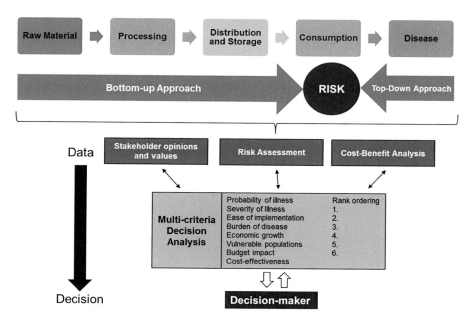

Fig. 6.5 Approaches to assessing risk. Adapted from EFSA Panel on Biological Hazards (BIOHAZ) (2012)

pathogen such as campylobacteriosis, making the top-down approach more appealing. EFSA has proposed a strategy to integrate top-down and bottom-up approaches in a scientific opinion about risk ranking (BIOHAZ 2015).

The outputs of risk assessment are used to inform risk management, where the goal is to control or limit the risks. Risk managers need to make decisions about the acceptable levels of risk and the selection and evaluation of intervention strategies: Is this a risk of public health concern? Should exposures be reduced? Should regulations be put into place? Should a material or substance be labeled or banned? Often resources are limited and priorities must be set; in these cases, risk-ranking exercises may be undertaken to aid prioritization. Ultimately, risk management decisions are often informed by other nonpublic health factors, such as economic, social, and political considerations; decision analysis, which is outside the purview of this chapter, can be used to identify and analyze decision alternatives in a transparent manner.

6.3 Burden Assessment

Burden of disease (or disease burden) refers to the total impact of a disease, including physical, social, and financial impacts, on society (population burden) and on the individual affected (individual burden). Burden of disease can be measured

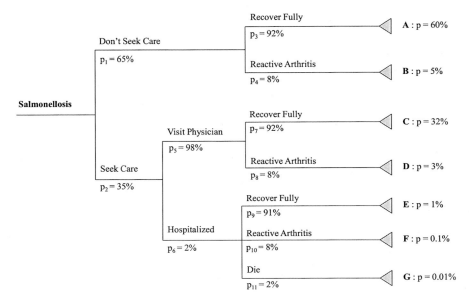

Fig. 6.6 Disease outcome tree for *Salmonella* spp. Based on Scallan et al. (2011) and Keithlin et al. (2015)

using a variety of metrics. Frequently, burden is estimated using the number of illnesses, hospitalizations, and deaths or the cost-of-illness (e.g., medical care, lost wages, and productivity). However, while these metrics provide a picture of the population-level occurrence of foodborne disease, they fail to account for the severity and duration of illness or the resulting disabilities and/or impacts on quality of life (Batz et al. 2012; Devleesschauwer et al. 2015; Mangen et al. 2010). Burden of disease is therefore increasingly quantified using summary measures of population health such as disability-adjusted life years or quality-adjusted life year losses.

The health impact of foodborne disease is defined by the health effects (health states) associated with infection with or exposure to the concerned foodborne hazard. The recommended approach is to design disease outcome trees to clearly define the potential outcomes associated with consuming food contaminated with a specific pathogen. A deterministic example using salmonellosis is illustrated in Fig. 6.6 based on Centers for Disease Control and Prevention estimates of probabilities associated with care seeking, hospitalization, and death (Scallan et al. 2011) and an estimate of the probability of reactive arthritis resulting from the acute infection (Keithlin et al. 2015). The probabilities associated with eight outcomes (A to G) can be assessed using this tree. For example, the probability of recovering fully without seeking care is 60% for an individual with salmonellosis ($Pr(A) = p_1 \times p_2$). Alternatively, the probability of being hospitalized and acquiring reactive arthritis as a sequela is only 0.1% ($Pr(F) = p_2 \times p_6 \times p_{10}$).

There are several things to note about the outcome tree in Fig. 6.6. First, the tree makes it clear that the sequela is assumed to be equally likely under all severity levels. Whether this is correct or the science just has not been able to discern these differences yet is unclear, but it is important to understand. Second, this is a relatively simple tree, even for salmonellosis. Other outcomes, such as whether the sick person provides a stool sample, is prescribed pharmaceuticals, misses work, or utilizes home health-care services, could be added. Depending on the economic technique used to examine the costs associated with illness, these additions to the model may be warranted. Third, uncertainty is not expressed in this tree, though most high-quality studies today do include uncertainty intervals, sensitivity analyses, or both.

Finally, the choice of values used often relies on the expertise of the modeler. For example, in Fig. 6.6, an estimate (8%) from a recent meta-analysis was used for the likelihood of reactive arthritis that focused on diagnoses made by specialists (Keithlin et al. 2015). Using the same meta-analysis, the value for all studies (6%), for those that had follow-ups within 90 days of the acute illness (12%), or for those studies involving more than 10,000 persons who had had salmonellosis (0.2%) could also have been used. Given the wide range of estimates available, modelers are forced to make judgments about which estimates to use and whether to include other estimates in sensitivity analyses.

Similarly, our choice to limit sequelae to reactive arthritis is not an easy one. There are many sequelae that have been associated with salmonellosis, including irritable bowel syndrome, inflammatory bowel disease, Crohn's disease, ulcerative colitis, Guillain-Barré syndrome, Miller Fisher syndrome, and hemolytic uremic syndrome (see Chap. 8 for further discussion). Generally speaking, health outcomes should be included in burden of illness estimates if causation between the acute illness and the outcomes can be sufficiently established. There are both empirical and theoretical criteria for demonstrating causation, including the Bradford Hill criteria (Hill 1965).

6.4 Quantifying the Health Impact of Foodborne Disease

As stated previously, the health impact of foodborne diseases may be defined based on the number of prevalent or incident cases or the number of deaths. However, these simple measures of population health do not provide a complete picture of the impact of foodborne diseases on human health (Batz et al. 2012; Devleesschauwer et al. 2015; Mangen et al. 2010). Indeed, these measures quantify the impact of either morbidity or mortality, thus prohibiting a comparative ranking of highly morbid but not necessarily fatal diseases (e.g., mild to moderate diarrhea) and diseases with a high case fatality (e.g., perinatal listeriosis). On the other hand, they only quantify occurrence of illness or death, thus treating each illness case, or each fatal case, alike. Foodborne diseases may however differ in clinical impact and duration of the concerned symptoms, such that the severity of different illness cases may differ. Likewise, fatal cases occurring at different ages will result in different numbers of potential life years lost, such that the impact of different fatal cases may differ.

To overcome the limitations of these simple measures, various summary measures of population health (SMPHs) have been developed as an additional source of information for measuring disease burden. What the wide range of proposed SMPHs all have in common is that they use time as a general unit of measure; they can further be divided into two broad families: health gaps (i.e., time not lived in good health) and health experiences of expectancies (i.e., time lived in good health) (Devleesschauwer et al. 2014a). The most powerful SMPHs allow combining information on mortality and nonfatal health outcomes, which requires weighting the time lived with disease or disability according to the health experienced or lost. Currently, the two most important SMPHs are the disability-adjusted life year (DALY) and the quality-adjusted life year (QALY).

The *DALY* belongs to the family of health gap measures and is currently the most widely used SMPH in epidemiological research. DALYs find their origin in the Global Burden of Disease (GBD) studies and are officially adopted by the WHO for reporting on health information (Murray et al. 2012; World Health Organization 2013).

DALYs measure the health gap from a life lived in perfect health and quantify this health gap as the number of healthy life years lost due to morbidity and mortality. A disease burden of 100 DALYs would thus imply a total loss of 100 healthy life years, irrespective of how these healthy life years were lost. Diseases, hazards, or risk factors accounting for more DALYs thus have a higher public health impact.

DALYs extend the notion of mortality gaps to include time lived in health states worse than ideal health (Devleesschauwer et al. 2014b). Specifically, they are the sum of years of life lost (YLL) due to premature mortality and years lived with disability (YLD), adjusted for severity:

$$DALY = YLL + YLD$$

YLLs are the product of the number of deaths (M) and the residual life expectancy (RLE) at the age of death:

$$YLL = M \times RLE$$

Two approaches exist for defining YLDs. Following an incidence perspective, YLDs are defined as the product of the number of incident cases (N), the duration until remission or death (D), and the disability weight (DW), which reflects the reduction in health-related quality of life on a scale from zero (full health) to one (death):

$$YLD_{inc} = N \times D \times DW$$

The incidence perspective assigns all health outcomes, including those in future years, to the initial event (e.g., *Campylobacter* infection). This approach therefore reflects the future burden of disease resulting from current events.

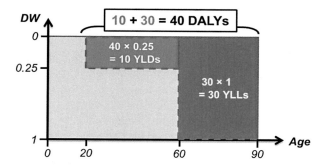

Fig. 6.7 Visual example of the disability-adjusted life year metric. *DW* disability weight, *YLD* years lived with disability, *YLL* years of life lost, *DALY* disability-adjusted life year

An alternative formula for calculating YLDs follows a prevalence perspective and defines YLDs as the product of the number of prevalent cases (P) with the disability weight (Murray et al. 2012):

$$\text{YLD}_{prev} = P \times \text{DW}$$

In this prevalence perspective, the health status of a population is assessed at a specific point in time, and prevalent diseases are attributed to events that happened in the past. This approach thus reflects the current burden of disease resulting from previous events. Although both perspectives are valid, the incidence perspective is more sensitive to current epidemiological trends (Murray 1994), including the effects of intervention measures, and therefore often preferred for assessing the burden of foodborne diseases (Devleesschauwer et al. 2015).

Figure 6.7 presents a theoretical example of calculating DALYs, following the incidence perspective. An individual is born with a perfect state of health. At age 20 years, a given event (e.g., foodborne disease) leads to a 25% decrease of his/her quality of life, and thereafter the person lives in this new health state for another 40 years, at which point he/she dies prematurely. The burden associated with this disease for this individual (total DALYs) is calculated by summing up the years lived with disability (YLD) with the years of life lost (YLL) due to premature death.

The recommended approach for quantifying the health impact of foodborne diseases is the *hazard-based* DALY calculation approach (Devleesschauwer et al. 2014c). This approach defines the burden of a specific foodborne disease as that resulting from all health states, i.e., acute symptoms, chronic sequelae, and death, which are causally related to the concerned hazard and which may become manifest at different time scales or have different severity levels (Mangen et al. 2013). The starting point for quantifying DALYs is therefore the construction of a disease model or outcome tree, such as Fig. 6.1 (Devleesschauwer et al. 2014c).

The *QALY* belongs to the family of health expectancies, and is a standard tool in health economic evaluations, and cost-utility analyses in particular. QALYs are healthy life years, obtained by weighting life years according to utility weights, or simply QALY weights, which reflect individual preferences for time spent in different health states. A number of methods are used to elicit QALY weights, including the standard gamble, time trade-off, and visual analog scale (Torrance 1986). Common to all methods is their use of a scale that measures health as being between 0 (death) and 1 (perfect health). The use of QALYs across multiple pathogens was made possible by the development of standardized QALY weights associated with multiple dimensions of well-being which has allowed for the generation of condition-specific QALY estimates without costly studies focused specifically on each pathogen in question (though expert opinion is needed to assign QALY weights in this case). For instance, in the EQ-5D multi-attribute utility scale, developed by the EuroQoL group, five dimensions of well-being are included (hence the acronym): mobility, self-care, usual activities, pain/discomfort, and anxiety/depression (Herdman et al. 2011).

The health impact of foodborne diseases may be quantified as QALY losses, i.e., the utility losses associated with foodborne disease that include both disability losses and pain and suffering losses. The measurement of QALY losses must account for the typical sufferer's initial QALY state, which is generally less than 1. Ideally, the initial state should be based on that of the typical person who gets a foodborne disease (who is older or younger and typically more immunocompromised than the average person), though the average population QALY level is typically used (Batz et al. 2014; Minor et al. 2015; Scharff 2015).

6.5   Quantifying the Economic Impact of Foodborne Disease

6.5.1   Costs Associated with Foodborne Disease

Decisions in a risk-based food safety system are driven by more than just public health impacts. Risk managers must also consider economic, social, and political factors in the decision-making process (Fig. 6.4). Therefore, it is important to understand the costs associated with foodborne disease: the individual who becomes sick from consuming tainted food, the retailer who sells the contaminated product, the food producer who allows contamination, and the government agencies that monitor, investigate, and regulate all incur costs from foodborne diseases. Figure 6.8, an adapted version of the taxonomy originally developed by the USDA Economic Research Service (Roberts 1989), illustrates these costs. Understanding each of these costs is important in a risk-based food safety system, though most efforts to measure economic cost have focused on household costs.

The *household* incurs costs whether or not an individual in the household has been made ill by their consumption of food. Specifically, consumers who are aware of risks associated with foods may face costs if they engage in self-protective efforts. For example, a consumer may choose to buy pasteurized products, avoid risky foods that he/she likes, or cook foods until any potential pathogens are destroyed (at an

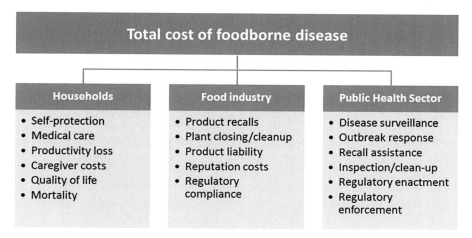

Fig. 6.8 Costs associated with foodborne disease

expense to taste). Each of these measures has a cost to the consumer, either monetary or through lost utility (well-being).

In the presence of illness, the costs include medical costs, productivity losses (to both sick persons and caregivers), pain and suffering losses, and mortality losses. Medical costs include costs for hospitalizations, physician services (both inpatient and outpatient), and drugs used (both prescription and over-the-counter). Ancillary medical services, such as tests of stool samples and urgent care/emergency room costs, are also included in this category.

Productivity losses occur when an individual is unable to perform productive tasks due to illness (either their own or someone they must care for, such as a child). Often this is measured as the costs of absenteeism from paid work. But some researchers have chosen to value the time of all ill persons at the average wage in the United States, regardless of their work status or age (the average wage is a proxy for the opportunity cost of the individual for time spent ill rather than engaging in his/her normal activities). There are also likely to be reductions in productivity for those that go to work sick, though these losses are likely to be significantly less than those for persons who stay home. Lost household production is also a cost of foodborne disease.

In some economic assessments, a monetary value is assigned to the intangible costs, i.e., the quality of life losses, associated with foodborne disease. The physical discomfort or pain associated with foodborne disease is one way an individual's quality of life is affected, but it is not the only way. Inability to engage in pleasurable activities (or reduced pleasure from those activities) also is an economic cost from foodborne disease. For example, if an individual with a mild case of illness decides not to go with friends to a concert because they do not want to deal with the consequences from a diarrheal illness in such a situation, their utility is reduced by an amount equal to the value they would have gotten from going to the concert while healthy minus the value they actually got from staying home with the illness.

Quality of life losses may also be borne by friends and family who must enduring seeing a loved one suffering. These may be quite high, especially when a parent is caring for a very ill child. Similar costs are borne when an individual dies due to a foodborne disease. In some instances, such as when chronic sequelae occur, other household cost categories, such as professional home health-care assistance, may be appropriate to include.

A number of costs accrue to *industry* as a result of foodborne diseases. First, if a firm determines that its product has the potential to make people sick, it is likely to institute a recall of the product. Costs associated with this effort include lost product sales equal to the market value of the recalled product and the cost of collecting and disposing of the product. If the recalled product has been in contact with processing and/or holding facilities, these facilities must conduct a thorough cleaning process, often entailing a lengthy closure of the operation. Next, if anyone was made ill due to the contaminated product, the firm responsible may be exposed to litigation and its attendant costs. Also, media coverage of outbreaks, litigation, and product recalls can have an effect on the reputation/value of the brand. Retailers and wholesalers may also suffer from costs associated with collecting and disposing of recalled product, as well as suffering from potential reputation costs if their customers perceive them as sourcing from unscrupulous suppliers. Finally, if the problem is not an isolated one and there are intervention measures that could remedy the problem, government may respond with costly regulation.

The *public health sector* also incurs costs as a result of foodborne disease. The various surveillance systems that track illnesses (see Chap. 13) are costly to maintain and often lead to the detection of outbreaks that are investigated and monitored. When recalls are initiated, government personnel are involved, whether or not illnesses have occurred. Inspections and assistance with cleaning up contaminated facilities are also activities that government funds. Finally, the promulgation of regulation involves costs, as do enforcement activities associated with the regulation.

6.5.2 Methods Used to Estimate the Costs

A number of methods have been developed to estimate the economic burden associated with foodborne disease. United States federal agencies that evaluate food safety interventions generally use benefit-cost analyses based on a cost-of-illness approach. Stated or revealed preference methods that generate willingness to pay/accept measures are used in some cases to supplement cost-of-illness studies and in others (primarily by academics) as a substitute for cost-of-illness. QALY losses or DALYs are in some cases monetized for use in cost-of-illness studies. Industry costs are often estimated using event studies using publically available data such as stock prices due to the proprietary nature of granular cost data. Attempts have been made to estimate recall and litigation costs, but these measures are generally very crude.

6.5.2.1   Cost-of-Illness

The cost-of-illness method is the most widely used approach among regulatory economists. The goal of this method is to calculate costs separately and aggregate them for presentation as a single cost number. The following equation illustrates a simple cost-of-illness formula for household costs:

$$\text{Cost}_i = \text{Medical}_i + \text{Productivity}_i + \text{QoL}_i + \text{Death}_i$$

The cost for individual i from an illness due to unanticipated risk is the sum of expected medical, productivity, quality of life (QoL), and death-related costs. Relevant industry and public health agency losses have also been added in by some, though not as much as they should be (perhaps due to the dearth of studies in this area). Total costs are defined as the sum of individual costs $\left(\sum_{i=1}^{n} \text{Cost}_i \right)$ and are often used by policymakers as a measure of problem scope, which can be used to set agency priorities and argue for expanded statutory authority. Cost per case measures is typically used in regulatory analyses as a means of demonstrating the economic value of an intervention. Costs per case is total costs divided by the number (n) made ill by the pathogen $\left(\sum_{i=1}^{n} \dfrac{\text{cost}_i}{n} \right)$. Cost per case is multiplied by number of cases averted by (or expected to be averted by) a given intervention to determine intervention effectiveness. The primary focus on household costs means that costs to industry and public health entities are often undervalued.

A number of valuation methods (and controversies) have arisen in response to the need for cost-of-illness estimates. Assuming the case of imperfect information (see Chap. 2), we explore the methods used to estimate medical costs, productivity losses, deaths, and lost utility below.

Medical costs are typically evaluated in one of two ways. Early efforts often relied on interviews or surveys of those that had been sickened in an outbreak. In this case, individuals report what they (or their insurance companies) spent on physician services, medication, hospital costs, and other costs. The primary problem with this approach is that the results may not be generalizable to the broader population outside of the outbreak area. That said, when other values are not available, estimates from outbreak reports can be useful.

An alternative means of estimating medical costs is by matching outcomes in a disease outcome tree with cost estimates from hospital and physician services databases. In the United States, for example, the National Inpatient Sample has cost data for hospitalizations and emergency room visits by ICD-9 classification. For example, in 2013 NIS has data on 6455 discharges with a primary diagnosis related to infection with *Salmonella*. The average cost was $9531 for an average of 5.1 days in the hospital, and 35 deaths were recorded. For physician services, there are references books, such as "Medical Fees" by PMIC, that catalog the costs of physician services and lab fees. Of course, these resources are only useful if the researcher has information about what services are expected to be used by persons made ill due to the pathogen of interest. Similarly, knowledge of prescription medicine costs is only useful if likelihood of use is known.

Productivity losses theoretically include lost work in both the paid and household sectors. Where work is compensated, costs include both wages and other compensation for the time away from work. Uncompensated work, or household production (Becker 1965), may also be lost, but it is unclear how much of this an ill person is able to do. A number of studies have looked at lost wages for persons who are ill (Scharff 2015; Buzby and Roberts 2009; Hoffmann et al. 2012, 2015; Scharff 2012). The most accurate of these have taken into account both wages and benefits using estimates for cost of compensation, rather than only wages. Also, some studies have included work loss due to caregiving for children (e.g., Scharff 2015). The availability of good surveillance data for some pathogens allows for the generation of age profiles for those made ill, which can be used to better predict work status and child care needs.

The inclusion of *quality of life losses* in cost-of-illness analyses is controversial. Originally, no cost-of-illness studies attempted to quantify pain and suffering. In the 1990s, however, the FDA began using a monetized QALY estimate for the value of lost quality of life, as suggested by Mauskopf and French (1991). Some have argued that the monetization of QALYs is not appropriate because it requires the imposition of a number of restrictive assumptions (Hammitt and Haninger 2007). Others have argued that the QALY is the best measure of welfare loss available (Adler 2006).

The monetization of QALYs typically involves obtaining the product of the average person's QALY losses from an illness and their value for a statistical life year (VSLY). VSLY is calculated using a value of statistical life (VSL) measure, a discount rate (r), and expected longevity (L): $VSLY = \dfrac{r \times VSL}{1 - (1 + r)^{-L}}$. Note that both QALY and VSLY values reflect annual losses, suggesting that resulting estimates need to be scaled for duration. For example, an individual who suffers a 0.3 QALY loss for 1 day of diarrheal illness and who faces a VSLY of \$300,000 would be calculated as losing $0.3 \times 1/365 \times \$300{,}000 \approx \$247$ from quality of life losses. It is important to note that productivity losses for the ill person are typically not included in cost-of-illness studies alongside QALY losses because QALYs account for utility losses due to loss of mobility, including internal productivity losses. External productivity losses may be included, however.

Losses from death due to foodborne disease are similar to quality of life losses in that there is a loss of utility and productivity from premature mortality. There are two methods used to assess these costs. First, some have simply used lost productivity for the remaining life span of the sick person, discounted appropriately. Alternatively, most policymakers in the United States now use a broader measure, the value of a statistical life (VSL).

The VSL measure generally used is based on labor market trade-offs between mortality risk and compensation (Viscusi and Aldy 2003). This is a revealed preference measure that essentially works as follows: if the typical individual is willing to accept (WTA) an increase in risk of 1/10,000 in exchange for $800, the implicit VSL is $8 million (VSL/10,000 = $800). Though the theoretically correct measure for a new risk reduction is a willingness to pay (WTP) measure, the revealed preference WTA measure is less likely to suffer from hypothetical biases that inflate stated preference WTP estimates because it is based on actual behavior rather than reported preferences (Murphy et al. 2005). In any case, it has been shown that for small changes in risk, such as those in most policy contexts, WTP and WTA are virtually identical (Kniesner et al. 2014). Note that VSL is not indexed by individual, illustrating the general use of an egalitarian assumption (all statistical lives are equally valuable). This assumption is typically used despite United States government guidance suggesting that VSL should be scaled to account for the population affected (U.S. Office of Management and Budget 2003). Specifically, research has demonstrated that the value for VSL varies by age, first increasing and then decreasing (Aldy and Smyth 2014). Given that many foodborne diseases have greatest incidence among the old and the young, this too suggests that government estimates of VSL are likely to be overestimates. USDA policymakers, however, have stricter food safety standards for the National School Lunch Program so that children are given stronger protection, based on the role of the state as a protector of children (Ollinger et al. 2014).

Despite the inclusion of many cost categories in cost-of-illness studies, some are not accounted for and others can be best seen as rough estimates. For example, the exclusion of self-protective actions and, often, quality of life losses leads to estimates that are likely to be underestimates of true cost, while the egalitarian assumption and assumption of uniform risk preferences may lead to values that are overestimates of true value. In response, some have suggested that the cost-of-illness approach leads to point estimates that give a false sense of precision. Though uncertainty intervals and sensitivity analyses are increasingly included in these analyses, these typically do not completely account for the structural deficiencies of the approach. Other approaches have been suggested as alternatives to the cost-of-illness approach.

6.5.2.2 Willingness to Pay

Foodborne disease cost-of-illness estimates have been criticized as being too limited, not including all of the losses to an individual who is made ill. An alternative is to assess the willingness to pay (WTP) to avoid foodborne disease. Theoretically, this is the most complete measure of utility loss for the affected individual because the individual is allowed to take into account all losses in making their assessment. The principal methods that have been used to elicit WTP for foodborne disease are experimental auctions and dichotomous choice experiments.

Early efforts to estimate WTP generally used experimental auction techniques (Hayes et al. 1995; Shin et al. 1992; Shogren et al. 1994). In these experiments, individuals bid to replace a product having a given risk with another that has a smaller (typically close to 0) risk. The winning bid pays the next highest bid to obtain the product (a mechanism designed to elicit accurate preferences and discourage gaming the auction). The best of these experiments are conducted using real products (and money exchanges) and are conducted using shoppers in a realistic setting (e.g., a grocery store). To be most meaningful, experimenters specify risks associated with the products in a manner that includes both probabilities of illness and likely consequences from becoming ill.

More recently, dichotomous choice experiments have been used in which individuals are asked to choose between two price/risk combinations for a given food product, where each person chooses between lower-risk/lower-price and higher-risk/higher-price options (Haninger and Hammitt 2011; Nayga et al. 2006; Teisl and Roe 2010). Experimenters vary the price/risk combinations across individuals and, in some cases, provide individuals with follow-up price/risk choices to more precisely assess WTP measures. Like auction experiments, these experiments are more likely to yield meaningful responses when the exchanges are not hypothetical, are conducted in realistic settings, and communicate risks in a meaningful way.

Despite the theoretical appeal of WTP measures, holistic WTP measures have not been used in policy settings for food safety. One reason for this is that the cost of conducting these experiments has led to the generation of estimates for a limited number of product/pathogen combinations. Second, WTP studies do not include external costs (e.g., costs to one's workplace from absenteeism, the costs to the insurance pool for claims made, and costs to family members for caregiving). Ideally, these costs would have to be assessed and added in. Perhaps most importantly, the values generated using these methods are not perceived as being plausible by some. This is because WTP estimates are routinely an order of magnitude higher than cost-of-illness estimates and are less sensitive to risk, duration, or consequences than would be expected. For example, Hammitt and Haninger (2007) found that people were implicitly willing to pay $8300 to avoid 1 day described as follows: "You will have an upset stomach and will feel tired, but these symptoms will not prevent you from going to work or from doing most of your regular activities." At the same time, the authors found that people were not willing to pay significantly more to avoid 3 days with the same symptoms and WTP increased less than proportionally with risk. This may be because biases such as the part-whole problem or yea-saying are at work. As a result, the linear extrapolation of individuals' WTP to reduce risk from a single meal or product in an experimental setting to a global WTP measure is likely to overestimate the value of the risk.

Though not used in a holistic fashion, WTP measures have been used to estimate VSL, which is used to place values on death and lost quality of life in some cost-of-illness studies. Many believe that VSL values are more reliable than most other food safety WTP measures because they are based on revealed preference measures derived from actual market behavior, rather than from an experimental setting.

6.5.2.3 Costs to Industry

Costs to industry are also important for both industry decision-makers and policy analysts. Though generalizable estimates of industry cost are not available, a number of event studies have been published. These studies look at effects on individual companies and industries as a result of food safety incidents.

Tangible costs accruing to companies implicated in food safety events include recalls and litigation. Recalls involve effort, destruction of product, and process changes, all of which are costly (Grocery Manufacturers Association 2010; Todd 1985). Resende-Filho and Buhr (2010) developed a model to assess recall costs and demonstrated how these costs decline significantly with the introduction of traceability into the system. Litigation is also a significant cost for those implicated in an outbreak. Buzby and Frenzen (1999) examined litigation associated with foodborne disease, providing both an overview of the system and estimates from litigated cases. The empirical estimates from this approach are of limited value, however, since, as the authors note, less than 0.01% of cases are litigated.

Perhaps the largest costs to companies implicated in a foodborne disease outbreak or recall are reputation costs. Several studies have found that food safety events can affect the stock prices of the firm implicated long after the outbreak is over (Seo et al. 2013, 2014), though this effect is not universally true for all recalls of tainted product (Salin and Hooker 2001). Researchers have also focused on changes in price and demand for products from implicated industries (Todd 1985; Arnade et al. 2009; Palma et al. 2010) finding significant industry spillover effects in some cases.

Though the literature has a number of event studies focused on costs to industry from foodborne disease recalls and outbreaks, peer-reviewed generalizable estimates are not available. Future research in this area would be beneficial.

6.6 Critical Appraisal of Foodborne Disease Burden Estimates

The preceding two sections have made it clear that there exist various methods for quantifying foodborne disease burden, which inevitably has led to large heterogeneity in published foodborne disease burden estimates (Haagsma et al. 2013). Furthermore, available foodborne disease burden studies may differ in their reference population and reference year and in their scope, i.e., the number and nature of foodborne hazards and corresponding sequelae included. Finally, it should be clear that when the underlying epidemiological and economic data are uncertain, the resulting foodborne disease burden estimates will inevitably also be uncertain. A realistic appraisal and quantification of this uncertainty should therefore be an integral part of every foodborne disease burden assessment.

Table 6.1 Comparison of the methods used to quantify the World Health Organization estimates of the global burden of foodborne disease (Devleesschauwer et al. 2015) and the Scharff estimates of the burden of foodborne disease in the United States (Scharff 2012, 2015)

	WHO/FERG	Scharff
Reference population	Global	United States
Reference year	2010	2017
What is valued	Health impact	Health-related economic impact
Metric	Disability-adjusted life years	US Dollars
Approach	Incidence-based Retrospective Top-down	Incidence-based Retrospective Mixed: Bottom-up/top-down
Number of pathogens included	31	30 pathogens + 1 set of unspecified agents
Inclusion of sequelae	Yes	Yes
Valuation of ill health	Disability weights (Salomon et al. 2015)	Dollar Values for: Medical costs Productivity losses Quality-adjusted life years Value of statistical life
Residual life expectancy	Highest UN projected life expectancy for 2050, with a life expectancy at birth of 92 years for both sexes (WHO 2013)	Age-invariant value of statistical life used
Time discounting	No	Yes: value of statistical life year based on discounted number of life years
Age weighting	No	No
Uncertainty propagation	Yes	Yes

Chapters 7 and 8 present two major efforts to quantify the burden of foodborne disease, i.e., the WHO initiative to estimate the global burden of foodborne disease and the Scharff estimates on the economic burden of foodborne disease in the United States. Table 6.1 compares the key characteristics of both studies.

6.7   Conclusion

A large body of research has developed to examine the burden of foodborne disease. This research is useful for researchers, policymakers, and industry professionals to support risk- and evidence-based food safety decision-making. The major metrics include summary measures of population health that quantify the intangible costs of foodborne disease and monetary metrics that quantify the costs to households, industry, and the public sector. There is increasing attention to include long-term health outcomes in economic evaluations of foodborne disease. Key uses of these evaluations are to support priority setting and evaluation of the costs and benefits of food safety interventions. As the literature continues to evolve, the efficiency of decisions made will improve, and all stakeholders will be better served.

References

Adler MD. QALY's and policy evaluation: a new perspective. Yale J Health Policy Law Ethics. 2006;6:1–92.

Aldy JE, Smyth SJ. Heterogeneity in the Value of Life (No. w20206). National Bureau of Economic Research, 2014.

Arnade C, Calvin L, Kuchler F. Consumer response to a food safety outbreak of *E. coli* O157:H7 linked to spinach. Appl Econ Perspect Pol. 2009;31:734–50.

Batz MB, Hoffmann S, Morris JG. Ranking the disease burden of 14 pathogens in food sources in the United States using attribution data from outbreak investigations and expert elicitation. J Food Prot. 2012;75:1278–91.

Batz M, Henke E, Kowalcyk B. Long-term consequences of foodborne infections. Infect Dis Clin N Am. 2013;27:599–616.

Batz M, Hoffmann S, Morris JG Jr. Disease-outcome trees, EQ-5D scores, and estimated annual losses of quality-adjusted life years (QALYs) for 14 foodborne pathogens in the United States. Foodborne Pathog Dis. 2014;11:395–402.

Becker GS. A theory of the allocation of time. Econ J (London). 1965;75:493–517.

Bouwknegt M, Knol AB, van der Sluijs JP, Evers EG. Uncertainty of population risk estimates for pathogens based on QMRA or epidemiology: a case study of *Campylobacter* in the Netherlands. Risk Anal. 2014;34:847–64.

Buzby JC, Frenzen PD. Food safety and product liability. Food Pol. 1999;24:637–51.

Buzby JC, Roberts T. The economics of enteric infections: human foodborne disease costs. Gastroenterology. 2009;136:1851–62.

Codex Alimentarius Commission. Codex Alimentarius Commission Procedural Manual. 16th ed. Joint FAO/WHO Food Standards Programme. 2006. ftp://ftp.fao.org/codex/Publications/ ProcManuals/Manual_16e.pdf. Accessed 08 May 2017.

Devleesschauwer B, Maertens de Noordhout C, Smit GSA, Duchateau L, Dorny P, Stein C, et al. Quantifying burden of disease to support public health policy in Belgium: opportunities and constraints. BMC Public Health. 2014a;14:1196.

Devleesschauwer B, Havelaar AH, Maertens de Noordhout C, Haagsma JA, Praet N, Dorny P, et al. Calculating disability-adjusted life years to quantify burden of disease. Int J Public Health. 2014b;59:565–9.

Devleesschauwer B, Havelaar AH, Maertens de Noordhout C, Haagsma JA, Praet N, Dorny P, et al. DALY calculation in practice: a stepwise approach. Int J Public Health. 2014c;59:571–4.

Devleesschauwer B, Haagsma JA, Angulo FJ, Bellinger DC, Cole D, Döpfer D, et al. Methodological framework for World Health Organization estimates of the global burden of foodborne disease. PLoS One. 2015;10:e0142498. https://doi.org/10.1371/journal.pone.0142498.

EFSA Panel on Biological Hazards (BIOHAZ). Scientific opinion on the development of a risk ranking framework on biological hazards. EFSA J. 2012;10:2724. https://doi.org/10.2903/j. efsa.2012.2724.

EFSA Panel on Biological Hazards (BIOHAZ). Scientific opinion on the development of a risk ranking toolbox for the EFSA BIOHAZ panel. EFSA J. 2015;13:3939.

Food and Agriculture Organization of the United Nations. Food safety risk analysis: a guide for national food safety authorities (food and nutrition paper 87). 2006. ftp://ftp.fao.org/docrep/ fao/009/a0822e/a0822e.pdf. Accessed 08 May 2017.

Grace D. Food safety in developing countries: an overview. Hemel Hempstead: Evidence on Demand; 2015. https://doi.org/10.12774/eod_er.oct2015.graced.

Grocery Manufacturers Association. Recall execution effectiveness: collaborative approaches to improving consumer safety and confidence. 2010. http://www.gmaonline.org/downloads/research-and-reports/WP_RecallExecution.pdf. Accessed 05 May 2017.

Grocery Manufacturers Association. Capturing recall costs: Measuring and recovering the losses. 2011. http://www.gmaonline.org/file-manager/images/gmapublications/Capturing_Recall_Costs_GMA_Whitepaper_FINAL.pdf. Accessed 05 May 2017.

Haagsma JA, Polinder S, Stein CE, Havelaar AH. Systematic review of foodborne burden of disease studies: quality assessment of data and methodology. Int J Food Microbiol. 2013;166:34–47. https://doi.org/10.1016/j.ijfoodmicro.2013.05.029.

Hammitt JK, Haninger K. Willingness to pay for food safety: sensitivity to duration and severity of illness. Am J Agric Econ. 2007;89:1170–5.

Haninger K, Hammitt JK. Diminishing willingness to pay per quality-adjusted life year: valuing acute foodborne illness. Risk Anal. 2011;31:1363–80.

Havelaar AH, Braunig J, Christiansen K, Cornu M, Hald T, Mangen MJ, et al. Towards an integrated approach in supporting microbiological food safety decisions. Zoonoses Public Health. 2007;54:103–17.

Havelaar AH, Kirk MD, Torgerson PR, Gibb HJ, Hald T, Lake RJ, et al. World Health Organization global estimates and regional comparisons of the burden of foodborne disease in 2010. PLoS Med. 2015;12:e1001923. https://doi.org/10.1371/journal.pmed.1001923.

Hayes DJ, Shogren JF, Shin SY, Kliebenstein JB. Valuing food safety in experimental auction markets. Am J Agric Econ. 1995;77:40–53.

Herdman M, Gudex C, Lloyd A, Janssen M, Kind P, Parkin D, Bonsel G, Badia X. Development and preliminary testing of the new five-level version of EQ-5D (EQ-5D-5L). Qual Life Res. 2011;20:1727–36. https://doi.org/10.1007/s11136-011-9903-x.

Hill AB. The environment and disease: association or causation? Proc R Soc Med. 1965;58:295.

Hoffmann S, Batz MB, Morris JG Jr. Annual cost of illness and quality-adjusted life year losses in the United States due to 14 foodborne pathogens. J Food Prot. 2012;75:1292–302.

Hoffmann S, Maculloch B, Batz M. Economic burden of major foodborne illnesses acquired in the United States. EIB-140, U.S. Department of Agriculture, Economic Research Service; 2015.

Hussain MA, Dawson CO. Economic impact of food safety outbreaks on food businesses. Foods. 2013;2:585–9.

Institute of Medicine. Improving food safety through a one health approach. Washington, D.C.: The National Academies Press; 2012.

Keithlin J, Sargeant JM, Thomas MK, Fazil A. Systematic review and meta-analysis of the proportion of non-typhoidal *Salmonella* cases that develop chronic sequelae. Epidemiol Infect. 2015;143:1333–51.

Kniesner TJ, Viscusi WK, Ziliak JP. Willingness to accept equals willingness to pay for labor market estimates of the value of a statistical life. J Risk Uncertain. 2014;48:187–205.

Koutsoumanis KP, Aspridou Z. Moving towards a risk-based food safety management. Curr Opin Food Sci. 2016;12:36–41. https://doi.org/10.1016/j.cofs.2016.06.008.

Mangen MJ, Batz MB, Käsbohrer A, Hald T, Morris JG, Taylor M, et al. Integrated approaches for the public health prioritization of foodborne and zoonotic pathogens. Risk Anal. 2010;30:782–97.

Mangen MJ, Plass D, Havelaar AH, Gibbons CL, Cassini A, Mühlberger N, BCoDE Consortium, et al. The pathogen- and incidence-based DALY approach: an appropriate [corrected] methodology for estimating the burden of infectious diseases. PLoS One. 2013;8:e79740.

Mauskopf JA, French MT. Estimating the value of avoiding morbidity and mortality from foodborne illnesses. Risk Anal. 1991;11:619–31.

Minor T, Lasher A, Klontz K, Brown B, Nardinelli C, Zorn D. The per case and total annual costs of foodborne illness in the United States. Risk Anal. 2015;35:1125–39.

Murphy JJ, Allen PG, Stevens TH, Weatherhead D. A meta-analysis of hypothetical bias in stated preference valuation. Environ Res Econ. 2005;30:313–25.

Murray CJ. Quantifying the burden of disease: the technical basis for disability-adjusted life years. Bull World Health Organ. 1994;72:429–45.

Murray CJ, Ezzati M, Flaxman AD, Lim S, Lozano R, Michaud C, et al. GBD 2010: design definitions and metrics. Lancet. 2012;380:2063–6.

National Research Council. Science and decisions: advancing risk assessment. Washington, DC: The National Academies Press; 2009.

National Research Council. Enhancing food safety: the role of the Food and Drug Administration. Washington, DC: The National Academies Press; 2010.

Nayga RM, Woodward R, Aiew W. Willingness to pay for reduced risk of foodborne illness: a nonhypothetical field experiment. Can J Agric Econ. 2006;54:461–75.

Ollinger M, Guthrie J, Bovay J. The food safety performance of ground beef suppliers to the national school lunch program. ERR-180, U.S. Department of Agriculture, Economic Research Service; 2014.

Palma MA, Ribera LA, Bessler D, Paggi M, Knutson RD. Potential impacts of foodborne illness incidences on market movements and prices of fresh produce in the US. J Agric Appl Econ. 2010;42:731–41.

Porter CK, Tribble DR, Aliaga PA, Halvorson HA, Riddle MS. Infectious gastroenteritis and risk of developing inflammatory bowel disease. Gastroenterology. 2008;135:781–6.

Pozo VF, Schroeder TC. Evaluating the costs of meat and poultry recalls to food firms using stock returns. Food Pol. 2016;59:66–77. https://doi.org/10.1016/j.foodpol.2015.12.007.

Resende-Filho MA, Buhr BL. Economics of traceability for mitigation of food recall costs. SSRN 995335. 2010; https://doi.org/10.2139/ssrn.995335.

Ribera LA, Plama MA, Paggi M, Knutson M, Masabni JG, Anciso J. Economic analysis of food safety compliance costs and foodborne illness outbreaks in the United States. HortTechnology. 2012;22:150–6.

Roberts T. Human illness costs of foodborne bacteria. Am J Agric Econ. 1989;71:468–74.

Roberts T, Kowalcyk B, Buck P, Blaser MJ, Frenkel JK, Lorber B, et al. The long-term health outcomes of selected foodborne pathogens. The Center for Foodborne Illness Research & Prevention. 2009. http://enhs.umn.edu/news/pdfs/CFIFinalReport.pdf. Accessed 05 May 2017.

Salin V, Hooker NH. Stock market reaction to food recalls. Rev Agric Econ. 2001;23:33–46.

Salomon JA, Haagsma JA, Davis A, de Noordhout CM, Polinder S, Havelaar AH, Cassini A, Devleesschauwer B, Kretzschmar M, Speybroeck N, Murray CJ, Vos T. Disability weights for the global burden of disease 2013 study. Lancet Glob Health. 2015;3:e712–23. https://doi.org/10.1016/S2214-109X(15)00069-8.

Scallan E, Hoekstra RM, Angulo FJ, Tauxe RV, Widdowson M-A, Roy SL, et al. Foodborne illness acquired in the United States—major pathogens. Emerg Infect Dis. 2011;17:7–15.

Scharff RL. Economic burden from health losses due to foodborne illness in the United States. J Food Prot. 2012;75:123–31.

Scharff RL. State estimates for the annual cost of foodborne illness. J Food Prot. 2015;78:1064–71.

Seo S, Jang SS, Miao L, Almanza B, Behnke C. The impact of food safety events on the value of food-related firms: an event study approach. Int J Hosp Manag. 2013;33:153–65.

Seo S, Jang SS, Almanza B, Miao L, Behnke C. The negative spillover effect of food crises on restaurant firms: did Jack in the box really recover from an E. coli scare? Int J Hosp Manag. 2014;39:107–21.

Shin SY, Kliebenstein J, Hayes DJ, Shogren JF. Consumer willingness to pay for safer food products. J Food Saf. 1992;13:51–9.

Shogren JF, Shin SY, Hayes DJ, Kliebenstein JB. Resolving differences in willingness to pay and willingness to accept. Am Econ Rev. 1994;84:255–70.

Teisl MF, Roe BE. Consumer willingness-to-pay to reduce the probability of retail foodborne pathogen contamination. Food Pol. 2010;35:521–30.

Todd EC. Economic loss from foodborne disease and non-illness related recalls because of mishandling by food processors. J Food Prot. 1985;48:621–33.

Torrance GW. Measurement of health state utilities for economic appraisal: a review. J Health Econ. 1986;5:1–30.

U.S. Food and Drug Administration. FDA's draft approach for designating high-risk foods as required by section 204 of FSMA. 2014. https://www.fda.gov/downloads/food/guidanceregulation/fsma/ucm380212.pdf. Accessed 08 May 2017.

U.S. Food and Drug Administration. FDA food safety modernization act (FSMA). 2017. https://www.fda.gov/Food/GuidanceRegulation/FSMA/default.htm. Accessed 08 May 2017.

U.S. Office of Management and Budget. Circular A-4: Regulatory analysis. Washington, D.C.: U.S. Office of Management and Budget; 2003.

Viscusi WK, Aldy JE. The value of a statistical life: a critical review of market estimates throughout the world. J Risk Uncertain. 2003;27:5–76.

World Health Organization. WHO methods and data sources for global burden of disease estimates 2000–2011. Global Health Estimates Technical Paper. WHO/HIS/HSI/GHE/2013.4. 2013. http://www.who.int/healthinfo/statistics/GlobalDALYmethods_2000_2011.pdf. Accessed 05 May 2017.

World Health Organization, Food and Agriculture Organization of the United Nations. Principles and methods for the risk assessment of chemicals in food. Geneva: World Health Organization; 2009. http://www.who.int/foodsafety/chem/dose_response.pdf. Accessed 08 May 2017.

Chapter 7
The Global Burden of Foodborne Disease

Brecht Devleesschauwer, Juanita A. Haagsma, Marie-Josée J. Mangen, Robin J. Lake, and Arie H. Havelaar

Abbreviations

DALY   Disability-adjusted life year
FBD   Foodborne diseases
FERG   Foodborne Disease Burden Epidemiology Reference Group
WHO   World Health Organization
YLD   Years lived with disability
YLL   Years of life lost

B. Devleesschauwer (✉)
Department of Public Health and Surveillance, Scientific Institute of Public Health (WIV-ISP), Brussels, Belgium
e-mail: brecht.devleesschauwer@wiv-isp.be

J. A. Haagsma
Department of Public Health, Erasmus MC, Rotterdam, The Netherlands
e-mail: j.haagsma@erasmusmc.nl

M.-J. J. Mangen
Centre for Infectious Disease Control, National Institute for Public Health and the Environment (RIVM), Bilthoven, The Netherlands
e-mail: marie-josee.mangen@rivm.nl

R. J. Lake
Institute of Environmental Science and Research, Christchurch, New Zealand
e-mail: rob.lake@esr.cri.nz

A. H. Havelaar
Department of Animal Sciences, Institute for Sustainable Food Systems, Emerging Pathogens Institute, University of Florida, Gainesville, FL, USA
e-mail: ariehavelaar@ufl.edu

© Springer International Publishing AG, part of Springer Nature 2018   107
T. Roberts (ed.), *Food Safety Economics*, Food Microbiology and Food Safety,
https://doi.org/10.1007/978-3-319-92138-9_7

7.1   Why Estimate the Global Burden of Foodborne Disease?

Foodborne diseases (FBD) present a constant threat to public health and a significant impediment to socioeconomic development worldwide. At the same time, food
safety remains a marginalized policy objective, especially in developing countries.
A major obstacle to adequately addressing food safety concerns is the lack of accurate data on the full extent, burden, and cost of FBD. Very few nations have assessed
their FBD burden, and information on the global burden of FBD has long been lacking. Several reasons may explain this knowledge gap. Although the potential threat
of FBD has long been recognized, epidemiological data on FBD remain scarce,
particularly in the developing world. Foodborne outbreaks may go unrecognized if
they are not connected to major public health or economic impact. Outbreaks are
only the tip of the iceberg; many more infections occur sporadically and often
remain unreported. Furthermore, the health effects of FBD are highly complex,
reaching far beyond acute gastroenteritis. Indeed, FBD may be caused by numerous
microbiological and chemical hazards and lead to a variety of health outcomes and
effects on different time scales. Certain diseases that may result from chronic exposure to contaminated food, such as cancer and kidney or liver failure, have multiple
causes, and the causal link is difficult to assess for individual cases. When taking a
global perspective, the sheer complexity of the problem becomes even more challenging, as the path from food production to food consumption across the globe is
highly diverse, and the range of potential contaminants in the food chain is astound-
ing and varies according to food type. Finally, to add to the complexity, food is not
the only transmission pathway of many food-related hazards, requiring a clear
delineation and quantification of the main transmission routes of food-related hazards. Figure 7.1 shows the complexity of transmission pathways that may exist for
a single hazard. It also illustrates the reservoir level and the exposure level as two
distinct potential points of attribution, each of which may be relevant depending on
where risk management is to be applied (Hald et al. 2016).

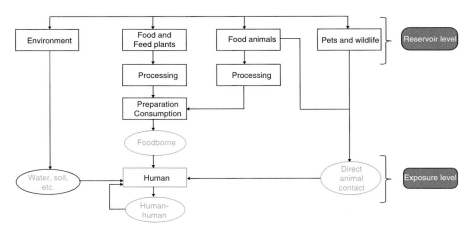

Fig. 7.1 Major transmission routes and points of attribution of human foodborne disease (Hald
et al. 2016)

To address these gaps, the World Health Organization (WHO) launched an initiative in 2006 to estimate the global burden of FBD. This initiative was carried forward by the Foodborne Disease Burden Epidemiology Reference Group (FERG), an expert group convened by WHO in 2007. In addition to providing estimates of the global burden of FBD by age, sex, and region, FERG was also tasked with strengthening country capacity to assess FBD burden, encouraging the use of FBD burden estimates to set evidence-informed policies, and increasing awareness and commitment to implement food safety standards. In 2015, FERG published the first-ever estimates of the global and regional burden of FBD (Havelaar et al. 2015; WHO 2015a).

In this chapter, we describe the methodological framework developed by FERG for estimating the global burden of FBD and present the key findings at a global and regional level.

7.2   Methodological Framework for WHO Estimates of the Global Burden of Foodborne Disease

FERG established five task forces focusing on groups of hazards (chemical, enteric, parasitic) or aspects of the methodology (source attribution, computation). The work of task force members was augmented by additional support from external resource advisors. The computational task force was responsible for integrating the work of the other task forces on DALY inputs and implementing FERG's methodological framework to generate DALY estimates (Fig. 7.2). This framework was structured around five distinct components leading to estimates of the global burden of FBD for the year 2010, expressed as disability-adjusted life years (DALYs): disease models and epidemiological data, imputation model, disability weights, probabilistic burden assessment, and source attribution.

In a first step, the hazard-specific task forces commissioned systematic reviews and other studies to provide the baseline epidemiological data needed to calculate burden estimates. This was done for 31 foodborne hazards that were chosen by each task force from a comprehensive list of hazards, taking into account presumed significance for the global burden of FBD and data availability. These 31 hazards included 11 diarrheal disease agents, 7 invasive disease agents, 10 helminths, and 3 chemicals and toxins (Table 7.1). For five hazards, including four bacterial toxins and one allergen, the data were found insufficient to generate global estimates, and burden estimates were presented for high-income regions only.

The epidemiological data were used to define and parameterize so-called disease models or outcome trees. These are schematic representations of the health states that are causally associated with the specific hazard. As a result, the burden of a foodborne hazard could be defined and quantified as the burden resulting from all related health states, including acute illness, chronic sequelae, and death. Across all considered hazards, 75 distinct health states were identified, highlighting the diverse nature of the health impact of FBD (Table 7.1). Where needed, the disease

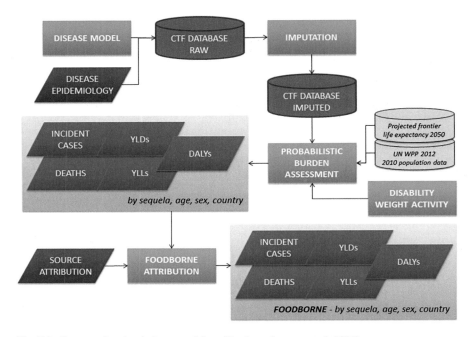

Fig. 7.2 Computational task force workflow (Devleesschauwer et al. 2015)

model further included an underestimation factor to correct the incidence data for underreporting and underascertainment (Gibbons et al. 2014). Finally, all retrieved information was compiled in a standardized spreadsheet database.

Even though all efforts were made to retrieve the best available epidemiological estimates, many data gaps remained, particularly for some of the world's most populous countries such as China, India, and Indonesia. FERG used statistical models to estimate these missing data from the available data and to quantify the associated uncertainties on a regional basis (Ezzati et al. 2002). Motivated by a strive for parsimony and transparency, a hierarchical Bayesian lognormal random effects model was adopted as the default model for imputing missing country-level incidence data (McDonald et al. 2015). After fitting this model to the available data, incidence values for countries with no data for a particular hazard were imputed based on the resulting posterior predictive distributions. For countries in a region where at least one of the other countries had data, the incidence was imputed as multiple random draws from a lognormal distribution reflecting a "random" country within the concerned region, with the uncertainty interval describing the variability within regions. For countries in a region where none of the countries had data, the incidence was imputed as multiple random draws from a lognormal distribution reflecting a "random" country within a "random" region, with the uncertainty interval describing the variability between and within regions. Of the 14 hazards to which

Table 7.1 Hazards and associated health states considered by the Foodborne Disease Burden Epidemiology Reference Group for quantifying the global burden of foodborne disease

Hazards	Health states
Diarrheal disease agents	
Viruses	
Norovirus	Diarrheal disease
Bacteria	
Campylobacter spp.	Diarrheal disease, Guillain-Barré syndrome
Enteropathogenic *E. coli*	Diarrheal disease
Enterotoxigenic *E. coli*	Diarrheal disease
Shiga toxin-producing *E. coli*	Diarrheal disease, hemolytic uremic syndrome, end-stage renal disease
Non-typhoidal *S. enterica*	Diarrheal disease, invasive salmonellosis
Shigella spp.	Diarrheal disease
Vibrio cholerae	Diarrheal disease
Protozoa	
Cryptosporidium spp.	Diarrheal disease
Entamoeba histolytica	Diarrheal disease
Giardia spp.	Diarrheal disease
Invasive infectious disease agents	
Viruses	
Hepatitis A virus	Hepatitis
Bacteria	
Brucella spp.	Acute brucellosis, chronic brucellosis, orchitis
Listeria monocytogenes	*Perinatal*: sepsis, central nervous system infection, neurological sequelae
	Acquired: sepsis, central nervous system infection, neurological sequelae
Mycobacterium bovis	Tuberculosis
Salmonella Paratyphi	Paratyphoid fever, liver abscesses, and cysts
Salmonella Typhi	Typhoid fever, liver abscesses, and cysts
Protozoa	
Toxoplasma gondii	*Congenital*: intracranial calcification, hydrocephalus, chorioretinitis early in life, chorioretinitis later in life, CNS abnormalities
	Acquired: chorioretinitis, acute illness, post-acute illness
Enteric intoxications	
Bacillus cereus[a]	Acute intoxication
Clostridium botulinum[a]	Moderate/mild botulism, severe botulism

(continued)

Table 7.1 (continued)

Hazards	Health states
Clostridium perfringens[a]	Acute intoxication
Staphylococcus aureus[a]	Acute intoxication
Helminths	
Cestodes	
Echinococcus granulosus	*Cases seeking treatment*: pulmonary cystic echinococcosis, hepatic cystic echinococcosis, central nervous system cystic echinococcosis
	Cases not seeking treatment: pulmonary cystic echinococcosis, hepatic cystic echinococcosis, central nervous system cystic echinococcosis
Echinococcus multilocularis	Alveolar echinococcosis
Taenia solium	Epilepsy, treated, seizure-free; epilepsy, treated, with recent seizures; epilepsy, severe; epilepsy, untreated
Nematodes	
Ascaris spp.	Ascariasis infestation, mild abdominopelvic problems due to ascariasis, severe wasting due to ascariasis
Trichinella spp.	Acute clinical trichinellosis
Trematodes	
Clonorchis sinensis	Abdominopelvic problems due to heavy clonorchiosis
Fasciola spp.	Abdominopelvic problems due to heavy fasciolosis
Intestinal flukes[b]	Abdominopelvic problems due to heavy intestinal fluke infections
Opisthorchis spp.	Abdominopelvic problems due to heavy opisthorchiasis
Paragonimus spp.	Central nervous system problems due to heavy paragonimiasis, pulmonary problems due to heavy paragonimiasis
Chemicals and toxins	
Aflatoxin	Hepatocellular carcinoma, diagnosis and primary therapy; hepatocellular carcinoma, metastatic; hepatocellular carcinoma, terminal phase with medication; hepatocellular carcinoma, terminal phase without medication
Cyanide in cassava	Konzo
Dioxin	Hypothyroid due to prenatal exposure, hypothyroid due postnatal exposure, male infertility
Peanut allergens[a]	Living with peanut-induced allergy

Adapted from Havelaar et al. (2015)
[a]Excluded from global burden assessments
[b]Includes *Echinostoma* spp., *Fasciolopsis buski*, *Heterophyes* spp., *Metagonimus* spp., and other foodborne intestinal trematode species (depending on data availability)

the random effects imputation model was applied, the Southeast Asian and Latin American regions were the ones for which most often no data could be identified. At a country level, at least one hazard had to be imputed for each country, while Cambodia had the highest number of data gaps, i.e., ten hazards with no data (Devleesschauwer et al. 2015). Figure 7.3 plots the number of data gaps per country.

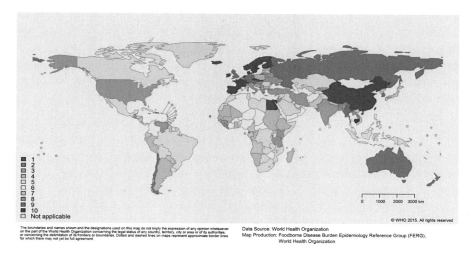

The boundaries and names shown and the designations used on this map do not imply the expression of any opinion whatsoever on the part of the World Health Organization concerning the legal status of any country, territory, city or area or of its authorities, or concerning the delimitation of its frontiers or boundaries. Dotted and dashed lines on maps represent approximate border lines for which there may not yet be full agreement.

Data Source: World Health Organization
Map Production: Foodborne Disease Burden Epidemiology Reference Group (FERG), World Health Organization

Fig. 7.3 Number of hazards requiring imputation per country (Devleesschauwer et al. 2015)

In a next step, the retrieved and imputed epidemiological data were translated into DALYs. DALYs combine years lived with disability (YLD) and years of life lost (YLL) due to premature mortality into a single estimate of healthy life-years lost. FERG used an incidence perspective for calculating YLDs, which defines YLDs as the product of the number of incident cases and the duration and severity of the health state. The estimates thus reflect the future health losses due to foodborne infections acquired in 2010. Compared to a prevalence perspective, which is for instance used in the recent iterations of the Global Burden of Disease study (2016), the incidence perspective was deemed to be more sensitive to current epidemiological trends and more consistent with the estimation of YLLs. To quantify the severity of health states, FERG adopted the disability weights used in the WHO Global Health Estimates. These, in turn, were largely based on the disability weights developed for the Global Burden of Disease 2010 study, which were based on population health equivalence and pairwise comparison surveys conducted face to face in Bangladesh, Indonesia, Peru, and Tanzania, telephone-based in the United States, and an open access web-based survey (Salomon et al. 2012). To estimate the YLLs due to premature mortality, FERG used as residual life expectancy table the highest United Nations projected life expectancy for 2050, with a life expectancy at birth of 92 years for both sexes. In line with current practice, age weighting and time discounting were not applied.

Many foodborne hazards are not exclusively transmitted by food; therefore, a separate effort was set up for the attribution of exposure to different sources, including food, the environment, and direct contact between humans or with animals. As many data are lacking for attribution, it was decided to apply structured expert elicitation to provide a consistent set of estimates. The global expert elicitation study involved 73 experts and 11 elicitors and was one of the largest, if not the largest study, of this kind ever undertaken (Hald et al. 2016). Due to the study constraints (remote elicitation instead of face-to-face meetings), individual experts' accuracies, elicited based on calibration questions, were generally lower than in other structured expert judgment studies. However, performance-based weighting, a key characteristic of Cooke's classical model, increased informativeness while retaining accuracy at acceptable levels (Aspinall et al. 2016).

All calculations were performed in a probabilistic framework, in which parameter, imputation and attribution uncertainties were propagated to the final foodborne DALY estimates by Monte Carlo simulation. The resulting uncertainty distributions were summarized by their median and 95% uncertainty interval. Estimates were presented per hazard, outcome, and age group (< or ≥5 years). Due to the limitations in data availability, FERG decided to present its estimates on a regional level only, even though all calculations were performed on a national level. The regional estimates are considered more robust as they build on data from several countries in most regions. It should however be noted that the regional estimates do not reflect the diversity of risks between countries in a region, or even within a country.

7.3   Global Estimates and Regional Comparisons of the Global Burden of Foodborne Disease

FERG estimated that in 2010, the 31 considered hazards caused 600 million foodborne illnesses, implying that roughly one out of every 10 people in the world would suffer from FBD annually. These illnesses were estimated to lead to 420,000 deaths and 33 million DALYs, making the global burden of FBD comparable to those of the major infectious diseases, HIV/AIDS, malaria, and tuberculosis (WHO 2015b) and comparable to certain other risk factors such as dietary risk factors, unimproved water and sanitation, and air pollution (GBD 2015 DALYs and HALE Collaborators 2016). Diarrheal disease agents accounted for more than 90% of all foodborne illnesses, but just over half of all foodborne deaths and DALYs – reflecting the fact that many diarrheal episodes are relatively benign (Table 7.2).

Table 7.2 Global burden of foodborne disease, 2010, by broad hazard groups

Hazard group	Foodborne illnesses (millions)	Foodborne deaths (thousands)	Foodborne disability-adjusted life years (millions)
All	600	420	33
Diarrheal disease agents	549	230	18
Invasive infectious disease agents	36	117	8
Helminths	13	45	6
Chemicals	0.2	19	0.9

Adapted from Havelaar et al. (2015)

Table 7.3 Major foodborne hazards contributing to the global burden of foodborne disease

#	Hazard	Estimate
Foodborne illnesses		
1	Norovirus	124,803,946
2	*Campylobacter* spp.	95,613,970
3	ETEC	86,502,735
4	NTS	78,707,591
5	*Shigella* spp.	51,014,050
6	*Giardia* spp.	28,236,123
7	*Entamoeba histolytica*	28,023,571
8	EPEC	23,797,284
9	Hepatitis A virus	13,709,836
10	*Ascaris* spp.	12,280,767
Foodborne deaths		
1	NTS	59,153
2	*Salmonella Typhi*	52,472
3	EPEC	37,077
4	Norovirus	34,929
5	*Taenia solium*	28,114
6	Hepatitis A virus	27,731
7	ETEC	26,170
8	*Vibrio cholerae*	24,649
9	*Campylobacter* spp.	21,374
10	Aflatoxin	19,455
Foodborne disability-adjusted life years		
1	NTS	4,067,929
2	*Salmonella Typhi*	3,720,565
3	EPEC	2,938,407
4	*Taenia solium*	2,788,426
5	Norovirus	2,496,078
6	*Campylobacter* spp.	2,141,926
7	ETEC	2,084,229
8	*Vibrio cholerae*	1,722,312
9	Hepatitis A virus	1,353,767
10	*Shigella* spp.	1,237,103

Adapted from Havelaar et al. (2015)
NTS non-typhoidal *Salmonella enterica*, *EPEC* enteropathogenic *Escherichia coli*, *ETEC* enterotoxigenic *Escherichia coli*

Table 7.3 shows the ten major foodborne hazards contributing to the global foodborne illnesses, deaths, and DALYs. The majority of foodborne illnesses were
caused by norovirus and other diarrheal disease agents, while non-typhoidal
Salmonella enterica was the major cause of foodborne deaths and DALYs. The
three included chemicals and toxins resulted in nearly 1 million foodborne DALYs,
a non-negligible share of the overall FBD burden. However, as there are many more

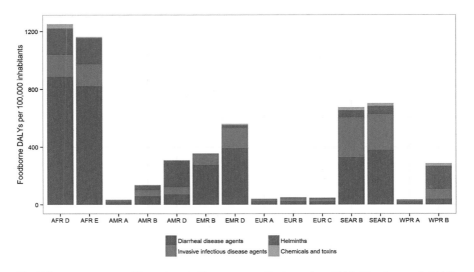

Fig. 7.4 Foodborne disability-adjusted life years (DALYs) by region, 2010 (Havelaar et al. 2015). *AFR* African Region, *AMR* Region of the Americas, *EMR* Eastern Mediterranean Region, *EUR* European Region, *SEAR* Southeast Asian Region, *WPR* Western Pacific Region; Strata A–E further subdivide the regions from low to high child and adult mortality, as documented by Ezzati et al. (2002)

chemical food contaminants beyond those included, the true disease burden of chemical foodborne hazards is expected to be considerably larger.

Figure 7.4 shows the estimated DALY rates per 100,000 person-years for the 14 considered regions, with a breakdown by four broad hazard groups, i.e., diarrheal disease agents, invasive infectious disease agents, helminths, and chemicals and toxins. There were considerable variations in disease burden across regions, confirming the close link between FBD and development. Indeed, while making up 41% of the world population, individuals living in low-income regions suffered from 53% of all foodborne illnesses, succumbed to 75% of all foodborne deaths, and bore 72% of the global foodborne DALYs. Specifically, the African regions were most affected (more than 1000 foodborne DALYs per 100,000 person-years), followed by the Southeast Asian regions (700 foodborne DALYs per 100,000 person-years). The European regions and the high-income American and Western Pacific regions on the other hand had the lowest foodborne disease burden, with 30–50 foodborne DALYs per 100,000 person-years. High-income countries have been largely successful in controlling foodborne deaths, partly by reducing exposure to hazards with high case-fatality rates but also because of better healthcare systems, leading to, e.g., much lower case-fatality rates for diarrheal disease. In contrast with these accomplishments, high-income countries have been less successful in controlling the incidence of FBD, which is only three- to four-folds lower than the global average (Table 7.4).

The pattern of contributing hazards also showed marked differences across regions. Bacterial agents were the dominant pathogens in most regions, i.e., nontyphoidal *S. enterica* in the African and European regions, *Salmonella Typhi* in

Table 7.4 Burden of foodborne disease in high-income regions, 2010

Metric (per 100,000)	Global average	AMR A (North America)	EUR A (Western Europe)	WPR A (Australia, New Zealand, Japan)
Incidence	8729	2577	2431	2798
Deaths	6	0.4	0.5	0.4
DALYs	477	35	41	36

Adapted from Havelaar et al. (2015)

Southeast Asian regions, and *Campylobacter* spp. in the eastern Mediterranean regions and the high-income American and Western Pacific regions. Parasites were the dominant pathogens in the remaining regions, i.e., the pork tapeworm (*Taenia solium*) in the middle- and low-income American regions and the lung fluke (*Paragonimus* spp.) in the middle-income Western Pacific region. Peanut allergy was a significant contributor to the foodborne disease burden in high-income regions, but data limitations did not allow generating estimates for other regions. Despite these differences, diseases caused by non-typhoidal *S. enterica*, *Campylobacter* spp., and *Toxoplasma gondii* were found to be a public health concern across the world.

Infants and young children are at particular risk of contracting and dying from common food-related diseases due to their immature immune system and their lack of protective immunity due to few past exposures. Even though children under the age of 5 make up only 9% of the world population, FERG estimated that they suffered from 38% of all foodborne illnesses, succumbed to 30% of all foodborne deaths, and bore 40% of global foodborne DALYs. The important contribution of children to the burden of FBD explains for a large part the relatively high burden of FBD in the African and Southeast Asian regions. Furthermore, at a global level, pre- and perinatal infections accounted for 21% of the burden of *Listeria monocytogenes* and for 32% of the burden of *Toxoplasma gondii*.

7.4 Discussion

The FERG estimates provide the first-ever comprehensive picture of the substantial global burden of FBD and address the lack of data to support food safety policy making. The estimates highlight significant differences between low- and high-income regions, suggesting that FBD are largely preventable by currently available methods. The WHO works with governments and stakeholders to implement effective food safety systems, which require preventive, risk-based and enabling methods, instead of reactive and repressive ones. These systems need to be complemented by effective laboratory-based surveillance networks at country, regional, and global levels, in order to monitor progress and detect emerging risks. In resource-poor settings, however, implementation of effective food safety systems may not receive sufficient priority. There is therefore an urgent need to develop cost-effective food

Fig. 7.5 World Health Organization's Five Keys to Safer Food (http://www.who.int/foodsafety/
areas_work/food-hygiene/5keys/en/)

hygiene interventions that can be implemented in such settings. High-income countries need to continue investing in food safety in order to maintain the current safety
levels. Hazards that remain of importance in these countries, such as *Salmonella*,
Campylobacter, and *Toxoplasma*, require novel control methods.

In addition to governments and food industries, consumers also play an important role in preventing FBD. The WHO calls on consumers and food handlers to
handle and prepare food safely, following the "Five Keys to Safer Food," i.e., keep
clean, separate raw and cooked, cook thoroughly, keep at safe temperatures, and use
safe water and raw materials (Fig. 7.5).

Even though the current FERG estimates show that the global burden of FBD is
considerable, the true FBD burden is expected to be even higher. Due to data
limitations and limited resources, only 31 foodborne hazards could be included.
The included microbiological hazards were the ones that were a priori deemed to

contribute most to the global burden and for which sufficient global data were available. A systematic review of the incidence of diarrheal illness commissioned by FERG was only able to attribute half of the incidence to the diarrheal disease agents included (Pires et al. 2015). A significant proportion of the unattributed incidence is likely to be due to foodborne pathogens, and so it is evident that the total foodborne burden including these remaining and unknown etiologies will be considerably higher.

Estimation of the burden of foodborne disease from chemical hazards presents specific challenges, particularly due to the lack of well-established methods for attributing disease incidence to chemical exposures. Due to model uncertainties (such as observed discrepancies between multiplicative and additive models) and a lack of data, global estimates could be generated for only three chemical hazards (aflatoxin, cassava cyanide, and dioxins) and for only few associated health states (liver cancer, konzo, hypothyroidism, and infertility, respectively)—despite the vast spectrum of chemical food contaminants. Indeed, heavy metals such as cadmium, lead, and methyl mercury are known risk factors for various metabolic disorders, while arsenic is associated with several cancers. Various food allergens and fish toxins may cause potentially fatal acute intoxications. Estimates of the burden for these chemicals would provide a much more comprehensive understanding of the impact that chemicals in the food supply have on the burden of disease (Gibb et al. 2015).

Further underestimation of the global burden of FBD resulted from the fact that not all endpoints could be considered for the included hazards, e.g., malnutrition and stunting due to diarrheal agents, post-infectious irritable bowel syndrome due to non-typhoidal *S. enterica*, and psychiatric consequences of *Toxoplasma gondii* infection. Finally, for non-typhoidal *S. enterica*, infections among the HIV-associated cases were excluded, even though non-typhoidal *S. enterica* infections in HIV positives are preventable by food safety interventions.

Data availability and data quality issues were encountered for all hazards across all regions, but particularly in low-income countries. To address these issues, there was a need for imputation and expert judgment, often resulting in large uncertainty intervals. Documenting these gaps and uncertainties would hopefully serve as an impetus for countries to conduct new epidemiological studies and to undertake national FBD studies, thereby adding to the evidence base that is required to generate an even better picture of the global burden of FBD. To support this goal and help countries develop capacity for national FBD studies, a sixth Country Studies Task Force was established by FERG (Lake et al. 2015). This task force developed a suite of tools to assist with the development of DALY estimates and conducted four pilot studies in individual countries. The availability of the FERG regional estimates provides an opportunity to address many of the data gaps faced by individual countries in developing national estimates. Currently the tools are being updated to incorporate the FERG results, and it is hoped that this resource will stimulate additional studies by individual countries.

The FERG methodological framework is to date the most comprehensive effort for generating comparable estimates of the global burden of FBD, but has some key limitations. First, the results were only presented at a regional level, even though

FBD burden may vary significantly between countries and even within countries. Second, the available data did not allow for modeling time trends in FBD burden. Third, comorbidities were not systematically taken into account, except for the possible associations between HIV and invasive salmonellosis or tuberculosis. Finally, the framework does not explicitly address the financial burden of FBD, but merely focuses on the intangible costs of illness and premature mortality expressed as DALYs. Although disease burden data for populations could be translated into economic metrics, additional financial costs related to illness such as healthcare costs, patient costs, and costs to other sectors, and particularly the value of lost production due to illness, are not included (e.g., Mangen et al. 2015; Scharff 2015), nor are the potentially substantial outbreak investigation and control costs that occur in the case of a community-acquired (food-related) outbreak (Suijkerbuijk et al. 2016). Nonetheless, FERG acknowledges that estimates of the economic burden of food-borne disease could have greater impact with those responsible for setting policy. It should be noted, however that, by providing regional estimates of the incidence of the multitude of health outcomes from foodborne disease, FERG has addressed one of the fundamental inputs into developing cost-of-illness estimates.

7.5   Conclusion

The global burden of FBD is considerable and of the same order as the major infectious diseases such as HIV/AIDS, malaria, and tuberculosis. It is also comparable to certain other risk factors such as dietary risk factors, unimproved water and sanitation, and air pollution. FBD affect everyone, but particularly children under the age of 5 and persons living in low-income regions of the world. Although reported data underestimate the true FBD burden and not all foodborne hazards have been included, the FERG estimates may be used by national and international stakeholders to support evidence-based priorities and contribute to improvements in food safety and population health.

FERG generated the first global and regional estimates of the burden of FBD, demonstrating that the global burden of FBD is considerable and of the same order as the major infectious diseases such as HIV/AIDS, malaria, and tuberculosis. It is also comparable to certain other risk factors such as dietary risk factors, unimproved water and sanitation, and air pollution. FBD affect individuals of all ages, but show a disproportionately high burden in children under the age of 5. Furthermore, a disproportionately high burden was established for the low-income regions of the world and for the African and Southeast Asian regions in particular. Although some hazards, such as non-typhoidal *Salmonella enterica*, *Campylobacter* spp., and *Toxoplasma gondii*, were found to be important causes of FBD in all regions of the world, others, such as *Salmonella Typhi*, *Taenia solium*, and *Paragonimus* spp., were of highly focal nature, resulting in high local burden and calling for context-specific policies.

By using these estimates to support evidence-based priorities, all stakeholders, both at national and international levels, can contribute to improvements in food safety and population health.

References

Aspinall WP, Cooke RM, Havelaar AH, Hoffmann S, Hald T. Evaluation of a performance-based expert elicitation: WHO global attribution of foodborne diseases. PLoS One. 2016;11:e0149817. https://doi.org/10.1371/journal.pone.0149817.

Devleesschauwer B, Haagsma JA, Angulo FJ, Bellinger DC, Cole D, Döpfer D, et al. Methodological framework for World Health Organization estimates of the global burden of foodborne disease. PLoS One. 2015;10:e0142498. https://doi.org/10.1371/journal.pone.0142498.

Ezzati M, Lopez AD, Rodgers A, Vander Hoorn S, Murray CJ. Comparative risk assessment collaborating group. Selected major risk factors and global and regional burden of disease. Lancet. 2002;360:1347–60. https://doi.org/10.1016/S0140-6736(02)11403-6.

GBD 2015 DALYs and HALE Collaborators. Global, regional, and national disability-adjusted life-years (DALYs) for 315 diseases and injuries and healthy life expectancy (HALE), 1990-2015: a systematic analysis for the global Burden of disease study 2015. Lancet. 2016;388:1603–58. https://doi.org/10.1016/S0140-6736(16)31460-X.

GBD 2015 Disease and Injury Incidence and Prevalence Collaborators. Global, regional, and national incidence, prevalence, and years lived with disability for 310 diseases and injuries, 1990-2015: a systematic analysis for the global Burden of disease study 2015. Lancet. 2016;388:1545–602. https://doi.org/10.1016/S0140-6736(16)31678-6.

Gibb H, Devleesschauwer B, Bolger PM, Wu F, Ezendam J, Cliff J, et al. World Health Organization estimates of the global and regional disease burden of four foodborne chemical toxins, 2010: a data synthesis. F1000Res. 2015;4:1393. https://doi.org/10.12688/f1000research.7340.1.

Gibbons CL, Mangen MJ, Plass D, Havelaar AH, Brooke RJ, Kramarz P, et al. Measuring underreporting and under-ascertainment in infectious disease datasets: a comparison of methods. BMC Public Health. 2014;14:147. https://doi.org/10.1186/1471-2458-14-147.

Hald T, Aspinall W, Devleesschauwer B, Cooke R, Corrigan T, Havelaar AH, et al. World Health Organization estimates of the relative contributions of food to the burden of disease due to selected foodborne hazards: a structured expert elicitation. PLoS One. 2016;11:e0145839. https://doi.org/10.1371/journal.pone.0145839.

Havelaar AH, Kirk MD, Torgerson PR, Gibb HJ, Hald T, Lake RJ, et al. World Health Organization global estimates and regional comparisons of the burden of foodborne disease in 2010. PLoS Med. 2015;12:e1001923. https://doi.org/10.1371/journal.pmed.1001923.

Lake RJ, Devleesschauwer B, Nasinyama G, Havelaar AH, Kuchenmüller T, Haagsma JA, et al. National studies as a component of the World Health Organization initiative to estimate the global and regional burden of foodborne disease. PLoS One. 2015;10:e0140319. https://doi.org/10.1371/journal.pone.0140319.

Mangen MJ, Bouwknegt M, Friesema IH, Haagsma JA, Kortbeek LM, Tariq L, et al. Cost-of-illness and disease burden of food-related pathogens in the Netherlands, 2011. Int J Food Microbiol. 2015;196:84–93. https://doi.org/10.1016/j.ijfoodmicro.2014.11.022.

McDonald SA, Devleesschauwer B, Speybroeck N, Hens N, Praet N, Torgerson PR, et al. Data-driven methods for imputing national-level incidence rates in global burden of disease studies. Bull World Health Organ. 2015;93:228–36. https://doi.org/10.2471/BLT.14.139972.

Pires SM, Fischer-Walker CL, Lanata CF, Devleesschauwer B, Hall AJ, Kirk MD, et al. Aetiology-specific estimates of the global and regional incidence and mortality of diarrhoeal diseases commonly transmitted through food. PLoS One. 2015;10:e0142927. https://doi.org/10.1371/journal.pone.0142927.

Salomon JA, Vos T, Hogan DR, Gagnon M, Naghavi M, Mokdad A, et al. Common values in assessing health outcomes from disease and injury: disability weights measurement study for the global Burden of disease study 2010. Lancet. 2012;380:2129–43. https://doi.org/10.1016/S0140-6736(12)61680-8.

Scharff RL. State estimates for the annual cost of foodborne illness. J Food Prot. 2015;78:1064–71. https://doi.org/10.4315/0362-028X.JFP-14-505.

Suijkerbuijk AW, Bouwknegt M, Mangen MJ, de Wit GA, van Pelt W, Bijkerk P, et al. The economic burden of a *Salmonella* Thompson outbreak caused by smoked salmon in the Netherlands 2012-2013. Eur J Pub Health. 2016; https://doi.org/10.1093/eurpub/ckw205.

World Health Organization. WHO estimates of the global burden of foodborne diseases. Foodborne Disease Burden Epidemiology Reference Group 2007–2015. Geneva: WHO Press; 2015a. http://apps.who.int/iris/bitstream/10665/199350/1/9789241565165_eng.pdf?ua=1. Accessed 08 May 2017

World Health Organization. Global Health estimates 2015: disease burden by cause, age, sex, by country and by region, 2000-2015. Geneva: World Health Organization; 2015b. http://www.who.int/healthinfo/global_burden_disease/estimates/en/index2.html. Accessed 08 May 2017

Chapter 8
The Economic Burden of Foodborne Illness in the United States

Robert L. Scharff

Abbreviations

CDC	Centers for Disease Control Research and Prevention/US
COI	Cost of illness
CPI	Consumer Price Index
FDA	Food and Drug Administration/US
ICD-9	The International Classification of Diseases, Ninth Revision
NIS	National Inpatient Sample
NDSS	National Notifiable Diseases Surveillance System
NORS	National Outbreak Reporting System
PFGE	Pulsed-field gel electrophoresis
PulseNet	National Molecular Subtyping Network for Foodborne Disease Surveillance
QALY	Quality-adjusted life year
STEC	Shiga-toxin *E. coli*
VSL	Value of a statistical life

8.1 Introduction

A number of methods for estimating the cost of foodborne illness have been employed both inside and outside the United States. In Chap. 6, many of these methods are described. In this chapter I present cost-of-illness estimates based on the methods employed by Scharff (2012, 2015). The Scharff approach integrates CDC estimates for incidence of illness (Scallan et al. 2011a, b) with estimates from alternative cost-of-illness models to illustrate the annual economic burden from foodborne illness in the United States. These measures are useful as metrics for prioritizing risk mitigation efforts, assessing whether given interventions are economically justified, and communicating the importance of the problem to the public.

R. L. Scharff (✉)
Department of Human Sciences, The Ohio State University, Columbus, OH, USA
e-mail: Scharff.8@osu.edu

© Springer International Publishing AG, part of Springer Nature 2018
T. Roberts (ed.), *Food Safety Economics*, Food Microbiology and Food Safety,
https://doi.org/10.1007/978-3-319-92138-9_8

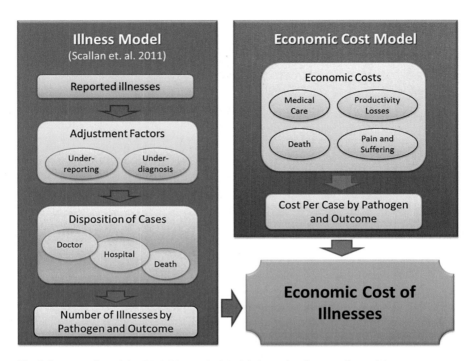

Fig. 8.1 Integration of the CDC Illness Model with the updated economic model

Figure 8.1 illustrates the general approach used to estimate economic costs associated with foodborne illnesses. Essentially, a full replication of the CDC illness model (preserving all measures of uncertainty) is combined with an updated version of the economic model developed by Scharff (2012, 2015) to produce economic burden of illness estimates. The illness model provides measures for illness incidence and likelihood of disease severity. The economic model provides values for the cost of illness associated with each disease endpoint, including costs associated with secondary conditions/complications.

Below I present the basic models, describe how the models are integrated with CDC illness estimates, reveal updated estimates, and discuss the usefulness and limitations of the estimates given.

8.2 Cost-of-Illness Modeling

The economic cost of a case of foodborne illness theoretically includes both monetary costs and utility losses to those directly impacted by the illness and, when the illness is part of an outbreak, to others (including industry and public health entities). Though industry and public health costs are not inconsequential, the approach used here follows other cost-of-illness studies by focusing on health-related costs.

For each identified pathogen, disease outcome trees are used to determine probabilistic illness outcomes (see Chap. 6). Disease outcomes vary by illness severity, requiring different levels of medical attention and implicating different types of secondary conditions/complications that may arise as a result of the initial acute illness. For example, one person made ill due to salmonellosis may have a routine self-limiting gastrointestinal illness that resolves in 3 or 4 days, while another is hospitalized for a week and is afflicted with reactive arthritis for months following the resolution of the initial acute illness. Costs are assessed for each end node of the tree and are aggregated based on probabilistic assessments of the likelihood of each outcome.

To make the model tractable, a limited number of outcomes and secondary conditions are evaluated. For acute conditions, potential outcomes are did not seek medical care, sought medical care/saw doctor, was hospitalized, and died. For secondary conditions/complications that arise, condition-specific outcome trees are employed. Specifically, costs are assessed for Guillain-Barré syndrome (*Campylobacter* spp.), hemolytic-uremic syndrome (*Escherichia coli* [STEC]), developmental disabilities (*Listeria monocytogenes*), and reactive arthritis (*Campylobacter* spp., *Salmonella* spp., *Shigella* spp., and *Yersinia enterocolitica*).

Potential measurable health-related costs from foodborne illnesses include medical costs (hospitalizations, physician services, and pharmaceuticals), lost productivity (for the person made ill or their caregiver), lost life expectancy, and lost quality of life. Though each of these cost categories is theoretically justified, many researchers have avoided using quality-of-life losses because the methods used to quantify them are controversial. For this reason, following Scharff (2012, 2015), I present two sets of estimates: one including quality-of-life losses and one excluding these losses. The first set of estimates is labeled as being from the "basic" model. The second set is from the "enhanced" model.

The basic cost-of-illness model measures the cost of a representative illness from pathogen p and is defined as

$$BCost_p = Hospital_p + Physician_p + Pharma_p + Prod_p + CProd_p + VSL_p \quad (8.1)$$

where, $Hospital_p$ is hospitalization costs, $Physician_p$ is physician costs (including lab fees), $Pharma_p$ is pharmaceutical costs, $Prod_p$ is productivity losses for the ill person, $CProd_p$ is caregiver productivity losses (for parents of ill children), and VSL_p is costs due to premature death (using the value of a statistical life).

The enhanced model adds a monetized value for quality-of-life losses:

$$ECost_p = Hospital_p + Physician_p + Pharma_p + CProd_p + VSL_p + QALY_p \quad (8.2)$$

where $QALY_p$ is quality-of-life losses. Note that productivity losses for ill persons are not in this model. This reflects the fact that QALY losses include lost utility due to functional limitations, which likely also reflects productivity losses. By omitting $Prod_p$ the potential for double counting is eliminated.

8.3 Methods for Estimating Costs

The methods for estimating costs for the models displayed in Eqs. (8.1) and (8.2) are described in this section. More detailed descriptions can be found in Scharff (2012, 2015) and linked appendices. All costs have been updated to reflect January 2017 dollars and are derived for 30 identified pathogens and the broader category of unspecified agents.

Medical costs are incurred when an individual sees a doctor, is hospitalized, or buys pharmaceuticals to treat their illness. Though the existence of private and public insurance means that only portion of these costs are paid directly by consumers, ultimately all of these costs fall on consumers due to a resulting rise insurance premiums and taxes (or deficits) to fund the expenses. As a result, all medical costs are included in the analysis. Estimates for these values are obtained from several sources.

Hospital costs are a combination of hospital services and inpatient physician services. Hospital service costs are taken from the Healthcare Cost and Utilization Project's National Inpatient Sample (NIS) (AHRQ 2016). Pathogen-specific costs are assessed by ICD-9 codes most closely related to the pathogen of interest. For example, ICD-9 code 003 specifically identifies hospitalizations due to *Salmonella* infections, allowing for direct assessment of salmonellosis hospitalization costs, while rarer hospitalizations resulting from infection with *Clostridium perfringens* are assumed to be captured by ICD-9 code 005.9 (food poisoning not otherwise specified). Costs are based on the most recent 5-year period of available data (generally 2009–2014) and are assumed to be uniformly distributed between the cost to the hospital (an underestimate of costs borne by patients/insurance) and hospital charges (an overestimate given that these charges are often negotiated down by insurance companies). Costs are updated to reflect January 2017 prices using the hospital services CPI (BLS 2017a).

Physician service costs are assessed for outpatient visits as well as emergency room and hospital inpatient visits. NIS data is used to determine the average length of hospital stays and the proportion of hospitalizations that utilize emergency room services (AHRQ 2016). Costs for each category of physician services (including lab work) are based on costs reported in a large annual physician survey (PMIC2017). Prescription drug costs reported in Scharff (2015) are updated to January 2017 prices using the prescription drug CPI (BLS 2017a).

When people become ill as a result of an infection with a foodborne illness, they are often unable to work. Productivity suffers as a result, and either wages are not paid or wages are paid to an absent employee. Either way, there is a cost to society. Costs associated with lost productivity are evaluated for workers who become ill and working parents of ill children aged 14 and younger. Following Scharff (2012, 2015) productivity losses are incurred when an adult misses work as a result of an illness and are equal to the cost of compensation for days of work missed (assuming that adults work on all weekdays except for 10 federal holidays and 10 vacation days). The percentage of adults employed reflects the most recent Bureau of Labor

Statistics estimates, and the hourly cost of compensation is based on December 2016 estimates (BLS 2016, 2017b). The proportion of illnesses attributable to adults and children are based on the most recent FoodNet (through 2015) and National Disease Surveillance system (NNDSS) (through 2014) data (CDC 2015, 2016, 2017).

When people die as a result of a foodborne illness, there is a utility loss for the person who dies. To measure this loss, an age-invariant value of statistical life (VSL) measure is used. This revealed preference measure is widely used and is derived from the trade-off between mortality risk and wages (Viscusi and Aldy 2003). Estimates from Scharff (2015) are revised to include nominal income growth from 2013 to 2015 and inflation experienced between 2015 and January 2017 (BLS 2017a, Census Bureau 2016).

Quality-of-life losses not associated with death are also experienced by those who become ill as a result of foodborne illness. These are legitimate economic costs but, because markets for these losses do not exist, are more difficult to place values on. The Food and Drug Administration (FDA) provides estimates for these losses by monetizing quality-adjusted life year (QALY) losses using value of statistical life year estimates, derived from the VSL (Minor et al. 2015). Though this method is controversial, it is also more complete than more standard cost-of-illness studies, as reflected by Eq. (8.1). In the enhanced model (Eq. 8.2), economic values for quality-of-life losses are included. Scharff (2015) QALY estimates are replaced in this analysis with more recent FDA pathogen-specific estimates (Minor et al. 2015), and VSLY estimates are revised to reflect updated VSL estimates, as described above.

For several pathogens, secondary conditions or complications (sequelae) may occur as a result of the initial acute illness. These conditions may be time-limited or chronic. Though many conditions have been examined by researchers, only a few are generally accepted as being definitively tied to foodborne illness. Sequelae for which costs are derived include Guillain-Barré syndrome (*Campylobacter* spp.), hemolytic-uremic syndrome (STEC), developmental disabilities (*Listeria monocytogenes*), and reactive arthritis (*Campylobacter* spp., *Salmonella* spp., *Shigella* spp., and *Yersinia enterocolitica*). Generally, Scharff (2012) estimates are used, with cost categories updated as described above for acute illnesses.

Estimates for the economic burden of illness have also been derived at the state level (Scharff 2015). Costs, at the state level, vary due to differences in illness incidence, medical costs, employment rates, and wages (which affect both productivity losses and VSL—through the effect of income on demand for risk reduction). Consequently, state-based estimates are valuable for local policymakers who are best served by making decisions based on local information. State-specific illness incidence is estimated for illnesses caused by several pathogens (*Brucella* spp., *Cryptosporidium* spp., *Cyclospora cayetanensis*, *Giardia intestinalis*, hepatitis A virus, *L. monocytogenes*, *Salmonella* spp., STEC, *Shigella* spp., and *Vibrio* spp.) based on illnesses reported to the CDC through the NNDSS by state health departments (CDC 2015, 2016). State-specific differences for physician services, hospitalizations, employment rates, wages, and household income are assessed for all pathogens.

The Scharff (2015) state cost model is updated here in the following ways. State medical costs are updated using the most recent published geographic adjustment factors (PMIC 2017). Productivity costs are updated by using the most recent state estimates for employment and wages (BLS 2016). Finally, VSL estimates are updated using the latest state estimates for household income (Census Bureau 2016).

Uncertainty is incorporated into the model by using @Risk 7.5 to perform a Monte Carlo analysis that incorporates hundreds of measures of uncertainty; described more fully in Scharff (2012, 2015).

8.4   Integration with the CDC Illness Model

The economic costs for each category of illness outcomes, as described above, are of little use without knowing how likely these events are when an illness occurs. Similarly, cost estimates in the absence of illness incidence rates are of limited usefulness. Fortunately, the CDC illness model generates these estimates for each of the 30 pathogens examined in this study and a separate, larger, category for illnesses from unspecified agents (Scallan et al. 2011a, b). As illustrated in Fig. 8.1, the integration of the illness model with the economic model provides all of the elements needed to produce economic cost estimates at both per case and national levels.

To preserve the uncertainty measures in the CDC model, a full replication of the model is conducted using data and methods provided in Scallan et al. (2011a, b) (and the papers' four technical appendices). Though this study uses @Risk 7.5 (to be compatible with the economic model), while Scallan used SAS for the empirical analyses; the resulting illness estimates are nearly identical.

CDC estimates of annual incidence of foodborne illness (generated through model replication and, where needed, adjusted to match CDC estimates) are presented in Table 8.1. More than half of all illnesses linked to a pathogen are caused by norovirus, followed by non-typhoidal *Salmonella* and *C. perfringens*. The presence of *C. perfringens* (a source of generally mild illnesses) near the top of the list of the most frequent sources of illness illustrates the value of economics as a means of providing severity weighted burden of illness estimates. As the results below demonstrate, the large number of *C. perfringens* illnesses are associated with a relatively modest economic cost. Severity differences are also apparent when illnesses are viewed in conjunction with hospitalizations and deaths. Though *Salmonella* is responsible for less than one fifth the number of illnesses associated with norovirus, *Salmonella* causes more hospitalizations and deaths than norovirus.

Significantly, of the almost 48 million foodborne illnesses that occur in the United States each year, most (over 38 million) are caused by unspecified agents. While most economic analyses have avoided placing costs on unspecified illnesses because of the difficulty in characterizing these illnesses (e.g., Hoffmann et al. 2012), the analysis used here includes values for these illnesses because failure to do so would lead to a gross mischaracterization of the health-related burden of foodborne illness in the United States. The cost estimates for these illnesses are based on the symp-

Table 8.1 Annual incidence of foodborne illness (CDC estimates)

Disease or agent	Illness	Hospitalizations	Deaths
Bacterial			
Bacillus cereus	63,400	20	0
Brucella spp.	839	55	1
Campylobacter spp.	845,024	8,463	76
Clostridium botulinum	55	42	9
Clostridium perfringens	965,958	438	26
STEC O157:H7	63,153	2,138	20
STEC non-0157	112,752	271	0
ETEC	17,894	12	0
Other diarrheagenic *E. coli*	11,982	8	0
Listeria monocytogenes	1,591	1,455	255
Salmonella spp., non-typhoidal	1,027,561	19,336	378
S. enterica serotype typhi	1,821	197	0
Shigella spp.	131,254	1,456	10
Staphylococcus aureus,	241,148	1,064	6
Streptococcus spp. group A,	11,217	1	0
Vibrio cholerae, toxigenic	84	2	0
Vibrio vulnificus	96	93	36
Vibrio parahaemolyticus	34,664	100	4
Vibrio spp., other	17,564	83	8
Yersinia enterocolitica	97,656	533	29
Parasitic			
Cryptosporidium spp.	57,616	210	4
Cyclospora cayetanensis	11,407	11	0
Giardia intestinalis	76,840	225	2
Toxoplasma gondii	86,686	4,428	327
Trichinella spp.	156	6	0
Viral			
Astrovirus	15,433	87	0
Hepatitis A	1,566	99	7
Norovirus	5,461,731	14,663	149
Rotavirus	15,433	348	0
Sapovirus	15,433	87	0
All specified pathogens	9,388,074	55,962	1,350
Unspecified agents	38,392,704	127,839	1,686
Total	47,780,778	183,801	3,036

toms of the gastrointestinal illnesses identified through the FoodNet Population Survey, which is the basis for the unspecified illness estimates in Scallan (2011b).

Probabilities for each of the four potential acute illness outcomes (did not seek medical care, sought medical care/saw doctor, was hospitalized, and died) are assigned using data from Scallan. Probabilities for secondary conditions or complications (sequelae) are obtained from other sources, as described in Scharff (2012). In many cases, more than one outcome occurs (e.g., hospitalization followed by death) leading the sum of probabilities to exceed one. The combination of outcome probabilities with outcome costs allows for the cost of a representative case to be assessed (cost per case).

Incidence estimates from Scallan are used to estimate the total economic burden of foodborne illness. For most pathogens, incidence values can be determined using the approach shown in Fig. 8.1; inflating reported illnesses to account for underdiagnosis and underreporting. Underdiagnosis occurs because many who are made ill do not visit a doctor, many of those who seek care do not submit a sample for testing, and some samples are false positives. Underreporting occurs when a patient is correctly determined to have an infection caused by a specific pathogen, but the case is not reported to the state health department or the CDC. Underreporting is generally lowest when active surveillance is used (ten pathogens), is higher in passive surveillance systems (ten pathogens), and is highest where outbreaks are the only source of data (five pathogens) (Scallan et al. 2011a). Five pathogens (including *Toxoplasma gondii* and four viral agents) are not covered by any form of surveillance. Various sources of data are used to produce estimates using top-down methods (Scallan et al. 2011a). A similar method is employed for unspecified agents (Scallan et al. 2011b).

It should be noted that the Scallan estimates for illness incidence have not been updated in this analysis despite the fact that up to 7 years of new data is available for many pathogens in many data categories, including reported illnesses. Though, in many cases, observed changes in reported illnesses may reflect actual changes in the incidence rate, these changes may also reflect, at least in part, changes in underdiagnosis and/or underreporting rates. Without evidence detailing how these rates have changed, any update to incidence rates would be speculative. As a result, the analysis here is based on dated, but defensible incidence estimates combined with updated cost estimates.

8.5 The Cost of Foodborne Illness in the United States

The cost per case of foodborne illness for a given pathogen is the product of costs associated with specific outcomes (e.g., hospitalization) and the probabilities that each outcome occurs. The resulting expected costs for each outcome are summed across all component categories to determine the total expected cost of a typical illness. In Table 8.2, expected costs for each major outcome are illustrated for each pathogen. There is substantial variability in costs for pathogens across all categories. As expected, mild illnesses, such as those from *C. perfringens*, are associated with minimal costs, while more serious illnesses, such as botulism, have high costs across all categories.

The total economic cost of foodborne illness for each pathogen is derived by combining the expected cost per case with number of illnesses for each pathogen.

Table 8.2 Expected cost per case of foodborne illness (Jan. $2017)

Pathogen or agent	Medical care	Productivity Loss		Quality of life	Death
		Ill person	Caregiver		
Bacterial					
Bacillus cereus	34	67	69	259	0
Brucella spp.	114	3,232	3,350	2,711	10,754
Campylobacter spp.	45	407	422	11,127	942
Clostridium botulinum	1,645	23,854	24,724	42,476	1,452,012
Clostridium perfringens	34	85	88	259	240
STEC O157:H7	85	462	479	6,627	11,090
STEC non-0157	36	462	479	2,288	0
ETEC	34	462	479	2,288	0
Other diarrheagenic *E. coli,*	34	462	479	786	0
Listeria monocytogenes	1,586	2,114	2,191	74,132	1,426,122
Salmonella, spp., non-typhoidal	51	752	666	10,772	3,274
S. enterica serotype typhi	140	1,085	1,010	11,745	0
Shigella spp.	40	645	668	12,140	678
Staphylococcus aureus	36	154	160	403	223
Streptococcus spp. group A	33	749	776	1,045	0
Vibrio cholerae, toxigenic	58	718	744	1,358	0
Vibrio vulnificus	702	985	1,021	38,156	3,336,694
Vibrio parahaemolyticus	34	581	603	1,229	1,036
Vibrio spp., other	34	581	603	1,260	4,033
Yersinia enterocolitica	31	1,026	1,063	11,137	2,642
Parasitic					
Cryptosporidium spp.	24	838	869	2,040	621
Cyclospora cayetanensis	37	513	532	4,573	0
Giardia intestinalis	23	1,231	1,276	6,449	230
Toxoplasma gondii	105	3,078	3,190	7,597	33,577
Trichinella spp.	74	5,027	5,211	15,683	0
Viral					
Astrovirus	33	413	428	676	0
Hepatitis A virus	158	1,078	1,117	4,555	39,840
Norovirus	32	142	147	403	243
Rotavirus	41	352	365	1,669	0
Sapovirus	32	352	365	417	0
All specified pathogens	37	296	294	2,987	1,345
Unspecified agents	32	279	290	388	391
Total	33	283	290	899	578

Total costs from all foodborne illnesses are the sum of costs for all 30 specified pathogens and the larger category of unspecified illnesses.

Means and 90% credible intervals for cost per case and total cost estimates are provided for the basic model in Table 8.3. The expected cost per case (which

Table 8.3 Economic cost of foodborne illness (Basic Model, Jan. $2017)

	Cost per case		Total cost ($millions)	
Disease or agent	Mean	(90% CI)	Mean	(90% CI)
Bacterial				
Bacillus cereus	183	(80–264)	12	(2–32)
Brucella spp.	20,548	(10,868–31,858)	17	(9–28)
Campylobacter spp.	2,210	(1,168–4,950)	1,867	(518–4,892)
Clostridium botulinum	1,619,234	(103,173–9,521,984)	90	(5–494)
Clostridium perfringens, foodborne	460	(209–1,628)	445	(51–1,710)
STEC O157:H7	13,003	(4,562–30,040)	821	(169–2,452)
STEC non-0157	1,046	(976–1,178)	118	(13–318)
ETEC	999	(973–1,064)	18	(<1–49)
Other diarrheagenic *E. coli*,	999	(973–1,064)	12	(<1–32)
Listeria monocytogenes	1,553,532	(100,980–4,740,267)	2,472	(118–8,164)
Salmonella spp., non-typhoidal	5,218	(1,867–12,188)	5,362	(1,782–13,093)
S. enterica serotype typhi	4,793	(2,777–7,891)	9	(<1–23)
Shigella spp.	2,355	(1,517–6,629)	309	(46–949)
Staphylococcus aureus, foodborne	639	(365–2,482)	154	(34–510)
Streptococcus spp. group A, foodborne	1,568	(1,557–1,581)	18	(<1–126)
Vibrio cholerae, toxigenic	1,710	(1,516–2,053)	0.14	(004–0.37)
Vibrio vulnificus	3,394,273	(724,596–6,436,726)	326	(67–658)
Vibrio parahaemolyticus	2,324	(1,261–5,701)	81	(34–202)
Vibrio spp., other	5,322	(1,859–12,110)	93	(33–218)
Yersinia enterocolitica	4,995	(2,206–19,984)	488	(80–2,006)
Parasitic				
Cryptosporidium spp.	2,431	(1,751–5,722)	140	(24–466)
Cyclospora cayetanensis	1,109	(1,078–1,246)	13	(<1–45)
Giardia intestinalis	2,818	(2,625–3,075)	217	(149–311)
Toxoplasma gondii	43,552	(16,397–75,925)	3,775	(1,376–6,942)
Trichinella spp.	11,087	(10,280–12,718)	2	(1–4)
Viral				
Astrovirus	983	(909–1,064)	15	(5–26)
Hepatitis A virus	43,757	(11,730–83,233)	69	(16–151)
Norovirus	631	(434–847)	3,446	(1,846–5,579)
Rotavirus	993	(866–1,133)	15	(6–27)
Sapovirus	813	(781–851)	13	(5–22)
All specified pathogens	2,175	(1,072–3,904)	20,415	(10,004–35,764)
Unspecified agents	1,056	(724–1,622)	40,524	(24,707–60,688)
Total	1,275	(805–1,970)	60,939	(37,221–90,820)

includes medical costs, productivity losses, and mortality costs) ranges from $183 for a typical illness resulting from *Bacillus cereus* to $3.4 million for the (often deadly) illnesses caused by *Vibrio vulnificus*. Across all pathogens (and unspecified agents), the average cost of an illness is $1275, though the cost is higher ($2175) for illnesses from identified pathogens.

Aggregated across all pathogens and other unspecified agents, the basic model estimates an annual economic cost from foodborne illness of $60.9 billion. Two thirds of these costs are attributable to unspecified agents. Of identified pathogens, *Salmonella* has the highest social cost ($5.4 billion) due to a large number of illnesses, a relatively lengthy illness duration (affecting productivity losses), and relatively high hospitalization and death rates. Despite causing five times as many illnesses, the cost of norovirus is lower ($3.4 billion) due to shorter illnesses and a lower probability of hospitalization or death. *C. perfringens*, with nearly as many illnesses as *Salmonella*, leads to costs more than an order of magnitude smaller than *Salmonella* ($0.4 billion) because of very low probabilities of costly outcomes.

Cost estimates derived using the enhanced model are presented in Table 8.4. The inclusion of a measure for quality of life increases average cost per case to $1887. The largest relative increase in costs resulted from the inclusion of a cost measure for quality-of-life losses due to reactive arthritis. Consequently, *Campylobacter*, *Salmonella*, *Shigella*, and *Yersinia* costs all are substantially larger when the enhanced model is used. Conversely, the cost per case for illnesses caused by unspecified agents is only marginally higher when the enhanced model is used, increasing by only $109 (compared to a $2666 increase in the cost per case for identified pathogens). Total cost estimates mirror these effects. Inclusion of quality-of-life losses increases the economic burden of illness by almost half, to $90.2 billion.

The relative effects of different measures of burden of illness are more clearly illustrated in Figs. 8.2 and 8.3. The contributions of specified disease-causing agents (and categories of agents) vary considerably depending on which metric of burden of illness is used.

As panel A of Fig. 8.2 illustrates, the largest category of costs in the basic model are generated by unspecified agents (66%), followed by bacterial (21%), parasitic (7%), and viral (6%) pathogens. Though bacterial pathogens represent a much larger portion of costs in the enhanced model (39%), the largest cost category is still unspecified agents. Note that, due to a lower cost per case, unspecified agents make up a larger share of illnesses (80%) than costs for both the basic (66%) and enhanced (50%) models.

When illnesses of interest are limited to those from identified pathogens (panel B), norovirus makes up the bulk (52%) of illnesses, though not the bulk of the costs. In the basic economic model *Salmonella* is responsible for the highest proportion of costs (26%), followed by *Toxoplasma gondii* (18%) and norovirus (17%). In the enhanced model, the valuation of quality-of-life losses from reactive arthritis increases the share of costs associated with illnesses from *Salmonella* (34%) and *Campylobacter* (24%).

Table 8.4 Economic cost of foodborne illness (Enhanced Model, Jan. $2017)

	Cost per case		Total cost ($millions)	
Disease or agent	Mean	(90% CI)	Mean	(90% CI)
Bacterial				
Bacillus cereus	375	(163–611)	24	(5–63)
Brucella spp.	20,027	(8,449–33,101)	17	(7–30)
Campylobacter spp.	12,759	(2,852–29,132)	10,782	(1,793–30,101)
Clostridium botulinum	1,637,856	(113,183–9,538,165)	91	(5–495)
Clostridium perfringens	634	(197–1,894)	613	(77–2,088)
STEC O157:H7	19,157	(5,532–39,137)	1,210	(225–3,642)
STEC non-0157	2,872	(1,034–5,127)	324	(30–1,017)
ETEC	2,825	(997–5,087)	51	(<1–158)
Other diarrheagenic *E. coli,*	1,323	(662–2,289)	16	(<1–46)
Listeria monocytogenes	1,584,975	(128,864–4,785,809)	2,522	(150–8,262)
Salmonella spp., non-typhoidal	15,238	(3,515–34,010)	15,658	(3,439–38,122)
S. enterica serotype typhi	15,454	(5,279–51,686)	28	(<1–102)
Shigella spp.	13,741	(2,953–31,933)	1,804	(168–6,408)
Staphylococcus aureus	887	(339–2,817)	214	(45–635)
Streptococcus spp. group A	1,864	(970–3,262)	21	(<1–143)
Vibrio cholerae, toxigenic	2,350	(1,223–3,719)	<1	(<1–1)
Vibrio vulnificus	3,431,444	(730,589–6,499,405)	329	(67–666)
Vibrio parahaemolyticus	2,971	(1,054–6,551)	103	(34–247)
Vibrio spp., other	6,001	(1,586–13,520)	105	(29–243)
Yersinia enterocolitica	14,996	(3,139–36,679)	1,464	(221–4,221)
Parasitic				
Cryptosporidium spp.	3,633	(1,265–7,902)	209	(28–705)
Cyclospora cayetanensis	5,169	(1,305–11,108)	59	(<1–237)
Giardia intestinalis	8,036	(2,413–16,700)	618	(179–1,370)
Toxoplasma gondii	48,072	(14,835–86,820)	4,167	(1,260–7,838)
Trichinella spp.	21,744	(8,491–42,140)	3	(<1–10)
Viral				
Astrovirus	1,246	(704–1,882)	19	(6–38)
Hepatitis A virus	47,235	(11,592–90,054)	74	(16–165)
Norovirus	892	(373–1,466)	4,871	(1,807–9,244)
Rotavirus	2,310	(967–3,940)	36	(9–77)
Sapovirus	879	(543–1,293)	14	(4–26)
All specified pathogens	4,841	(1,375–9,830)	45,446	(12,679–92,039)
Unspecified agents	1,165	(546–1,989)	44,714	(19,846–76,534)
Total	1,887	(720–3,492)	90,159	(34,244–161,752)

In most cases the 90% credible intervals for cost estimates have wide distributions, reflecting a large amount of uncertainty for several important model parameters. The largest single source of uncertainty in the economic model is the estimate for the value of statistical life (VSL) which is responsible for mortality costs of $27.6 billion (90% CI, $5.5–$55.0 billion). The VSL plays a large role in driving

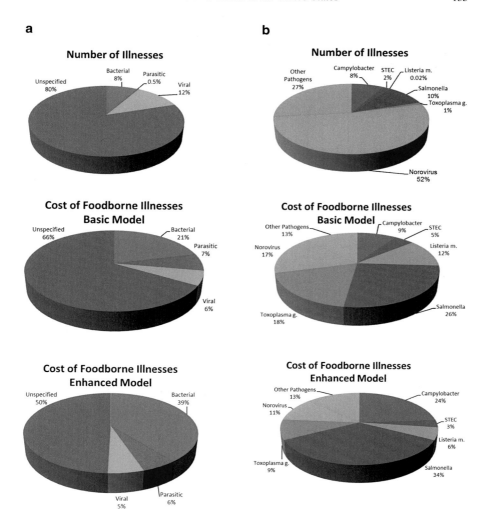

Fig. 8.2 Distribution of burden of illness by agent. Panel A: burden of illness by etiology. Panel B: burden of illness for identified pathogens

uncertainty, both because it is the source of a large portion of costs (used both in death cost and quality-of-life estimates) and because the credible interval for VSL is large, with each statistical death valued at $8.9 million (90% CI, $1.8–$16.1 million). VSL is especially influential because it is assumed that there is one true value for the parameter that, if discovered, would be applied uniformly across all pathogens. At the pathogen level, the number of illnesses is also highly variable for several sources of illness including STEC O157 (63,153; 90% CI, 17,587–149,631), *Campylobacter* (845,024; 90% CI, 337,031–1,611,083), and *Shigella* (13,254; 90% CI, 24,511–374,789) (Scallan et al. 2011a). Though these sources of uncertainty affect cost dis-

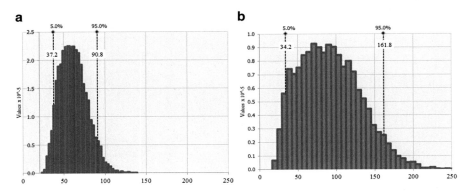

Fig. 8.3 Distribution of the total cost of foodborne illness (2017 $billion). Panel A: basic model. Panel B: enhanced model

tributions for individual pathogens, the effect on the total cost of illness is muted by the fact that these distributions are independent from each other (e.g., it is less likely that high values in multiple distributions will be selected concurrently).

The relative role of uncertainty in the total cost estimates derived using the basic and enhanced models, respectively, is illustrated in Fig. 8.3. In panel A, the basic model total cost distribution that results from a Monte Carlo analysis of hundreds of uncertain parameter estimates is relatively narrow, with 90% of generated total cost estimates between $37.2 and $90.8 billion. For the enhanced model, in panel B, greater reliance on uncertain VSL numbers (though the use of VSLY to monetize QALYs) leads to a wider distribution of cost estimates, with 90% of estimates between $34.2 and $161.8 billion.

Efforts have also been made to evaluate costs at the state level (Scharff 2015). These estimates account for differences in incidence of illness due to differences in consumption patterns, regulatory regimes, and environmental conditions. Cost differences are also reflected in state-based figures. Tables 8.5 and 8.6 present state-specific cost-of-illness estimates using the basic and enhanced models, respectively. These updated estimates demonstrate that costs vary significantly across the states. Under the basic (enhanced) model, the average cost per case ranges from $933 ($1370) in West Virginia to $1981 ($2527) in Washington D.C. The average cost per case across all states is $1293 ($1917). Consequently, the estimates for total burden of foodborne illness are also higher when costs are first aggregated at the state level. Recognition of these cost differences allows states to tailor illness response efforts toward the needs of their residents.

Table 8.7 provides a sensitivity analysis for national burden of illness estimates. The primary approach, labeled as the "national" estimate is described in detail above. In addition, estimates based on initial aggregation at the state level ("state-based") are provided. These have the advantage of not assuming uniform costs across the states. These are more difficult to derive, however, without having a large effect on aggregate costs. The final set of estimates ("population-adjusted") is based

Table 8.5 State-level cost of foodborne illness (Basic Model, Jan. $2017)

	Cost per case ($)	Total cost ($million)
US total	*1,293*	*61,796*
Alabama	1,049	759
Alaska	1,461	158
Arizona	1,161	1,186
Arkansas	1,021	460
California	1,451	8,360
Colorado	1,410	1,149
Connecticut	1,670	881
D.C.	1,981	198
Delaware	1,379	194
Florida	1,207	3,747
Georgia	1,192	1,834
Hawaii	1,609	344
Idaho	1,081	270
Illinois	1,345	2,535
Indiana	1,112	1,081
Iowa	1,246	586
Kansas	1,195	514
Kentucky	991	646
Louisiana	1,092	765
Maine	1,205	236
Maryland	1,614	1,434
Massachusetts	1,677	1,695
Michigan	1,161	1,686
Minnesota	1,425	1,169
Mississippi	999	453
Missouri	1,181	1,068
Montana	1,071	165
Nebraska	1,251	356
Nevada	1,132	485
New Hampshire	1,516	298
New Jersey	1,623	2,134
New Mexico	1,067	329
New York	1,465	4,263
North Carolina	1,134	1,705
North Dakota	1,299	145
Ohio	1,168	1,996
Oklahoma	1,088	634
Oregon	1,240	748
Pennsylvania	1,300	2,449
Rhode Island	1,405	219
South Carolina	1,105	819
South Dakota	1,186	155

(continued)

Table 8.5 (continued)

	Cost per case ($)	Total cost ($million)
Tennessee	1,055	1,034
Texas	1,213	5,000
Utah	1,241	558
Vermont	1,339	124
Virginia	1,439	1,783
Washington	1,414	1,515
West Virginia	933	250
Wisconsin	1,285	1,109
Wyoming	1,304	113

Table 8.6 State-level cost of foodborne illness (Enhanced Model, Jan. $2017)

	Cost per case ($)	Total cost ($million)
US Total	*1,917*	*91,600*
Alabama	1,622	1,174
Alaska	2,206	239
Arizona	1,732	1,770
Arkansas	1,647	741
California	2,091	12,047
Colorado	2,014	1,641
Connecticut	2,415	1,274
D.C.	2,527	253
Delaware	2,083	293
Florida	1,905	5,913
Georgia	1,852	2,849
Hawaii	2,571	550
Idaho	1,585	396
Illinois	1,974	3,721
Indiana	1,612	1,568
Iowa	1,883	885
Kansas	1,783	766
Kentucky	1,468	958
Louisiana	1,732	1,213
Maine	1,691	331
Maryland	2,477	2,200
Massachusetts	2,458	2,485
Michigan	1,630	2,368
Minnesota	2,090	1,714
Mississippi	1,682	764
Missouri	1,731	1,565
Montana	1,570	242
Nebraska	1,861	529

(continued)

Table 8.6 (continued)

	Cost per case ($)	Total cost ($million)
Nevada	1,636	701
New Hampshire	2,232	439
New Jersey	2,424	3,187
New Mexico	1,612	497
New York	2,087	6,075
North Carolina	1,693	2,544
North Dakota	1,864	208
Ohio	1,685	2,878
Oklahoma	1,693	987
Oregon	1,735	1,046
Pennsylvania	1,881	3,543
Rhode Island	1,994	310
South Carolina	1,742	1,290
South Dakota	1,912	251
Tennessee	1,565	1,534
Texas	1,849	7,621
Utah	1,892	850
Vermont	1,927	179
Virginia	2,160	2,678
Washington	2,010	2,154
West Virginia	1,370	367
Wisconsin	1,903	1,642
Wyoming	1,968	171

Table 8.7 Sensitivity analysis: alternative estimates for the economic cost of foodborne illness in the United States (Jan. $2017)

Method	Cost per case		Total cost ($millions)	
	Mean	(90% CI)	Mean	(90% CI)
Basic model				
National	1,275	(805–1,970)	60,939	(37,221–90,820)
State-based	1,293	(813–2,000)	61,796	(37,699–92,291)
Population-adjusted	1,275	(805–1,970)	65,857	(40,224–98,149)
Enhanced model				
National	1,887	(720–3,492)	90,159	(34,244–161,752)
State-based	1,917	(733–3,552)	91,600	(34,833–164,210)
Population-adjusted	1,887	(720–3,492)	97,435	(37,007–174,805)

on a relaxation of the assumption that illness estimates in Scallan et al. (2011a, b) are still true today. Instead, this scenario assumes that incidence *rates,* not illness *numbers* have remained steady in the decade following the data collection for the CDC study. Consequently, illnesses are assumed to grow proportionately with the US population, yielding estimates of aggregate cost that are $5 billion to $7 billion greater than "national" estimates.

8.6   The Use of Cost-of-Illness Estimates

Estimates of the economic cost of foodborne illness can be used in efforts to mitigate harms from foods contaminated with pathogens and other harmful agents. Risk managers seeking to improve social welfare can use these measures to help prioritize food safety efforts, evaluate interventions, and educate consumers. Essentially, cost estimates provide valuable information in an atmosphere where information is often scarce.

All food safety risk managers must make choices about where to employ their limited resources. This is true whether the manager is a decision-maker at a regulatory agency, an educator in an extension program, or a food safety manager in industry. As an aid in these efforts, managers often rely on burden of illness measures to assess where harms are greatest. This, presumably, provides some insight about where mitigation efforts are likely to have the biggest impact. In the absence of information about economic costs, burden of illness estimates are difficult to compare. For example, how might a manager compare the risk posed by *C. perfringens* (causing 965,958 illnesses, 438 hospitalizations, and 26 deaths) with *Listeria m.* (causing 1591 illnesses, 1455 hospitalizations, and 255 deaths)? Though it might appear that *C. perfringens* poses the larger threat due to the sheer number of illnesses, economic cost estimates suggest the opposite. The 1591 illnesses due to *Listeria m.* impose costs of $2.5 billion, compared to only $0.4 billion in costs from *C. perfringens.* Essentially, economics provides an objective means of completing the difficult task of weighting burdens based on illness severity.

The role that these estimates play in prioritizing resources is also apparent at higher levels. Legislators, department heads, and industry leaders all must decide what levels of resources to target toward food safety, as opposed to other goals. The finding that foodborne illnesses impose health-related social costs of up to $90 billion is powerful evidence that resources should be directed toward the mitigation of food safety risks. Nevertheless, it is important not to overstate the value of these estimates. While aggregate economic burden of illness estimates are useful as a means of highlighting the importance of the problem, they provide little guidance regarding whether a particular intervention is justified.

Though burden of illness estimates are insufficient as a means of evaluating intervention effectiveness, the cost per case estimates provided here can play an important role. Specifically, these estimates are used in benefit cost analyses to determine whether the costs of an intervention (or set of interventions) are justified by corresponding benefits. The estimates generated here can help improve benefit cost analyses that are often poorly calculated for major federal rules (Hahn and Dudley 2007).

A risk manager attempting to maximize social welfare would try to solve the following problem:

$$\max_{ip} \text{CostperCase}_p \times \text{AvertedIll}_{ip} - \text{InterventionCost}_{ip} \qquad (8.3)$$

Essentially, the manager will choose a set of interventions with the goal of maximizing net benefits for the mix of interventions (i) used to reduce illnesses from pathogen (p). Benefits from the intervention are estimated to be the cost per case for the pathogen (CostperCase_p) times the expected number of illnesses averted by the set of interventions (AvertedIll_{ip}). As long as these benefits exceed the costs of the intervention, it will improve social welfare, though, from an economist's perspective, society will be best off when the difference between benefits and costs is greatest. In practice, the simultaneous evaluation of all potential interventions is impossible. Instead, policymakers will often be interested in knowing benefit cost ratios $\left(\dfrac{\text{CostperCase}_p \times \text{AvertedIll}_{ip}}{\text{InterventionCost}_{ip}} \right)$ for a single intervention (or set of interventions) as a rough metric of return on investment. See Chap. 4 for an example of benefit cost analysis related to HACCP requirements for meat and poultry.

Economic cost estimates can also be used to educate consumers. In their roles as household risk managers, information about the economic burden of foodborne illness may influence consumers to take more care to prepare foods safely and buy foods from trusted sources. In their roles as voters, they can use the information to decide whether they support elected officials' food safety efforts.

8.7   Conclusion

Foodborne illness imposes a substantial burden on the American public. Under alternative models and scenarios, the health-related cost of foodborne illness ranges from $60.9 billion to $97.4 billion or $1275 to $1917 per case. Cost per case and total cost estimates vary significantly, however, based on the pathogen causing the illness and the population affected. Total cost estimates are useful as a means of describing the burden of illness, which can be used to prioritize scarce resources. The cost per case estimates can be used as a tool for evaluating specific interventions. Nevertheless, some caution is advised when using these numbers. First, not all economic consequences are included. Costs to public health authorities and industry are omitted. Second, the illness model used is based on data that is more than a decade old. It is unclear how the incidence of foodborne illness has changed in intervening years. Third, the method used to measure quality-of-life losses (for the enhanced model) is controversial. This model is included because these types of losses are an important and theoretically justified, but the measurement technique is not universally accepted. Finally, modeling of uncertainty leads to large credible intervals for most estimates. This can lead to difficult decisions for risk managers who have to decide whether to adopt interventions that lead to potentially large, but uncertain, benefits or interventions leading to more modest, but certain, benefits. Future research aimed at ameliorating these issues would be of great value.

References

Agency for Healthcare Research and Quality (AHRQ). HCUPnet, Healthcare Cost and Utilization Project, Rockville. 2016. https://hcupnet.ahrq.gov. Accessed 15 Feb 2017.

Bureau of Labor Statistics (BLS). May 2015 National occupational employment and wage estimates: United States, Washington. 2016. https://www.bls.gov/cpi. Accessed 17 Feb 2017.

Bureau of Labor Statistics (BLS). Consumer Price Index—all urban consumers, Washington. 2017a. http://www.bls.gov/cpi. Accessed 24 Mar 2017.

Bureau of Labor Statistics (BLS). Employer costs for employee compensation news release, Washington. 2017b. https://www.bls.gov/news.release/ecec.nr0.htm. Accessed 24 Mar 2017.

Census Bureau (Census). Current population survey, annual social and economic supplements, Washington. 2016. https://www.census.gov/did/www/saipe/data/model/info/cpsasec.html. Accessed 24 Mar 2017.

Centers for Disease Control and Prevention. (CDC). MMWR: Summary of Notifiable Diseases, Atlanta. 2015. https://www.cdc.gov/mmwr/mmwr_nd/index.html. Accessed on 12 February 2017.

Centers for Disease Control and Prevention (CDC). MMWR: Summary of notifiable diseases, Atlanta. 2016. https://www.cdc.gov/mmwr/mmwr_nd/index.html. Accessed 12 Feb 2017.

Centers for Disease Control and Prevention (CDC). FoodNet 2015 Surveillance Report (Final Data), Atlanta. 2017. https://www.cdc.gov/foodnet/reports/annual-reports-2015.html. Accessed 10 Mar 2017.

Hahn RW, Dudley PM. How well does the US government do benefit-cost analysis? Rev Environ Econ Pol. 2007;1(2):192–211.

Hoffmann S, Batz MB, Morris JG Jr. Annual cost of illness and quality-adjusted life year losses in the United States due to 14 foodborne pathogens. J Food Prot. 2012;75(7):1292–302.

Minor T, Lasher A, Klontz K, Brown B, Nardinelli C, Zorn D. The per case and total annual costs of foodborne illness in the United States. Risk Anal. 2015;35(6):1125–39.

Practice Management Information Corporation (PMIC). Medical fees 2017. Los Angeles: PMIC; 2017.

Scallan E, Hoekstra RM, Angulo FJ, Tauxe RV, Widdowson MA, Roy SL, Jones JL, Griffin PM. Foodborne illness acquired in the United States—major pathogens. Emerg Infect Dis. 2011a;17(1):7–15.

Scallan E, Griffin PM, Angulo FJ, Tauxe RV, Hoekstra RM. Foodborne illness acquired in the United States—unspecified agents. Emerg Infect Dis. 2011b;17(1):16.

Scharff RL. Economic burden from health losses due to foodborne illness in the United States. J Food Prot. 2012;75(1):123–31.

Scharff RL. State estimates for the annual cost of foodborne illness. J Food Prot. 2015; 78(6):1064–71.

Viscusi WK, Aldy JE. The value of a statistical life: a critical review of market estimates throughout the world. J Risk Uncertain. 2003;27(1):5–76.

Chapter 9
Improving Burden of Disease and Source Attribution Estimates

Barbara B. Kowalcyk, Sara M. Pires, Elaine Scallan, Archana Lamichhane, Arie H. Havelaar, and Brecht Devleesschauwer

Abbreviations

AF	Aflatoxin
CD	Crohn's disease
CDC	US Centers for Disease Control and Prevention
CeD	Celiac disease
CIDT	Culture-independent diagnostic test
DALY	Disability-adjusted life year

B. B. Kowalcyk (✉)
Department of Food Science and Technology, The Ohio State University, Columbus, OH, USA

RTI International, Research Triangle Park, NC, USA

Center for Foodborne Illness Research and Prevention, Raleigh, NC, USA
e-mail: kowalcyk.1@osu.edu

S. M. Pires
National Food Institute, Technical University of Denmark, Lyngby, Denmark
e-mail: smpi@food.dtu.dk

E. Scallan
Colorado School of Public Health, University of Colorado, Aurora, CO, USA
e-mail: Elaine.scallan@ucdenver.edu

A. Lamichhane
RTI International, Research Triangle Park, NC, USA
e-mail: alamichhane@rti.org

A. H. Havelaar
Department of Animal Sciences, Institute for Sustainable Food Systems, Emerging Pathogens Institute, University of Florida, Gainesville, FL, USA
e-mail: ariehavelaar@ufl.edu

B. Devleesschauwer
Department of Public Health and Surveillance, Scientific Institute of Public Health (WIV-ISP), Brussels, Belgium
e-mail: brecht.devleesschauwer@wiv-isp.be

© Springer International Publishing AG, part of Springer Nature 2018
T. Roberts (ed.), *Food Safety Economics*, Food Microbiology and Food Safety, https://doi.org/10.1007/978-3-319-92138-9_9

ECDC	European Centre for Disease Prevention and Control
EED	Environmental enteric dysfunction
EFSA	European Food Safety Authority
ETEC	Enterotoxigenic *Escherichia coli*
ExPEC	Extraintestinal pathogenic *E. coli*
FAO	Food and Agriculture Organization of the United Nations
FBD	Foodborne disease
FERG	WHO Foodborne Disease Burden Epidemiology Reference Group
FGD	Functional gastrointestinal disorders
FUTI	Foodborne urinary tract infection
GBS	Guillain-Barré syndrome
GMI	Global Microbial Identifier
HUS	Hemolytic uremic syndrome
IBD	Inflammatory bowel disease
IBS	Inflammatory bowel syndrome
JECFA	Joint FAO/WHO Expert Committee on Food Additives
LTHO	Long-term health outcomes
OR	Odds ratio
PAF	Population attributable fraction
PCR	Polymerase chain reaction
PI-IBS	Post-infectious irritable bowel syndrome
QALY	Quality-adjusted life year
ReA	Reactive arthritis
RR	Relative risk
STEC	Shiga toxin-producing *E. coli*
UC	Ulcerative colitis
UTI	Urinary tract infection
WGS	Whole genome sequencing
WHO	World Health Organization

9.1   Introduction

Disease burden estimates provide the foundation for evidence-informed policy making and are critical to public health priority setting around food safety. Several efforts have recently been undertaken to better quantify the burden of foodborne disease, as presented in Chaps. 7 and 8, but there is still much work to be done. This chapter outlines areas of improvement that would lead to improved estimates such as enhancing foodborne disease surveillance infrastructure and improving our understanding of the burden of chronic sequelae associated with foodborne disease. We also give an overview of attribution studies that will increase the usefulness of disease burden estimates by identifying the most important (groups of) foods or reservoirs that contribute to the disease burden.

9.2 Foodborne Disease Surveillance

Many studies use data from public health surveillance to estimate the overall burden of foodborne disease (Flint et al. 2005; Haagsma et al. 2013; Scallan et al. 2011a, b). Public health surveillance systems for foodborne disease are largely passive and often require a laboratory confirmed diagnosis; therefore, only a relatively small number of cases are actually reported to public health agencies (Fig. 9.1). To estimate the overall burden of foodborne disease using data from public health surveillance, investigators must have a good understanding of how many cases of illness are lost at each stage of the surveillance pyramid due to underdiagnosis (i.e., medical care seeking, specimen submission, laboratory testing practices, laboratory test sensitivity) or underreporting (i.e., diagnosed illness not reported to surveillance). By estimating the degree of underdiagnosis (e.g., only 20% people seek medical care) and underreporting (e.g., only 90% of diagnoses illnesses were reported to public health), investigators adjust for undercounts by creating multipliers (e.g., the inverse of the proportion (1/0.20) equates to a multiplier of 5 for medical care seeking) to scale up the number of illnesses reported in public health surveillance to estimate the overall number of illnesses in the community. An example is provided in Fig. 9.2.

Surveys of the general population have been used to estimate the number of people with a diarrheal illness that seek medical care and submit stool sample for testing (Jones et al. 2007; Scallan et al. 2005). Limitations of these retrospective surveys include the fact that they are based on self-report and people with a diarrheal illness in the community may not be representative of those with an enteric infection reported to surveillance, given that those with more severe symptoms may be more likely to seek medical care and submit a stool sample for testing (O'Brien et al. 2010; Scallan et al. 2006). Some investigators have tried to account for severity by estimating medical care seeking and stool sample submission separately for those with mild and severe illness, using symptoms such as bloody diarrhea or duration of illness as a marker for severity (Haagsma et al. 2013; Scallan et al. 2011a, b; Kirk et al. 2014). Most surveys estimating the rate of medical care seeking and specimen submission focus on diarrheal illness, so estimates of underdiagnosis are often lacking for foodborne diseases that do not have diarrhea as a primary symptom (e.g., brucellosis, listeriosis) or that are not associated with diarrhea (e.g., toxoplasmosis and most diseases associated with foodborne chemical hazards (Gibb et al. 2015)).

Because laboratory confirmation is often required for a foodborne disease to be diagnosed and reported to public health agencies, investigators must determine how often laboratories routinely test for specific pathogens as well as the sensitivity and specificity of the laboratory test that was used. Laboratory test sensitivity can be challenging to estimate as it encapsulates more than just sensitivity of the test in a controlled setting. Rather it is meant to capture the "real-world" laboratory test sensitivity which includes reductions in sensitivity caused by issues with transportation and transport media, timeliness of specimen collection and testing, and other

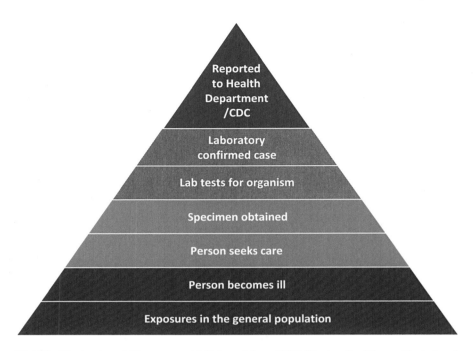

Fig. 9.1 The burden of illness pyramid (Adapted from CDC (Centers for Disease Control and Prevention (CDC) 2015))

factors. Studies have derived estimates of laboratory test sensitivity from a variety of sources including quality assurance surveys (Hall et al. 2008), outbreaks (Chalker and Blaser 1998), and expert opinion (Ingram et al. 2013).

The increased use of culture-independent diagnostic testing (CIDT) for foodborne pathogens poses a number of challenges for accurately estimating the burden of foodborne disease (Cronquist et al. 2012). CIDTs for bacterial enteric pathogens include nucleic acid amplification tests (such as PCR) and antigen-based methods (such as enzyme immunoassays and lateral flow assays) and are being increasingly used by clinical laboratories. While there are many advantages to CIDTs, including more rapid diagnosis and testing for pathogens not previously tested for routinely (e.g., Enterotoxigenic *E. coli*), any changes in laboratory test or practices will require investigators estimating the burden of foodborne disease to reassess the multipliers used to adjust for laboratory testing and laboratory test sensitivity when estimating total illnesses. The sensitivity and specificity of CIDTs is different from culture which has been the standard for many decades, and there is a lot of variation in test performance across different tests. In addition, the demographic characteristics of patients with detected infections have also shifted, suggesting that testing practices have changed with the introduction of new tests. To account for the increased use of CIDTs, more information is needed on laboratory testing practices, sensitivity and specificity, and changes in clinician testing practices.

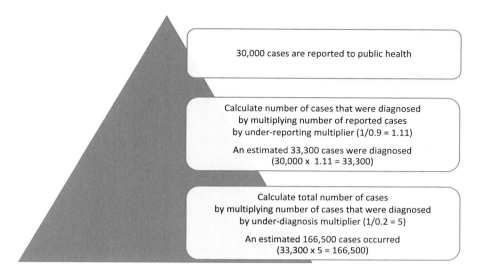

Fig. 9.2 Example of the use of multipliers in estimating the number of illnesses

An alternative approach to obtaining population-level incidence estimates of diarrheal disease and attribution to specific pathogens is through (1) prospective cohort studies with community and etiologic components and (2) cross-sectional surveys with or without supporting targeted studies (Flint et al. 2005). Prospective cohort studies invite patients in the general population and/or presenting at general practices to provide detailed information on their health status during a pre-defined follow-up period. Patients meeting a case definition of acute gastroenteritis are invited to submit stool specimens for pathogen detection and to complete questionnaires on health, risk factors, and other relevant factors. Healthy controls may be invited to strengthen etiologic and risk factor analysis. Such prospective cohort studies are relatively expensive and complex and have been organized by only a few countries. Yet, these studies have the advantage of providing community incidence rates that are pathogen-specific. Key examples are the IID-1 and IID-2 studies in the United Kingdom (Tam et al. 2012; Wheeler et al. 1999) and the Sensor/NIVEL studies in the Netherlands (De Wit et al. 2001a, b, c). Prospective cohort studies have also implemented in major, recent international studies on the incidence and etiology of enteric disease in low and middle income countries, although these studies typically included patients presenting to health care and therefore do not provide population-based incidence estimates. Key examples are the GEMS (Kotloff et al. 2013) and MAL-ED studies (The MAL-ED Network Investigators 2014; Platts-Mills et al. 2015).

Cross-sectional surveys, which are also known as prevalence studies, examine the association between a risk factor(s) and a disease by collecting data on both exposures and outcomes at a specific point in time rather than by following a group of patients over time, as is done in a prospective cohort study. In food safety, cross-sectional surveys are typically based on random-dialing telephone surveys, and provide information on (self-reported) incidence of gastrointestinal illness and, depending on questionnaire design, other variables of interest for burden estimation and risk factor analysis. While these types of studies are faster and cheaper to conduct than prospective cohort studies, they cannot prove causality, and, as such, etiological information often must be obtained from other sources. Flint et al. (Flint et al. 2005) provide examples of studies implemented in different high-income countries, including the population surveys used to estimate the number of people with a diarrheal illness that seek medical care and submit stool sample for testing that were discussed previously.

Many foodborne pathogens are not routinely captured as part of routine surveillance and may only be reported to public health agencies as part of a recognized outbreak. Therefore, outbreak reports may provide the only source of data for some pathogens. Because only a fraction of diagnosed cases are associated with an outbreak, studies apply an "outbreak multiplier" (in addition to any adjustments for underdiagnosis) to estimate the number outbreak of cases that would have been reported had all outbreak cases been reported to routine disease surveillance. Studies have derived an outbreak multiplier by comparing the number of cases reported to national surveillance with the number of cases reported as part of outbreak for the given pathogen or pathogens with both types of data available (e.g., *Salmonella*) (Scallan et al. 2011a, b; Kirk et al. 2014); however, it is not clear how representative these extrapolations are.

Outbreak reports also provide information on the routes of transmission and the foods responsible for illness, and these data have been used to attribute the burden of illness to specific sources (Adak et al. 2005; Painter et al. 2013). While data from outbreak reports can provide extremely valuable information on foods, there are several limitations. First, it is not known how representative outbreak-associated cases are of all cases of illness with regard to the implicated product. For example, chicken is thought to be the most important cause of *Campylobacter* infections, but most detected *Campylobacter* outbreaks have been linked to unpasteurized milk (Adak et al. 2005; Painter et al. 2013). Second, many outbreaks do not implicate a food vehicle as part of the outbreak investigations, so information may be missing or food vehicles may be reported as a "complex food" (e.g., lasagna) without a clear ingredient being identified as the culprit. Finally, outbreak data may be lacking for some pathogens of interest; for example, *Campylobacter* is rarely associated with outbreaks but causes a significant number of illnesses annually.

Public health surveillance and outbreak reports are important sources of data for estimating the overall burden of disease and attributing the burden of illness to specific sources. More complete surveillance data accompanied by supplemental studies that illuminate different points in the surveillance pyramid increase the accuracy of burden of disease estimates based on public health surveillance data. In particular, more work is needed to understand the surveillance pyramid for non-diarrheal foodborne pathogens. Understanding laboratory testing practices, laboratory test sensitivity and specificity, and changes in physician testing practices in the age of CIDTs is also of critical importance. Outbreak reports provide critical information on pathogens not routinely reported to public health surveillance and provide data needed to attribute illness to specific foods. This underscores the importance of investigating outbreaks, identifying the causative pathogen and implicating a food vehicle, and systematically collecting these data in a central location.

9.3 Disease Burden of Chronic Sequelae

Traditionally, burden of disease estimates have focused on the incidence of acute foodborne illness, hospitalization, and death (Scallan et al. 2011a, 2011b; Mead et al. 1999). However, foodborne illness has been associated with several chronic diseases, including functional gastrointestinal disorders, renal dysfunction, reactive arthritis, neurologic disorders, cognitive and developmental deficits (Table 9.1) (Batz et al. 2013; Keithlin et al. 2014a, b, 2015; Kowalcyk et al. 2013; Roberts et al. 2009), and increased long-term mortality (Helms et al. 2003). These long-term health outcomes (LTHO), which are described below, are major drivers of disease burden and cost (Havelaar et al. 2012; Mangen et al. 2014), but few long-term follow-up studies of FBD have been conducted, and most that have been conducted have significant limitations that restrict their generalizability (Roberts et al. 2009). As a result, there are significant gaps in our understanding of the strength and consistency of effect, temporality, dose response, burden of disease, and clinical management of the LTHOs associated with foodborne illness (Deising et al. 2013). Due to the lack of data and conclusive evidence on causality, many chronic sequelae associated with FBD have not been systematically included in burden of disease estimates. For example, Scharff (Scharff 2012) included the burden associated with irritable bowel syndrome (IBS) but did not include the burden associated with reactive arthritis (ReA), while other researchers included ReA but excluded IBS from their burden estimates (Batz et al. 2012). When such discrepancies exist, it is difficult to compare burden estimates and/or make recommendations to decision-makers. Research is needed to address these important epidemiologic research gaps, which would lead to improved burden of disease estimates.

9.3.1 Functional Gastrointestinal Disorders and Inflammatory Bowel Disease

Exposure to foodborne pathogens has been associated with several functional gastrointestinal disorders (FGDs) that cause chronic or recurrent gastrointestinal symptoms. While the biological mechanism for this is not fully understood, it is hypothesized that exposure to the foodborne pathogen alters the gut flora, alters intestinal permeability and/or motility, and increases the number of intraepithelial lymphocytes, lamina propria T cells, and mast cells, triggering an immune response (Barbara et al. 2009; Marshall et al. 2004; Dunlop et al. 2003;

DuPont 2008; Smith and Bayles 2007). Post-infectious irritable bowel syndrome (PI-IBS) has been associated with exposure to *Campylobacter*, *Salmonella*, Shiga toxin-producing *E. coli* (STEC), *Shigella*, *Yersinia*, *Giardia*, *Trichinella*, and norovirus, with the incidence varying by pathogen from 3%–36% (Dai and Jiang 2012; Halvorson et al. 2006; Ilnyckyj et al. 2003; Marshall et al. 2006; Pitzurra et al. 2011; Porter et al. 2011, 2013a; Thabane et al. 2007). For example, patients from the 2000 Walkerton, Ontario, waterborne outbreak of *E. coli* O157:H7 and *Campylobacter* had an increased risk of PI-IBS (odds ratio (OR): 3.12; 95% confidence interval (CI): 1.99–5.04) 8 years following the outbreak when compared to controls (Marshall et al. 2010). Increased risk of Crohn's disease (CD) and ulcerative colitis (UC) has also been associated with acute gastroenteritis generally (Garcia Rodriguez et al. 2006; Gradel et al. 2009; Jess et al. 2011; Porter et al. 2008; Ternhag et al. 2008) as well as with specific enteric pathogens, such as *Campylobacter* and *Salmonella* (Gradel et al. 2009; Jess et al. 2011; Ternhag et al. 2008). A meta-analysis of nine studies found a twofold increase in risk of developing functional dyspepsia (FD) following infectious gastroenteritis (Pike et al. 2013). Celiac disease (CeD), an autoimmune disorder triggered by the protein epitopes of gluten, has been associated with *Campylobacter*, but the epidemiologic evidence is limited (Riddle et al. 2012, 2013). Identified risk factors for developing FGDs following acute gastroenteritis vary by FGD but generally include family history, age, gender, severity of acute infections, prior antibiotic use, smoking,

Table 9.1 Selected health outcomes associated with foodborne pathogens (Batz et al. 2013)

Health outcome	Foodborne pathogen
Celiac disease	*Campylobacter*
Chronic diarrhea	*Campylobacter*, *Cryptosporidium*, *Giardia lamblia*, *Salmonella*, *Yersinia enterocolitica*
Diabetes	*E. coli* O157:H7, *Shigella*
Dyspepsia	*Campylobacter*, *E. coli* O157:H7, *Salmonella*, *Norovirus*
Inflammatory bowel disease (IBD)	*Campylobacter*, *Giardia lamblia*, *Salmonella*, *Shigella*
Irritable bowel syndrome (IBS)	*Campylobacter*, *E. coli* O157:H7, *Giardia lamblia*, *Norovirus*, *Salmonella*, *Shigella*
Gastroesophageal reflux disease	*Norovirus*
Guillain-Barré syndrome (GBS)	*Campylobacter*
Hemolytic uremic syndrome (HUS)	*E. coli* O157:H7, *Salmonella*, *Shigella*
Multiple sclerosis	*Clostridium perfringens*
Neurological disorders	*Cryptosporidium*, *E. coli* O157:H7, *Giardia lamblia*, *Listeria monocytogenes*, *Shigella*, *Vibrio vulnificus*
Reactive arthritis (ReA)	*Campylobacter*, *Cryptosporidium*, *E. coli* O157:H7, *Giardia lamblia*, *Salmonella*, *Shigella*, *Yersinia enterocolitica*
Renal impairment	*E. coli* O157:H7, *Shigella*
Schizophrenia, depression	*Toxoplasma gondii*

education level, psychosocial factors (e.g., stress, neuroses, hypochondrias), and health-care seeking behaviors (Dunlop et al. 2003; Riddle et al. 2012; Gwee et al. 1999; Locke 3rd et al. 2000; Marshall et al. 2007; Neal et al. 1997; Nicholl et al. 2008; Ruigómez et al. 2007).

9.3.2 Autoimmune Disorders

Exposure to foodborne pathogens can also cause autoimmune responses, such as reactive arthritis (ReA) and Guillain-Barré syndrome (GBS). Several studies have found an association between infectious gastroenteritis and ReA, a painful form of inflammatory arthritis that is triggered by an infection in another part of the body (Keithlin et al. 2014a, b; Ajene et al. 2013; Hannu 2011; Pope et al. 2007; Porter et al. 2013b). For example, a review of 14 cohort studies estimated the weighted mean incidence of ReA following *Campylobacter*, *Salmonella*, and *Shigella* infection to be 9, 12, and 12 per 100,000, respectively (Ajene et al. 2013). Estimates, however, vary across studies and reviews; this is likely due to the variability in measuring exposure and outcomes and/or differences in host/pathogen factors. Similarly, several studies and reviews have found an association between GBS, a rare but serious autoimmune disorder that causes paralysis and is fatal in 4–15% of patients, and infectious pathogens such as *Cytomegalovirus*, Epstein-Barr virus, Zika virus, *Salmonella*, and *Campylobacter* (Keithlin et al. 2015; Esan et al. 2017; Frenzen 2008; McCarthy and Giesecke 2001; McGrogan et al. 2009; Moore et al. 2005; Mori et al. 2000; Tam et al. 2006, 2007; Winer 2001). *Campylobacter jejuni* infection, in particular, has been identified in 20–40% of GBS cases, making it the predominant antecedent infection for GBS (Hughes and Cornblath 2005; Nyati and Nyati 2013; Poropatich et al. 2010). GBS patients often develop long-term chronic sequelae with 31% showing moderate to severe neurological sequelae and 38% and 44% reporting changes in their work situation and leisure activities, respectively, 2.6–6.4 years post-GBS (Bernsen et al. 2002).

9.3.3 Hemolytic Uremic Syndrome

Hemolytic uremic syndrome (HUS) is a severe and potentially fatal complication characterized by acute hemolytic anemia (destruction of red blood cells), nephropathy (kidney failure), and thrombocytopenia (reduced platelets) that can occur during or immediately following the acute phase of foodborne illness and is most commonly associated with Shiga toxin-producing *E. coli* (STEC), although cases following *Shigella* and *Salmonella* infection have been reported (Keithlin et al. 2014a; Garg et al. 2003; Karpman et al. 1998; Mayer et al. 2012; Siegler and Oakes 2005). The proportion of cases that develop HUS varies by pathogen species, serotype, and virulence factors. A meta-analysis of 82 studies found that 4.2–17.2% of

E. coli O157:H7 cases develop HUS (Keithlin et al. 2014a) which is consistent with data collected through national surveillance systems in the United States and the United Kingdom (Byrne et al. 2015; Gould et al. 2013, 2009), with an estimated 3–5% of cases being fatal (Andreoli et al. 2002; Mody et al. 2015; Siegler 1995). Children below 5 years of age and the elderly are at higher risk of developing HUS following STEC infection (Wong et al. 2012); bloody diarrhea, fever, treatment with β-lactam antibiotics, and serotypes with class 2 Shiga toxin and eae (intimin encoding) virulence genes are additional risk factors (Brandal et al. 2015; Ethelberg et al. 2004; Gianantonio et al. 1968; Launders et al. 2016; Siegler et al. 1994; Werber et al. 2003). Chronic sequelae, including renal impairment, hypertension, diabetes mellitus, cardiovascular disease, and neurological sequelae such as seizures, hemiparesia, epilepsy, and developmental delay, have been associated with HUS (Garg et al. 2003; Bale Jr. et al. 1980; Bauer et al. 2014; Magnus et al. 2012; Nathanson et al. 2010; Buder et al. 2015; Clark et al. 2010; Eriksson et al. 2001; Gagnadoux et al. 1996; Kelles et al. 1994; Suri et al. 2005). The Walkerton Health Study found that, 8 years after the Walkerton outbreak, patients with moderate to severe acute symptoms were significantly more likely to develop hypertension, renal impairment, and self-reported cardiovascular disease than asymptomatic or mildly ill cases (Clark et al. 2010). A large meta-analysis found similar results with 25% of HUS cases suffering renal sequelae (95% CI: 20–30%), 10% hypertension (95% CI: 8–12%), and 15% proteinuria (95% CI: 10–20%) (Garg et al. 2003). Another meta-analysis found that 3.2% (95% CI: 1.3–5.1%) of children with HUS develop diabetes during the acute illness and 38% (95% CI: 24–55) develop persistent diabetes (Suri et al. 2005). It is important to note that most studies of the long-term health impacts of HUS are retrospective, have small sample sizes and/or short follow-up, and often do not include neurological sequelae. Consequently, renal impairment is commonly the only chronic sequela included in burden estimates.

9.3.4 Neurological Dysfunction

Foodborne disease has been associated with neurological sequelae such as impaired/delayed cognitive development, motor impairment, seizures, palsies, and vision/hearing loss (Roberts et al. 2009). While several pathogens—such as *Salmonella*, *Campylobacter*, *Shigella*, *Brucella*, and several parasites, including *Taenia solium*, *Trichinella*, *Echinococcus*, *Diphyllobothrium*, *Paragonimus*, *Spirometra*, and *Toxocara* – have been associated with neurological sequelae, the most notable are *Listeria monocytogenes* and *Toxoplasma gondii* (Batz et al. 2013; Schlech 3rd 2000). *L. monocytogenes* can cause sepsis, meningoencephalitis, or acute respiratory distress syndrome, particularly in fetuses, newborns, the elderly, and those with compromised immune systems—resulting in residual neurological deficits and sometimes death (Lomonaco et al. 2015; Mylonakis et al. 1998, 2002; Swaminathan and Gerner-Smidt 2007). In a meta-analysis of 87 studies, 43.8% of perinatal and 13.7% of non-perinatal listeriosis cases with central nervous system infections subsequently developed neurological sequelae, including long-term hearing/vision loss and stroke outcomes (Maertens de Noordhout et al. 2014). While most infected individuals experience no symptoms, *T. gondii* can cause serious illness and significant neurological sequelae, particularly in fetuses, newborns, and individuals with compromised immune systems. Congenital toxoplasmosis is usually more severe than acquired toxoplasmosis and has been associated with vision/hearing impairment, cognitive impairment, psychomotor deficiencies, and seizures (Havelaar et al. 2007a, 2007b).

9.3.5 Psychological Disorders

There is emerging evidence that infection with foodborne pathogens may increase risk of psychological disorders, such as depression, chronic fatigue, anxiety, bipolar disorder, schizophrenia, and post-traumatic stress disorder (Bolton and Robertson 2016). For example, a follow-up study of 389 patients sickened in the 2011 *E. coli* O104 outbreak in Germany found that 6 months after the infection, 43% of patient had clinically relevant fatigue and 3% of patients suffered from post-traumatic stress syndrome (Löwe et al. 2014). Of all the foodborne pathogens, the links between psychological disorders and *T. gondii* have been the most comprehensively studied. A meta-analysis of 50 case-control studies found significant differences in seroprevalence of *T. gondii* between healthy controls and patients with schizophrenia (OR: 1.81; CI: 1.51–2.16; p-value (p) < 0.001), bipolar disorder (OR: 1.52; CI, 1.06–2.18; $p = 0.02$), addiction (OR: 1.91; 95% CI, 1.49–2.44; $p < 0.001$), and obsessive-compulsive disorder (OR: 3.4; 95% CI: 1.73–6.68; $p < 0.001$) (Sutterland et al. 2015). It has also been suggested that exposure to infectious agents could be associated through gut-brain interactions with autism spectrum disorder although, as with many of the chronic sequelae discussed here, more research is needed to establish a conclusive link (Bolton and Robertson 2016). The mechanisms by which bacterial and parasitic pathogens affect mental health are not well understood, but the hypothesis is that the pathogens directly infect the brain, as with *T. gondii*, or indirectly impact the brain by activating the peripheral nervous system (Sutterland et al. 2015; Torrey and Yolken 2003). More research is needed in this emerging area that may greatly contribute to the burden of foodborne disease.

9.3.6 Urinary Tract Infections

The association between urinary tract infections (UTIs) and extraintestinal pathogenic *E. coli* (ExPEC) is well established, but there is emerging evidence that UTIs may also be associated with foodborne pathogens (Sutterland et al. 2015; Nordstrom et al. 2013; Toval et al. 2014). In one study, isolates of *E. coli* strains from retail meats and ready-to-eat foods were found to be genetically related to strains from women with UTIs, suggesting that foodborne transmission may play a role (Vincent et al. 2010). In another study, women with UTIs caused by antimicrobial resistant *E. coli* reported consuming poultry and pork more frequently than women with UTIs caused by fully susceptible *E. coli*, suggesting meat as a potential reservoir (Manges et al. 2007). Based on this evidence, it has been hypothesized that, in foodborne UTIs (FUTIs), the patient is exposed to ExPEC through food, the gut is colonized, and the pathogen is subsequently transferred to the urinary tract (Nordstrom et al. 2013); however, additional studies are needed.

9.3.7 Malnutrition and Growth Impairment

Childhood growth impairment is a topic of big concern given the high prevalence of stunted children under 5 years of age; a 2017 report estimated that 115 million children worldwide are stunted (UNICEF, WHO, World Bank Group 2017). Previous published findings suggest that childhood stunting is associated with poor cognitive development (Grantham-McGregor et al. 2007; Walker et al. 2011; Prendergast and Humphrey 2014; Prendergast et al. 2015), increased morbidity and mortality from infectious and chronic diseases (Caulfield et al. 2004; De Boer et al. 2012; Guerrant et al. 2008; Prendergast and Humphrey 2014), as well as reduced incomes throughout life (Prendergast and Humphrey 2014). However, the pathogenesis of childhood stunting is poorly understood (Owino et al. 2016). In the last several decades, various epidemiological or intervention studies have extensively explored the relationships of malnutrition and growth/stunting and infection/diarrheal disease and growth/stunting (Bhutta et al. 2013; Dewey and Adu-Afarwuah 2008; Richard et al. 2013, 2014). However, the modest relationships with stunting suggest that, while nutrition and diarrheal disease are important factors for linear growth, they are not the only factors. This increased realization has encouraged researchers to delve more into potential pathways such as chronic gut injury with systemic inflammation and immunostimulation that can ultimately impair growth (Campbell et al. 2003; Mbuya and Humphrey 2016). Of particular interest is the hypothesis that exposure to poor sanitation and hygiene causes enteropathy in the gut that leads to stunting (Humphrey 2009). This enteropathy, which has been recently termed environmental enteric dysfunction (EED), has been associated with increased intestinal permeability, impaired gut immune function, recurrent/persistent diarrhea, nutrient malabsorption, and stunting (Owino et al. 2016; Crane et al. 2015; Keusch et al. 2013; McCormick and Lang 2016). Multiple factors seem to contribute to EED including nutritional deficiencies, (asymptomatic) colonization by enteric pathogens, and environmental toxins such as mycotoxins (Prendergast et al. 2015). However, the relative contribution of each of the factors is unknown (Prendergast et al. 2015; Kelly et al. 2004). Recently, the MAL-ED study identified a high *Campylobacter* prevalence in primarily asymptomatic children in eight low-resource settings being associated with a lower length-for-age Z score, increased intestinal permeability, and intestinal and systemic inflammation at 24 months of age (Amour et al. 2016).

Many chemical hazards are assumed to increase the risk for chronic diseases, including malnutrition and growth impairment. For example, mycotoxin contamination can cause various health issues and economic losses worldwide. Mycotoxins are toxic secondary metabolites produced by fungi that commonly contaminate foods such as maize, peanuts, and cereal grains (Wu 2013); 25% of the world's crop are contaminated with mycotoxins (Reddy et al. 2010), and high levels are reported for sub-Saharan Africa, Asia, and Central America. Developing countries with tropical climates (high temperature and humidity) are particularly impacted by mycotoxin contamination (Reddy et al. 2010), and over 4.5 billion people are at risk for chronic aflatoxin exposure through food (Centers for Disease Control and Prevention (CDC) 2012). Despite the significant public health impact (Wild and Gong 2010), very few epidemiological studies have explored the longitudinal relationships of mycotoxin exposure on health outcomes and, particularly, childhood growth impairment. Current findings suggest exposure to mycotoxins— including aflatoxins and fumonisins—is associated with several serious health outcomes, including adverse birth outcomes, childhood stunting, impaired nutrient absorption, immune suppression, mental impairment, liver disease, and cancer (Wu 2013; Alborzi et al. 2006; Food and Drug Administration (FDA) 2012; International Agency for Research in Cancer (IARC) 1993; Shuaib et al. 2010; Smith et al. 2012; Turner et al. 2007; Turner 2013). Potential biological mechanisms/pathways related to mycotoxins exposure and child growth impairment are less well understood. Hence, well-characterized epidemiological studies with multiple exposures/biomarkers and in multi-country settings (such as MAL-ED) can provide valuable insights into the contribution of mycotoxins and EED, along with various factors, in the pathogenesis of childhood stunting and burden of disease calculation.

9.4 Exploring the Association of Health Outcomes

As Sect. 9.3 has shown, there are a large number of health outcomes that are potentially associated with foodborne hazards. Establishing this association is, however, not always straightforward. In this section, we describe and discuss the methods that are used for establishing such associations and argue for scenario analyses to evaluate potential uncertainties and knowledge gaps.

The first, and most straightforward, method for causal attribution is *categorical attribution*. This approach can be used when a foodborne hazard results in an outcome (death or a specific symptom) that is identifiable as caused by the hazard (and only the hazard) in individual cases (Devleesschauwer et al. 2015). For instance, an individual diarrhea case may be attributed to *Salmonella* based on laboratory confirmation, or an anaphylactic reaction may be attributed to peanut exposure based on anamnesis.

When the foodborne hazard elevates the risk of an outcome that occurs from other causes as well, causal attribution can no longer be made on a case-by-case basis but, only statistically, at a population level (Devleesschauwer et al. 2015). For instance, *T. gondii* is reported to increase the risk of schizophrenia and other psychological disorders (Sutterland et al. 2015), but it is not possible to attribute an individual case of schizophrenia to *T. gondii* infection. Likewise, aflatoxin may increase the risk of hepatocellular carcinoma, but it is not possible to specify that a specific liver cancer case was caused by aflatoxin since (1) there is a long latency period between the exposure and the development of cancer and (2) many other exposures and/or genetic risk factors could have caused the liver cancer. In this situation, the standard approach for calculating the burden of foodborne disease is to use a *counterfactual analysis* in which the current disease outcomes with current exposure are statistically compared to the disease outcomes under an alternate exposure (a minimum risk exposure which could be zero or some accepted background level) (Prüss-Üstün et al. 2003). This allows calculation of the relative risk and population attributable fraction, which are population-level metrics of the association between the foodborne hazard and the associated outcome. Specifically, the relative risk is defined as the ratio of the outcome incidence among exposed individuals and the outcome incidence among non-exposed individuals. The population attributable fraction is a function of the relative risk and the exposure distribution and is defined as the proportion of incident cases that would be prevented in a population if exposure could be reduced to the minimum risk exposure level. However, these metrics are generally obtained through observational studies, which demonstrate association, but not necessarily causation. Information on the causal attribution between the concerned hazard-outcome pairs is therefore often limited. Furthermore, estimation of the relative risk, and thus the population attributable fraction, may be done under the competing assumptions of an additive versus a multiplicative model. The additive model assumes that RR_{AB}, the expected RR for a person experiencing risk factor A and risk factor B, equals $RR_A + RR_B - 1$, while the multiplicative model assumes that RR_{AB} equals $RR_A \times RR_B$. Both assumptions can lead to widely varying estimates, thus resulting in important methodological uncertainty. Finally, to calculate the burden of the concerned hazard-outcome pair, the population attributable fraction must be multiplied with the all-cause burden estimates for the relevant disease outcome (the so-called burden envelope); for instance, the burden of *T. gondii*-associated schizophrenia is obtained by multiplying an all-cause schizophrenia burden estimate with the *T. gondii* population attributable fraction. The counterfactual approach is, therefore, not only dependent on estimates of the population attributable fraction but also on the availability and quality of the concerned burden envelopes.

In cases where there are insufficient data for categorical attribution and counterfactual analysis (considered top-down approaches)—this is the case for many foodborne chemical hazards, *risk assessment* approach (considered a bottom-up approach) is often used (Devleesschauwer et al. 2015). The risk assessment approach is the standard methodology applied to assess the safety of human exposure to foodborne chemicals and increasingly used for microbial risks. In this approach, the incidences of the hazard-associated outcomes (e.g., diarrhea due to *Salmonella* exposure or liver cancer due to aflatoxin exposure) are estimated by combining exposure and dose-response data. The dose-response model may, for instance, define the probability of illness at a given exposure level, which can then be translated into an estimate of the number of incident cases expected to occur in the exposed population (Prüss-Üstün et al. 2003)). As this approach does not involve burden attribution, it does not necessarily ensure consistency with existing health statistics. Furthermore, the risk assessment approach is often limited by uncertainty on the dose-response relationship. For instance, when dose-response data are extracted from animal models, a tenfold correction factor is generally included to account for the potential differences between animals and humans and another tenfold factor for the difference between humans. This strategy is relevant when estimating maximum allowable intake levels but might lead to overestimation when the aim is to assess true disease burden. Even when human dose-response data are used, these are not necessarily representative for the general population that is of interest in burden of disease studies. For example, when dose-response relationships are based only on data from high-exposure events, there may remain important uncertainty in the lower end of the dose-response curve, which may be most relevant for the general population (Teunis and Havelaar 2000). For instance, Teunis et al. (Teunis et al. 2012) developed a dose-response model for *Trichinella* spp. in humans based on published outbreaks of human trichinellosis; likewise, Crump et al. (Crump et al. 2003) developed a dose-response model for dioxin and cancer based on data from three occupationally exposed cohorts. Since these dose-response models were developed using data from high-exposure events, they may overestimate risk at lower exposure levels that may be more representative of exposure in the general population. When microbial dose-response relationships are based on data from human or animal feeding trials, the virulence and pathogenicity of the applied isolates or their physiological state may not be representative for that of the isolates circulating in foods. For example, Teunis et al. (Teunis et al. 2002) explored the strain differences in available *Cryptosporidium* dose-response models. Chen et al. (Chen et al. 2006) demonstrated that fresh (animal passaged) isolates of *Campylobacter jejuni* showed higher colonization potential in chickens and less within isolate variation than isolates that had been repeatedly subcultured in the laboratory.

In addition to the methodological issues that arise when modeling the association between foodborne hazards and health outcomes, causal attribution may also be hampered by ethical controversies. For instance, whether or not to include miscarriage and stillbirth in burden of disease calculations implies ethical and moral discussions on how the life, and death, of an embryo or fetus compares to that of a human after birth (Jamison et al. 2006; Phillips and Millum 2015). For this reason, many burden estimates exclude miscarriages and stillbirths.

Given the various sources of methodological and structural uncertainty regarding the association of health outcomes to foodborne hazards, a valid approach would be to generate estimates based on different, well-defined scenarios. Such scenario analyses would allow the reader to assess the impact of alternating methodological and structural choices and to adopt the estimates that correspond to what is deemed the most acceptable scenario. For instance, estimates could be generated using both a counterfactual and risk assessment approach, to assess the impact of different methodological approaches (Jakobsen et al. 2015). Likewise, estimates could be generated that either include or exclude an uncertain health outcome, allowing the reader to assess the impact of this uncertainty. For instance, Smit et al. (Smit et al. 2017) showed that the disease burden of congenital toxoplasmosis in Belgium would be twice as high if fetal losses at ≥22 weeks of gestational age would be included.

9.5 Attributing the Burden of Foodborne Diseases to Specific Foods, Food Groups, or Reservoirs

While burden of disease estimates are crucial to raising awareness of foodborne diseases, estimating their public health impact, and ranking diseases according to their importance, they may be insufficient for policy making. To identify and prioritize food safety intervention strategies to prevent and reduce the burden of diseases in a population, knowledge on the most important sources of the causative foodborne hazards is needed.

Several source attribution methods are available, including approaches based on the analysis of data from occurrence of hazards in foods and humans, epidemiological studies, intervention studies, and expert elicitations. All methods present both advantages and limitations, and their utility and applicability depend on the public health questions being addressed and on characteristics and distribution of the hazard (Table 9.2). As examples, epidemiological studies may be useful for source attribution of disease by microbiological hazards, which lead mostly to acute disease and thus enable an association of exposure to specific contaminated foods with the onset of symptoms; on the contrary, they are usually insufficient to attribute disease by chemical hazards, which is typically chronic and appears a long time after exposure. Additionally, methods have different data requirements and attribute human illness at either the point of production (reservoir) or of exposure to the food, and therefore their utility will vary depending on the hazard and/or the country or region in question (Pires 2013).

9.5.1 Overview of Source Attribution Methods

Approaches to source attribution can be grouped broadly into four categories: microbiological, epidemiological, expert elicitation, and intervention studies (Batz et al. 2005; Pires et al. 2009). Methods in all categories have been used to estimate the sources of several pathogens in different subpopulations (e.g., *Salmonella*, *Campylobacter*, *L. monocytogenes*). For chemical hazards, source attribution has

Table 9.2 Strengths and limitations of source attribution methods (adapted from (Pires 2013))

SA approach	Strengths	Limitations
Occurrence approaches		
Subtyping approach	• Identifies the most important reservoirs of the hazard and therefore (1) is useful to prioritize interventions at production level and (2) reduces uncertainty due to cross-contamination or spread to accidental sources	• Limited to hazards heterogeneously distributed among the reservoirs • No information on transmission pathways from reservoirs to humans • Data intensive, requiring a collection of representative isolates from all (major) sources
Comparative exposure assessment	• Accounts for different transmission routes from the same reservoir • Easily updated	• Often limited by lack of data, resulting in large uncertainties
Epidemiological studies		
Case-control studies (including systematic review)	• Able to identify variety of risk factors, including exposure routes, predisposing, behavioral, or seasonal factors • A systematic review of published studies can be useful for regional analysis and may detect temporal and geographical variations • Can identify a wide range of known and unknown risk factors	• Misclassification due to immunity may reduce attributable risk or suggest protection • Most studies only explain a small fraction of all cases • Cases may reflect a mixture of possible sources of exposure • Misclassification due to recall bias may lead to an underestimation of the attribution proportion
Analysis of data from outbreaks	• Documentation that a hazard was transmitted to humans via a specific food item can be available • Data may capture the effect of contamination at multiple points from the farm-to-consumption chain • Wide variety of foods represented, including uncommon foods • Most readily available information for source attribution in some countries or regions	• Quality of evidence varies • Large outbreaks, outbreaks associated with point sources, outbreaks with short incubation periods, or more severe are more likely to be investigated • Investigated cases may not be representative of all foodborne illnesses • Certain pathogens and foods are more likely to be associated with reported outbreaks, which can lead to an overestimation of the attribution proportion

(continued)

Table 9.2 (continued)

SA approach	Strengths	Limitations
Intervention studies		
	• Allows for a direct measure of the impact of a source on the number of infections, avoiding accounting for the effect of external factors	• Interpretation of data from "large-scale" interventions is difficult, since usually several interventions are implemented at the same time • Complex and resource demanding
Expert elicitations		
	• Allows for attribution to main transmission routes • Useful tool when data is lacking • May be the only available method for source attribution	• Conclusions are based on the individual experts' judgment, which may be misinformed or biased

been done mostly unintentionally, i.e., as a part of methods applied for risk assessment or burden of disease studies.

Microbiological approaches for source attribution include the subtyping approach and the comparative exposure assessment approach. Both involve the use of data on the occurrence of foodborne hazards in animal, food, and/or environmental sources. These data are ideally available from surveillance or monitoring programs in a country but may also be obtained through, e.g., targeted projects or literature review. The subtyping approach was designed to attribute human cases to the reservoir level, i.e., the closest possible to the origin of the pathogen, and gives no information on the relative contribution of different exposure routes to humans. On the contrary, the comparative exposure assessment approach estimates the relative importance of different routes for exposure, including several routes from the same reservoir.

Epidemiological approaches comprise case-control studies of sporadic and analyses of data from outbreak investigations. Case-control studies are useful to identify sources and risk factors for a disease, as well as the fraction of human cases that can be attributable to these (by estimating population attributable fractions, PAF). Even if case-control studies are not often conducted and are insufficient to extrapolate source attribution estimates at national level, a meta-analysis of several case-control studies (i.e., combining studies conducted in several countries) can be used to estimate the number of illnesses attributable to each exposure at regional and global level. In contrast, foodborne outbreak data are widely available from most world regions. Outbreak investigations are often able to identify the contaminated source or ingredient that caused infections, and an analysis of these data can show the relative contribution of the most important sources of disease. These analyses can be done at national, regional, and global levels, and, despite the limitations of assuming that outbreak data are representative of all cases in the population (i.e., also of sporadic cases of disease), outbreak attribution analyses are useful evidence for source prioritization.

Expert elicitations can be used to estimate the proportion of illnesses that are attributed to foodborne, environmental, contact with animals, environmental, or human-to-human transmission pathways (Hald et al. 2016).

Source attribution can take place at different points along the food chain (points of attribution), including at the origin of the pathogen, i.e., the point of reservoir, such as the animal production stage, or at the point of exposure, such as the food consumption stage. The different source attribution methods attribute disease at different points and will as mentioned depend on the availability of data and on the risk management question being addressed.

9.5.2 Attribution to Main Types of Transmission

The first step in the source attribution process is to estimate the overall proportion of the burden of disease that can be attributed to the four main transmission routes, i.e., foodborne, environmental, direct contact to animals, and person-to-person. For most foodborne hazards, data-driven methods, based, for example, on surveillance and monitoring data, would require an exhaustive review and inclusion of all potential sources and pathways within these main routes and consequently are not the most appropriate tool for this initial step when applied individually. A combination of epidemiological methods could provide a more adequate picture of the relative importance of the types of transmission, namely, a combination of an analysis of outbreak data and of studies of sporadic cases. For hazards that are transmitted through a limited number of routes (e.g., *Brucella* spp.), the application of one epidemiological approach for source attribution may be sufficient. Alternatively, two methods are currently available to attribute disease to these main routes: expert elicitations and intervention studies.

Attribution of foodborne disease to food and other transmission routes could be undertaken for individual foodborne hazards or for syndromic groups, e.g., diarrheal disease. In both cases, expert elicitations can be conducted at a country or regional level, whereas interventions are optimally designed as small scale population-based studies. The latter are additionally expensive and difficult to apply.

The WHO-FERG has undertaken a large-scale expert elicitation to attribute disease by 19 foodborne hazards to main transmission groups at a global, regional, and subregional level (Hald et al. 2016; Havelaar et al. 2015). The study applied structured expert judgment using Cooke's Classical Model (Cooke 1991) to obtain estimates for the relative contributions of different transmission pathways for 11 diarrheal diseases, 7 other infectious diseases, and 1 chemical (lead). Experts were selected based on their experience including international working experience and included in ten global panels or nine subregional panels. This study presented the first worldwide estimates of the proportion of specific diseases attributable to food and other major transmission routes. Other expert elicitations have been conducted to deliver similar estimates but at a national level, specifically in the Netherlands and in Canada (Davidson et al. 2011; Havelaar et al. 2008; Lake et al. 2010; Vally et al. 2014). Similar country-specific initiatives will be useful to improve estimates and reduce uncertainties.

9.5.3   Attribution to Specific Foods and Exposure Routes

As mentioned before, the risk management question, the characteristics of the hazard causing the disease, and the data available influence the utility of source attribution methods. When more than one source attribution method proves useful, the final choice of method will be determined by the question that needs answering and will be influenced by the analytical capacity in a country and the level of data sharing between agencies.

The type of reservoir of the hazard will influence the applicability of some source attribution methods, particularly the subtyping approach. This approach applies to hazards with one or more animal reservoirs, to which disease can be traced back and where the hazard can potentially be controlled. All other approaches are, in principle, applicable regardless of the origin of the hazard, since they focus on routes of transmission or the point of exposure.

There may also be differences in the utility of methods for regional or national level. In general, epidemiological approaches, specifically analysis of outbreak data and systematic review and meta-analysis of case-control studies of sporadic infections, are useful for source attribution at a regional level when data are not available a country level.

The applicability and usefulness of the source attribution methods vary for enteric, parasitic, and chemical hazards. The subtyping approach is appropriate to attribute human disease for an enteric pathogen if that pathogen has mainly an animal reservoir, can be subtyped by appropriate discriminatory methods, and subtyping data are available. This has been verified for only two pathogens (*Salmonella* spp. and *Campylobacter* spp.) (Pires 2013). For the majority of the remaining enteric hazards, source attribution by an analysis of data from outbreak investigations is appropriate. The comparative exposure assessment approach has been shown to be useful for attributing infections by pathogens that are mostly transmitted by a limited number of food routes, namely, STEC, *L. monocytogenes*, and *Brucella* (Food and Drug Administration (FDA) 2003; Kosmider et al. 2010); it has also been applied to other pathogens, e.g., *Campylobacter* (Evers et al. 2008; Pintar et al. 2017). A systematic review of epidemiological studies of sporadic infections can be useful for enteric hazards that have been extensively studied throughout the world (Domingues et al. 2012a, 2012b).

For chemical hazards, the comparative exposure assessment approach is the most appropriate method to attribute disease and is also often done as part of the method applied to estimate the burden of disease caused by exposure to the hazard through multiple food routes. Given the availability of data, this approach is of simple application. Epidemiological studies, particularly cohort studies, have been undertaken for some of these chemicals, and a review of these could be useful for source attribution. However, because disease caused by chemicals often appears a long time after exposure, epidemiological studies may have challenges identifying cases and sources.

9.5.4   Challenges and Future Directions in Source Attribution

Controlling foodborne diseases and thus improving food safety requires efforts at several levels. All research and risk management initiatives, including the ones relying on source attribution studies, are dependent on efficient surveillance, which has been the target for improvements and investments throughout the world, either through national, regional, or capacity building initiatives. Multinational organizations such as WHO at the international level and the European Food Safety Authority (EFSA) and the European Centre for Disease Prevention and Control (ECDC) at the regional level play an increasingly important role in the harmonization of surveillance statuses across countries and will be crucial to encourage countries to invest in the integration of food safety components.

In developed countries, improvements in surveillance have been largely focused on the development and use of sophisticated typing methods (e.g., molecular techniques), which have substantially increased the opportunities for research and the production of scientific evidence for interventions. Recently, whole genome sequencing (WGS) has opened yet another spectrum of possibilities, providing new and faster ways to diagnose, monitor, and track foodborne pathogens. We are now witnessing extensive research on the applications of these methods, particularly on how to best use WGS in surveillance and how to translate these data into useful epidemiological evidence.

Several factors have favored the use of such techniques in foodborne disease surveillance: (1) WGS has become mature and has been increasingly introduced in routine laboratories; (2) the price of WGS has been falling dramatically, in some cases, below the price of traditional identification; (3) the availability of a vast amount of IT resources and a fast Internet; and (4) the idea that, via a One Health approach, infectious diseases could be better controlled and prevented (Global Microbial Identifier (GMI) 2013). In this context, initiatives to harmonize methodologies and data collection and sharing are crucial. An example is the Global Microbial Identifier, a genomic epidemiological database for global identification of microorganisms which is a platform for storing WGS data of microorganisms, for the identification of relevant genes, and for the comparison of genomes to detect outbreaks and emerging pathogens (http://www.globalmicrobialidentifier. org).

Traditional microbiological foodborne disease surveillance systems have relied on the collection of samples at different stages of the food production chain, isolation and quantification of foodborne pathogens in these samples, and typing of these with different methods of phenotypic or genotypic characterization. The recent development of molecular typing methods is changing the way surveillance systems work. These changes may be particularly relevant in developing countries where surveillance of foodborne diseases is still behind with regards to their ability to diagnose/identify specific causes of disease. In these countries where systems are not yet entrenched, affordable WGS may represent a significant technological shortcut.

In the context of burden of disease and source attribution, opportunities are immense but are still to be explored. Along with pathogen characterization techniques and surveillance, the scientific methods available to produce evidence for food safety interventions are also likely to change. This will require extensive research. A major challenge of using data generated from molecular typing methods, and in particular WGS, will be to define meaningful subtypes to provide appropriate level of discrimination for source attribution models (European Food Safety Authority Panel on Biological Hazards (EFSA) 2013). Such research will also depend on the accessibility to potential enormous amounts of data that needs to be compiled, analyzed, and shared among the scientific community. Developing such a coordinated system is timely and should be carried out at a global level.

9.6   Conclusion

In a world of limited resources, policy makers are constantly being asked to prioritize the allocation of resources to efforts. Should they allocate resources to preventing this disease or another one? Which intervention strategies should they invest in? Burden of disease estimates provide policy makers a quantitative measurement of the impact on public health, while source attribution estimates provide information on where to intervene. Significant advancements have recently been made in understanding the burden and sources of foodborne illness, but there is still room for improvement. Public health surveillance is an important source of data for disease burden and attribution studies, but few countries have the infrastructure needed to reliably provide such data. Even in countries that do have strong surveillance systems, there are still significant gaps in understanding and a need for constant improvement as clinical and laboratory practices evolve (e.g., CIDTs, WGS). Important epidemiologic gaps also remain about the burden of foodborne disease, particularly for chemical foodborne hazards and the long-term health impact of all foodborne pathogens. As a result, these health impacts are often not included in estimates, leading to underestimates of the burden of disease. There are significant opportunities to improve the ability of policy makers to effectively allocate resources by expanding our understanding of the burden and sources of foodborne disease, but this will require substantial investments in surveillance and research.

References

Adak GK, Meakins SM, Yip H, Lopman BA, O'Brien SJ. Disease risks from foods, England and Wales, 1996-2000. Emerg Infect Dis. 2005;11(3):365–72. https://doi.org/10.3201/eid1103.040191.

Ajene AN, Fischer Walker CL, Black RE. Enteric pathogens and reactive arthritis: a systematic review of *Campylobacter*, *Salmonella* and *Shigella*-associated reactive arthritis. J Health Popul Nutr. 2013;31(3):299–307.

Alborzi S, Pourabbas B, Rashidi M, Astaneh B. Aflatoxin M1 contamination in pasteurized milk in Shiraz (south of Iran). Food Control. 2006;17:582–4.

Amour C, Gratz J, Mduma E, Svensen E, Rogawski ET, McGrath M, et al. Epidemiology and impact of *Campylobacter* infection in children in 8 low-resource settings: results from the MAL-ED study. Clin Infect Dis. 2016;63(9):1171–9.

Andreoli SP, Trachtman H, Acheson DW, Siegler RL, Obrig TG. Hemolytic uremic syndrome: epidemiology, pathophysiology, and therapy. Pediatr Nephrol. 2002;17:293–8.

Bale JF Jr, Brasher C, Siegler RL. CNS manifestations of the hemolyticuremic syndrome. Relationship to metabolic alterations and prognosis. Am J Dis Child. 1980;134(9):869–72.

Barbara G, Cremon C, Pallotti F, De Giorgio R, Stanghellini V, Corinaldesi R. Postinfectious irritable bowel syndrome. J Pediatr Gastroenterol Nutr. 2009;48(Suppl 2):S95–7.

Batz MB, Doyle MP, Morris JG, Painter J, Singh R, Tauxe RV, Taylor MR, DMA LFW, The Food Attribution Working Group. Attributing illness to food. Emerg Infect Dis. 2005;11:993–9.

Batz MB, Hoffmann S, Morris JG. Ranking the disease burden of 14 pathogens in food sources in the United States using attribution data from outbreak investigations and expert elicitation. J Food Prot. 2012;75(7):1278–91.

Batz M, Henke E, Kowalcyk B. Long-term consequences of foodborne infections. Infect Dis Clin North Am. 2013;27(3):599–616.

Bauer A, Loos S, Wehrmann C, Horstmann D, Donnerstag F, Lemke J, Hillebrand G, Löbel U, Pape L, Haffner D, Bindt C, Ahlenstiel T, Melk A, Lehnhardt A, Kemper MJ, Oh J, Hartmann H. Neurological involvement in children with *E. coli* O104:H4-induced hemolytic uremic syndrome. Pediatr Nephrol. 2014;29:1607–15.

Bernsen RA, de Jager AE, Schmitz PI, van der Meché FG. Long-term impact on work and private life after Guillain-Barré syndrome. J Neurol Sci. 2002;201(1–2):13–7.

Bhutta ZA, Das JK, Rizvi A, Gaffey MF, Walker N, Horton S, Webb P, Lartey A, Black RE, Lancet Nutrition Interventions Review Group, The Maternal and Child Nutrition Study Group. Evidence-based interventions for improvement of maternal and child nutrition: what can be done and at what cost? Lancet. 2013;382(9890):452–77.

Bolton DJ, Robertson LJ. Mental health disorders associated with foodborne pathogens. J Food Prot. 2016;79(11):2005–17.

Brandal LT, Wester AL, Lange H, Lobersli I, Lindstedt BA, Vold L, Kapperud G. Shiga toxin-producing *Escherichia coli* infections in Norway, 1992–2012: characterization of isolates and identification of risk factors for haemolytic uremic syndrome. BMC Infect Dis. 2015;15:324.

Buder K, Latal B, Nef S, Neuhaus TJ, Laube GF, Spartà G. Neurodevelopmental long-term outcome in children after hemolytic uremic syndrome. Pediatr Nephrol. 2015;30(3):503–13.

Byrne L, Jenkins C, Launders N, Elson R, Adak GK. The epidemiology, microbiology and clinical impact of Shiga toxin-producing *Escherichia coli* in England, 2009–2012. Epidemiol Infect. 2015;143(16):3475–87.

Campbell DI, Elia M, Lunn PG. Growth faltering in rural Gambian infants is associated with impaired small intestinal barrier function, leading to endotoxemia and systemic inflammation. J Nutr. 2003;133(5):1332–8.

Caulfield LE, de Onis M, Blossner M, Black RE. Undernutrition as an underlying cause of child associated with diarrhea, pneumonia, malaria, and measles. Am J Clin Nutr. 2004;80(1):193–8.

Centers for Disease Control and Prevention (CDC), Atlanta, GA: U.S. Centers for Disease Control and Prevention (updated 13 January 2012). Aflatoxin. http://www.cdc.gov/nceh/hsb/chemicals/aflatoxin.htm. Accessed Aug 11, 2015.

Centers for Disease Control and Prevention (CDC), Atlanta, GA: U.S. Centers for Disease Control and Prevention (updated 7 December 2015). FoodNet Surveillance. https://www.cdc.gov/foodnet/surveillance.html. Accessed 30 Oct 2017.

Chalker RB, Blaser MJA. Review of human salmonellosis: III. Magnitude of *Salmonella* infection in the United States. Rev Infect Dis. 1998;10(1):111–24.

Chen L, Geys H, Cawthraw S, Havelaar A, Teunis P. Dose response for infectivity of several strains of *Campylobacter jejuni* in chickens. Risk Anal. 2006;26:1613–1.

Clark WF, Sontrop J, Macnab JJ, Salvadori M, Moist L, Suri R, Garg AX. Long term risk for hypertension, renal impairment, and cardiovascular disease after gastroenteritis from drinking water contaminated with *Escherichia coli* O157:H7: a prospective cohort study. BMJ. 2010;341:c6020. https://doi.org/10.1136/bmj.c6020.

Cooke RM. Experts in uncertainty—opinion and subjective probability in science. Environmental ethics and science policy series. Oxford: Oxford University Press; 1991.

Crane RJ, Kelsey DJJ, Berkley JA. Environmental enteric dysfunction: an overview. Food Nutr Bull. 2015;36(10):S76–87.

Cronquist AB, Mody RK, Atkinson R, Besser J, D'Angelo MT, Hurd S, Robinson T, Nicholson C, Mahon BE. Impacts of culture-independent diagnostic practices on public health surveillance for bacterial enteric pathogens. Clin Infect Dis. 2012;54(Suppl 5):S432–9. https://doi.org/10.1093/cid/cis267.

Crump KS, Canady R, Kogevinas M. Meta-analysis of dioxin cancer dose response for three occupational cohorts. Environ Health Perspect. 2003;111:681–7.

Dai C, Jiang M. The incidence and risk factors of post-infectious irritable bowel syndrome: a meta-analysis. Hepatogastroenterology. 2012;59(113):67–2. https://doi.org/10.5754/hge10796.

Davidson VJ, Ravel A, Nguyen TN, Fazil A, Ruzante JM. Food-specific attribution of selected gastrointestinal illnesses: estimates from a Canadian expert elicitation survey. Foodborne Pathog Dis. 2011;8(9):983–5.

De Boer MD, Lima AA, Oriá RB, Scharf RJ, Moore SR, Luna MA, Guerrant RL. Early childhood growth failure and the developmental origins of adult disease: do enteric infections and malnutrition increase risk for the metabolic syndrome? Nutr Rev. 2012;70(11):642–53.

De Wit MAS, Koopmans MPG, Kortbeek LM, Van Leeuwen NJ, Bartelds AIM, van Duynhoven YT. Gastroenteritis in sentinel general practices, The Netherlands. Emerg Infect Dis. 2001a;7:82–1.

De Wit MA, Koopmans MP, Kortbeek LM, Van Leeuwen NJ, Vinje J, van Duynhoven YT. Etiology of gastroenteritis in sentinel general practices in the Netherlands. Clin Infect Dis. 2001b;33:280–8.

De Wit MA, Koopmans MP, Kortbeek LM, Wannet WJ, Vinje J, Van Leusden F, Bartelds AI, van Duynhoven YT. Sensor, a population-based cohort study on gastroenteritis in the Netherlands: incidence and etiology. Am J Epidemiol. 2001c;154:666–74.

Deising A, Gutierrez RL, Porter CK, Riddle MS. Postinfectious functional gastrointestinal disorders: a focus on epidemiology and research agendas. Gastroenterol Hepatol. 2013;9(3):145–7.

Devleesschauwer B, Haagsma JA, Angulo FJ, Bellinger DC, Cole D, Döpfer D, Fazil A, Fèvre EM, Gibb HJ, Hald T, Kirk MD, Lake RJ, Maertens de Noordhout C, Mathers CD, McDonald SA, Pires SM, Speybroeck N, Thomas MK, Torgerson PR, Wu F, Havelaar AN, Praet N. Methodological framework for World Health Organization estimates of the global burden of foodborne disease. PLoS One. 2015;10(12):e0142498. https://doi.org/10.1371/journal.pone.0142498.

Dewey KG, Adu-Afarwuah S. Systematic review of the efficacy and effectiveness of complementary feeding interventions in developing countries. Matern Child Nutr. 2008;4(Suppl 1):24–85.

Domingues AR, Pires SM, Halasa T, Hald T. Source attribution of human salmonellosis using a meta-analysis of case-control studies of sporadic infections. Epidemiol Infect. 2012a;140(6):959–69.

Domingues AR, Pires SM, Halasa T, Hald T. Source attribution of human campylobacteriosis using a meta-analysis of case-control studies of sporadic infections. Epidemiol Infect. 2012b;140(6):970–81.

Dunlop SP, Jenkins D, Spiller RC. Distinctive clinical, psychological, and histological features of postinfective irritable bowel syndrome. Am J Gastroenterol. 2003;98(7):1578–83.

DuPont AW. Postinfectious irritable bowel syndrome. Clin Infect Dis. 2008;46(4):594–9.

Eriksson KJ, Boyd SG, Tasker RC. Acute neurology and neurophysiology of haemolytic-uraemic syndrome. Arch Dis Child. 2001;84(5):434–5.

Esan OB, Pearce M, van Hecke O, Roberts N, Collins DRJ, Violato M, McCarthy N, Perera R, Fanshawe TR. Factors associated with sequelae of *Campylobacter* and non-typhoidal *Salmonella* infections: a systematic review. EBioMedicine. 2017;15:100–11.

Ethelberg S, Olsen KE, Scheutz F, Jensen C, Schiellerup P, Enberg J, Petersen AM, Olesen B, Gerner-Smidt P, Molbak K. Virulence factors for hemolytic uremic syndrome, Denmark. Emerg Infect Dis. 2004;10(5):842–7.

European Food Safety Authority Panel on Biological Hazards (EFSA). Scientific opinion on the evaluation of molecular typing methods for major food-borne microbiological hazards and their use for attribution modelling, outbreak investigation and scanning surveillance: part 1 (evaluation of methods and applications). EFSA J. 2013;11(12):3502. https://doi.org/10.2903/j.efsa.2013.3502. Available online: www.efsa.europa.eu/efsajournal

Evers EG, Van Der Fels-Klerx HJ, Nauta MJ, Schijven JF, Havelaar AH. *Campylobacter* source attribution by exposure assessment. Int J Risk Assess Manag. 2008; https://doi.org/10.1504/IJRAM.2008.016151.

Flint JA, van Duynhoven YT, Angulo FJ, DeLong SM, Braun P, Kirk M, Scallan E, Fitzgerald M, Adak GK, Sockett P, Ellis A, Hall G, Gargouri N, Walke H, Braam P. Estimating the burden of acute gastroenteritis, foodborne disease, and pathogens commonly transmitted by food: an international review. Clin Infect Dis. 2005;41(5):698–704. https://doi.org/10.1086/432064.

Food and Drug Administration (FDA). Quantitative assessment of the relative risk to public health from foodborne *Listeria monocytogenes* among selected categories of ready-to-rat foods. Center for Food Safety and Applied Nutrition (FDA) and Food Safety Inspection Service (USDA), 2003.

Food and Drug Administration (FDA). Bad bug book: foodborne pathogenic microorganisms and natural toxins. 2nd ed. Center for Food Safety and Applied Nutrition (CFSAN) of the Food and Drug Administration (FDA), US Department of Health and Human Services, 2012.

Frenzen PD. Economic cost of Guillain-Barré syndrome in the United States. Neurology. 2008;71(1):21–7.

Gagnadoux MF, Habib R, Gubler MC, Bacri JL, Broyer M. Long-term (15-25 years) outcome of childhood hemolytic-uremic syndrome. Clin Nephrol. 1996;46:39–41.

Garcia Rodriguez LA, Ruigomez A, Panes J. Acute gastroenteritis is followed by an increased risk of inflammatory bowel disease. Gastroenterology. 2006;130:1588–94.

Garg AX, Suri RS, Barrowman N, Rehman F, Matsell D, Rosas-Arellano MP, Salvadori M, Haynes RB, Clark WF. Long-term renal prognosis of diarrhea associated hemolytic uremic syndrome: a systematic review, meta-analysis, and meta-regression. JAMA. 2003;290:1360–70.

Gianantonio CA, Vitacco M, Mendilaharzu F, Gallo G. The hemolytic-uremic syndrome. Renal status of 76 patients at long-term follow-up. J Pediatr. 1968;72(6):757–65.

Gibb H, Devleesschauwer B, Bolger PM, Wu F, Ezendam J, Cliff J, Zeilmaker M, Verger P, Pitt J, Baines J, Adegoke G, Afshari R, Liu Y, Bokkers B, van Loveren H, Mengelers M, Brandon E, Havelaar AH, Bellinger D. World Health Organization estimates of the global and regional disease burden of four foodborne chemical toxins, 2010: a data synthesis. F1000Res. 2015;4:1393.

Global Microbial Identifier (GMI). 2013. http://www.globalmicrobialidentifier.org/about-gmi/vision-and-objectives. Assessed June 2017.

Gould LH, Demma L, Jones TF, Hurd S, Vugia DJ, Smith K, Shiferaw B, Segler S, Palmer A, Zansky S, Griffin PM. Hemolytic uremic syndrome and death in persons with *Escherichia coli* O157:H7 infection, foodborne diseases active surveillance network sites, 2000-2006. Clin Infect Dis. 2009;49:1480–5.

Gould LH, Mody RK, Ong KL, Clogher P, Cronquist AB, Garman KN, Lathrop S, Medus C, Spina NL, Webb TH, White PL, Wymore K, Gierke RE, Mahon BE, Griffin PM, Emerging Infections Program Foodnet Working Group. Increased recognition of non-O157 Shiga toxin-producing

Escherichia coli infections in the United States during 2000–2010: epidemiologic features and comparison with *E. coli* O157 infections. Foodborne Pathog Dis. 2013;10(5):453–60.

Gradel KO, Nielsen HL, Schønheyder HC, Ejlertsen T, Kristensen B, Nielsen H. Increased short- and long-term risk of inflammatory bowel disease after *Salmonella* or *Campylobacter* gastroenteritis. Gastroenterology. 2009;137:495–501.

Grantham-McGregor S, Cheung YB, Cueto S, Glewwe P, Richter L, Strupp B, International Child Development Steering Group. Developmental potential in the first 5 years for children in developing countries. Lancet. 2007;369(9555):60–70.

Guerrant R, Oriá RB, Moore SR, Oriá MO, Lima AA. Malnutrition as an enteric infectious disease with long-term effects on child development. Nutr Rev. 2008;66(9):487–505.

Gwee KA, Leong YL, Graham C, McKendrick MW, Collins SM, Walters SJ, Underwood JE, Read NW. The role of psychological and biological factors in postinfective gut dysfunction. Gut. 1999;44(3):400–6.

Haagsma JA, Geenen PL, Ethelberg S, Fetsch A, Hansdotter F, ansen A, Korsgaard H, O'Brien SJ, Scavia G, Spitznagel H, Stefanoff P, Tam CC, Havelaar AH, Med-Vet-Net Working Group. Community incidence of pathogen-specific gastroenteritis: reconstructing the surveillance pyramid for seven pathogens in seven European Union member states. Epidemiol Infect. 2013;141(8):1625–39. https://doi.org/10.1017/S0950268812002166.

Hald T, Aspinall W, Devleesschauwer B, Cooke R, Corrigan T, Havelaar AH, Gibb HJ, Torgerson PR, Kirk MD, Angulo FJ, Lake RJ, Speybroeck N, Hoffmann S. World Health Organization estimates of the relative contributions of food to the Burden of disease due to selected foodborne hazards: a structured expert elicitation. PLoS One. 2016;11(1):e0145839. https://doi.org/10.1371/journal.pone.0145839

Hall G, Yohannes K, Raupach J, Becker N, Kirk M. Estimating community incidence of *Salmonella, Campylobacter*, and Shiga toxin-producing *Escherichia coli* infections, Australia. Emerg Infect Dis. 2008;14(10):1601–9. https://doi.org/10.3201/eid1410.071042.

Halvorson H, Schlett C, Riddle M. Postinfectious irritable bowel syndrome—a meta-analysis. Am J Gastroenterol. 2006;101:1894–9.

Hannu T. Reactive arthritis. Best Pract Res Clin Rheumatol. 2011;25:347–57.

Havelaar AH, Bräunig J, Christiansen K, Cornu M, Hald T, Mangen MJ, Mølbak K, Pielaat A, Snary E, Van Pelt W, Velthuis A, Wahlström H. Towards an integrated approach in supporting microbiological food safety decisions. Zoonoses Public Health. 2007a;54(3–4):103–17.

Havelaar AH, Kemmeren JM, Kortbeek LM. Disease burden of congenital toxoplasmosis. Clin Infect Dis. 2007b;44:1467–74.

Havelaar AH, Galindo AV, Kurowicka D, Cooke RM. Attribution of foodborne pathogens using structured expert elicitation. Foodborne Pathog Dis. 2008;5:649–59.

Havelaar AH, Haagsma JA, Mangen MJ, Kemmeren JM, Verhoef LP, Vijgen SM, Wilson M, Friesema IH, Kortbeek LM, van Duynhoven YT, van Pelt W. Disease burden of foodborne pathogens in the Netherlands, 2009. Int J Food Microbiol. 2012;156(3):231–8.

Havelaar AH, Kirk MD, Torgerson PR, Gibb HJ, Hald T, Lake RJ, Praet N, Bellinger DC, de Silva NR, Gargouri N, Speybroeck N, Cawthorne A, Mathers C, Stein C, Angulo FJ, Devleesschauwer B, World Health Organization Foodborne Disease Burden Epidemiology Reference Group. World Health Organization global estimates and regional comparisons of the burden of foodborne disease in 2010. PLoS Med. 2015;12:e1001923.

Helms M, Vastrup P, Gerner-Smidt P, Mølbak K. Short and long term mortality associated with foodborne bacterial gastrointestinal infections: registry based study. BMJ. 2003;326:357.

Hughes RA, Cornblath DR. Guillain-Barré syndrome. Lancet. 2005;366:1653–66.

Humphrey JH. Child undernutrition, tropical enteropathy, toilets and handwashing. Lancet. 2009;374:1032–5.

Ilnyckyj A, Balachandra B, Elliott L, Choudhri S, Duerksen DR. Post-traveler's diarrhea irritable bowel syndrome: a prospective study. Am J Gastroenterol. 2003;98:596–9.

Ingram M, St John J, Applewhaite T, Gaskin P, Springer K, Indar L. Population-based estimates of acute gastrointestinal and foodborne illness in Barbados: a retrospective cross-sectional study. J Health Popul Nutr. 2013;31(4 Suppl 1):81–97.

International Agency for Research in Cancer (IARC). Ochratoxin A. In some naturally occurring substances; food items and constituents, heterocyclic aromatic amines and mycotoxins. In: IARC monographs on the evaluation of carcinogenic risks to humans, vol. 56. Lyon: IARC; 1993. p. 489–521.

Jakobsen LS, Nauta M, Knudsen VK, Pires SM, Poulsen M. Burden of disease estimates of cancer caused by dietary exposure to acrylamide: how methodological choices affect the outcome. Toxicol Lett. 2015;238(2):115. https://doi.org/10.1016/j.toxlet.2015.08.371

Jamison DT, Shahid-Salles SA, Jamison J, Lawn JE, Zupan J. chap. Incorporating deaths near the time of birth into estimates of the global burden of disease. In: Global burden of disease and risk factors. Washington: World Bank Publications; 2006.

Jess T, Simonsen J, Nielsen NM, Jørgensen KT, Bager P, Ethelberg S, Frisch M. Enteric *Salmonella* or *Campylobacter* infections and the risk of inflammatory bowel disease. Gut. 2011;60:318–24.

Jones TF, McMillian MB, Scallan E, Frenzen PD, Cronquist AB, Thomas S, Angulo FJA. Population-based estimate of the substantial burden of diarrhoeal disease in the United States; FoodNet, 1996-2003. Epidemiol Infect. 2007;135(2):293–301. https://doi.org/10.1017/S0950268806006765.

Karpman D, Håkansson A, Perez MT, Isaksson C, Carlemalm E, Caprioli A, Svanborg C. Apoptosis of renal cortical cells in the hemolytic-uremic syndrome: in vivo and in vitro studies. Infect Immun. 1998;66:636–44.

Keithlin J, Sargeant J, Thomas MK, Fazil A. Chronic sequelae of *E. coli* O157: systematic review and meta-analysis of the proportion of *E. coli* O157 cases that develop chronic sequelae. Foodborne Pathog Dis. 2014a;11(2):79–95.

Keithlin J, Sargeant J, Thomas MK, Fazil A. Systematic review and meta-analysis of the proportion of *Campylobacter* cases that develop chronic sequelae. BMC Public Health. 2014b;14:1203.

Keithlin J, Sargeant JM, Thomas MK, Fazil A. Systematic review and meta-analysis of the proportion of non-typhoidal *Salmonella* cases that develop chronic sequelae. Epidemiol Infect. 2015;143(7):1333–51.

Kelles A, Van Dyck M, Proesmans W. Childhood haemoltyic uraemic syndrome: long-term outcome and prognostic features. Eur J Pediatr. 1994;153:38–42.

Kelly P, Menzies I, Crane R, Zulu I, Nickols C, Feakins R, Mwansa J, Mudenda V, Katubulushi M, Greenwald S, Farthing M. Responses of small intestinal architecture and function over time to environmental factors in a tropical population. Am J Trop Med Hyg. 2004;70(4):412–9.

Keusch GT, Rosenberg IH, Denno DM, Duggan C, Guerrant RL, Lavery JV, Tarr PI, Ward HD, Black RE, Nataro JP, Ryan ET, Bhutta ZA, Coovadia H, Lima A, Ramakrishna B, Zaidi AKM, Hay Burgess DC, Brewer T. Implications of acquired environmental enteric dysfunction for growth and stunting in infants and children living in low- and middle-income countries. Food Nutr Bull. 2013;34(3):357–64.

Kirk M, Ford L, Glass K, Hall G. Foodborne illness, Australia, circa 2000 and circa 2010. Emerg Infect Dis. 2014;20(11):1857–64. https://doi.org/10.3201/eid2011.131315.

Kosmider RD, Nally P, Simons RLR, Brouwer A, Cheung S, Snary EL, Wooldridge M. Attribution of human VTEC O157 infection from meat products: a quantitative risk assessment approach. Risk Anal. 2010;30:753–65.

Kotloff KL, Nataro JP, Blackwelder WC, Nasrin D, Farag TH, Panchalingam S, Wu Y, Sow SO, Sur D, Breiman RF, Faruque AS, Zaidi AK, Saha D, Alonso PL, Taboura B, Sanogo D, Onwuchekwa U, Manna B, Ramamurthy T, Kanungo S, Ochieng JB, Omore R, Oundo JO, Hossain A, Das SK, Ahmed S, Qureshi S, Quadri F, Adegbola RA, Antonio M, Hossain MJ, Akinsola A, Mandomando I, Nhampossa T, Acácio S, Biswas K, O'Reilly CE, Mintz ED, Berkeley LY, Muhsen K, Sommerfelt H, Robins-Browne RM, Levine MM. Burden and aetiology of diarrhoeal disease in infants and young children in developing countries (the global enteric multicenter study, GEMS): a prospective, case-control study. Lancet. 2013;382(9888):209–22.

Kowalcyk B, Smeets H, Succop P, DeWit N, Havelaar A. Relative risk of irritable bowel syndrome following acute gastroenteritis and associated risk factors. Epidemiol Infect. 2013;13:1–10.

Lake RJ, Cressey PJ, Campbell DM, Oakley E. Risk ranking for foodborne microbial hazards in New Zealand: burden of disease estimates. Risk Anal. 2010;30(5):743–52.

Launders N, Byrne L, Jenkins C, Harker K, Charlett A, Adak GK. Disease severity of Shiga toxin-producing *E. coli* O157 and factors influencing the development of typical haemolytic uraemic syndrome: a retrospective cohort study, 2009–2012. BMJ Open. 2016;6(1):e009933.

Locke GR 3rd, Zinsmeister AR, Talley NJ, Fett SL, Melton LJ. Risk factors for irritable bowel syndrome: role of analgesics and food sensitivities. Am J Gastroenterol. 2000;95(1):157–65.

Lomonaco S, Nucera D, Filipello V. The evolution and epidemiology of *Listeria monocytogenes* in Europe and the United States. Infect Genet Evol. 2015;35:172–83.

Löwe B, Andresen V, Faedrich K, Gapprnayer K, Wedgscheider K, Treszl A, Riegel B, Rose M, Lohse AW, Broicher W. Psychological outcome, fatigue, and quality of life after infection with Shiga toxin-producing *Escherichia coli* O104. Clin Gastroenterol Hepatol. 2014;12:1848–55.

Maertens de Noordhout C, Devleesschauwer B, Angulo FJ, Verbeke G, Haagsma J, Kirk M, Havelaar A, Speybroeck N. The global burden of listeriosis: a systematic review and meta-analysis. Lancet Infect Dis. 2014;14:1073–82.

Magnus T, Rother J, Simova O, Meier-Cillien M, Repenthin J, Möller F, Gbadamosi J, Panzer U, Wengenroth M, Hagel C, Kluge S, Stahl RK, Wegscheider K, Urban P, Eckert B, Glatzel M, fiehler J, Gerloff C. The neurological syndrome in adults during the 2011 northern German *E. coli* serotype O104:H4 outbreak. Brain. 2012;135:1850–9.

Mangen MJ, Bouwknegt M, Friesema IH, Haagsma JA, Kortbeek LM, Tariq L, Wilson M, van Pelt W, Havelaar AH. Cost-of-illness and disease burden of food-related pathogens in the Netherlands, 2011. Int J Food Microbiol. 2014;196:84–93.

Manges AR, Smith SP, Lau BJ, Nuval CJ, Eisenberg JN, Dietrich PS, Riley LW. Retail meat consumption and the acquisition of antimicrobial resistant *Escherichia coli* causing urinary tract infections: a case-control study. Foodborne Pathog Dis. 2007;4:419–31.

Marshall JK, Thabane M, Garg AX, Clark W, Meddings J, Collins SM, WEL Investigators. Intestinal permeability in patients with irritable bowel syndrome after a waterborne outbreak of acute gastroenteritis in Walkerton, Ontario. Aliment Pharmacol Ther. 2004;20(11–12):1317–22.

Marshall JK, Thabane M, Garg AX, Clark WF, Salvadori M, Collins SM, Walkerton Health Study Investigators. Incidence and epidemiology of irritable bowel syndrome after a large waterborne outbreak of bacterial dysentery. Gastroenterology. 2006;131:445–50.

Marshall JK, Thabane M, Borgaonkar MR, James C. Postinfectious irritable bowel syndrome after a food-borne outbreak of acute gastroenteritis attributed to a viral pathogen. Clin Gastroenterol Hepatol. 2007;5(4):457–60.

Marshall J, Thabane M, Garg AX, Clark WF, Moayyedi P, Collins SM, Walkerton Health Study Investigators. Eight year prognosis of post infectious irritable bowel syndrome following waterborne bacterial dysentery. Gut. 2010;59(5):605–11.

Mayer CL, Leibowitz CS, Kurosawa S, Sterns-Kurosawa DJ. Shiga toxins and the pathophysiology of hemolytic uremic syndrome in humans and animals. Toxins (Basel). 2012;4:1261–87.

Mbuya MN, Humphrey JH. Preventing environmental enteric dysfunction through improved water, sanitation and hygiene: an opportunity for stunting reduction in developing countries. Matern Child Nutr. 2016;12(Suppl 1):106–20.

McCarthy N, Giesecke J. Incidence of Guillain-Barré syndrome following infection with *Campylobacter jejuni*. Am J Epidemiol. 2001;153(6):610–4.

McCormick BJJ, Lang DR. Diarrheal disease and enteric infections in LMIC communities: how big is the problem? Trop Dis Travel Med Vaccines. 2016;2:11. https://doi.org/10.1186/s40794-016-0028-7.

McGrogan A, Madle GC, Seaman HE, de Vries CS. The epidemiology of Guillain-Barré syndrome worldwide. A systematic literature review. Neuroepidemiology. 2009;32(2):150–63.

Mead PS, Slutsker L, Dietz V, McCaig LF, Bresee JS, Shapiro C, Griffin PM, Tauxe RV. Food-related illness and death in the United States. Emerg Infect Dis. 1999;5(5):607–25.

Mody RK, Gu W, Griffin PM, Jones TF, Rounds J, Shiferaw B, Tobin-D'Angelo M, Smith G, Spina N, Hurd S, Lathrop S, Palmer A, Boothe E, Luna-Gierke RE, Hoekstra RM. Postdiarrheal hemolytic uremic syndrome in United States children: clinical spectrum and predictors of in-hospital death. J Pediatr. 2015;166:1022–9.

Moore JE, Corcoran D, Dooley JS, Fanning S, Lucey B, Matsuda M, McDowell DA, Megraud F, Millar BC, O'Mahony R, O'Riordan L, O'Rourke M, Rao JR, Rooney PJ, Sails A, Whyte P. *Campylobacter*. Vet Res. 2005;36(3):351–82.

Mori M, Kuwabara S, Miyake M, Noda M, Kuroki H, Kanno H, Ogawara K, Hattori T. *Haemophilus influenzae* infection and Guillain-Barré syndrome. Brain. 2000;123(Pt 10):2171–8.

Mylonakis E, Hohmann EL, Calderwood SB. Central nervous system infection with *Listeria monocytogenes*. 33 years' experience at a general hospital and review of 776 episodes from the literature. Medicine. 1998;77:313–36.

Mylonakis E, Paliou M, Hohmann EL, Calderwood SB, Wing EJ. Listeriosis during pregnancy: a case series and review of 222 cases. Medicine. 2002;81:260–9.

Nathanson S, Kwon T, Elmaleh M, Charbit M, Launay EA, Harambat J, Brun M, Ranchin B, Bandin F, Cloarec S, Bourdat-Michel G, Piètrement C, Champion G, Ulinski T, Deschênes G. Acute neurological involvement in diarrhea-associated hemolytic uremic syndrome. Clin J Am Soc Nephrol. 2010;5(7):1218–28.

Neal K, Hebden J, Spiller R. Prevalence of gastrointestinal symptoms six months after bacterial gastroenteritis and risk factors for development of the irritable bowel syndrome: postal survey of patients. BMJ. 1997;314(7083):779–82.

Nicholl BI, Halder SL, Macfarlane GJ, Thompson DG, O'Brien S, Musleh M, McBeth J. Psychosocial risk markers for new onset irritable bowel syndrome—results of a large prospective population-based study. Pain. 2008;137(1):147–55.

Nordstrom L, Liu CM, Price LB. Foodborne urinary tract infections: a new paradigm for antimicrobial-resistant foodborne illness. Front Microbiol. 2013;4:29.

Nyati KK, Nyati R. Role of *Campylobacter jejuni* infection in the pathogenesis of Guillain-Barré syndrome: an update. Biomed Res Int. 2013;2013:852195. https://doi.org/10.1155/2013/852195.

O'Brien SJ, Rait G, Hunter PR, Gray JJ, Bolton FJ, Tompkins DS, McLauchlin J, Letley LH, Adak GK, Cowden JM, Evans MR, Neal KR, Smith GE, Smyth B, Tam CC, Rodrigues LC. Methods for determining disease burden and calibrating national surveillance data in the United Kingdom: the second study of infectious intestinal disease in the community (IID2 study). BMC Med Res Methodol. 2010;10:39. https://doi.org/10.1186/1471-2288-10-39.

Owino V, Ahmed T, Freemark M, Kelly P, Loy A, Manary M, Loechi C. Environmental enteric dysfunction and growth failure/stunting in global child health. Pediatrics. 2016;138(6) https://doi.org/10.1542/peds.2016-0641.

Painter JA, Hoekstra RM, Ayers T, Tauxe RV, Braden CR, Angulo FJ, Griffin PM. Attribution of foodborne illnesses, hospitalizations, and deaths to food commodities by using outbreak data, United States, 1998-2008. Emerg Infect Dis. 2013;19(3):407–15. https://doi.org/10.3201/eid1903.111866.

Phillips J, Millum J. Valuing stillbirths. Bioethics. 2015;29(6):413–23.

Pike BL, Porter CK, Sorrell TJ, Riddle MS. Acute gastroenteritis and the risk of functional dyspepsia: a systematic review and meta-analysis. Am J Gastroenterol. 2013;108(10):1558–63.

Pintar KDM, Thomas KM, Christidis T, Otten A, Nesbitt A, Marshall B, Pollari F, Hurst M, Ravel A. A comparative exposure assessment of *Campylobacter* in Ontario, Canada. Risk Anal. 2017;37(4):677–715. https://doi.org/10.1111/risa.12653.

Pires SM. Assessing the applicability of currently available methods for attributing foodborne disease to sources, including food and food commodities. Foodborne Pathog Dis. 2013;10:206–13.

Pires SM, Evers EG, van Pelt W, Ayers T, Scallan E, Angulo FJ, Havelaar A, Hald T, Med-Vet-Net Workpackage 28 Working Group. Attributing the human disease burden of foodborne infections to specific sources. Foodborne Pathog Dis. 2009;6(4):417–24.

Pitzurra R, Fried M, Rogler G, Rammert C, Tschopp A, Hatz C, Steffen R, Mutsch M. Irritable bowel syndrome among a cohort of European travelers to resource-limited destinations. J Travel Med. 2011;18:250–6.

Platts-Mills JA, Babji S, Bodhidatta L, Gratz J, Haque R, Havt A, McCormick BJ, McGrath M, Olortegui MP, Samie A, Shakoor S, Mondal D, Lima IF, Hariraju D, Rayamajhi BB, Qureshi S, Kabir F, Yori PP, Mufamadi B, Amour C, Carreon JD, Richard SA, Lang D, Bessong P, Mduma

E, Ahmed T, Lima AA, Mason CJ, Zaidi AK, Bhutta ZA, Kosek M, Guerrant RL, Gottlieb M, Miller M, Kang G, Houpt ER, MAL-ED Network Investigators. Pathogen-specific burdens of community diarrhoea in developing countries: a multisite birth cohort study (MAL-ED). Lancet Glob Health. 2015;3(9):e564–75.

Pope JE, Krizova A, Garg AX, Thiessen-Philbrook H, Ouimet JM. *Campylobacter* reactive arthritis: a systematic review. Semin Arthritis Rheum. 2007;37:48–55.

Poropatich KO, Walker CL, Black RE. Quantifying the association between *Campylobacter* infection and Guillain-Barré syndrome: a systematic review. J Health Popul Nutr. 2010;28:545–52.

Porter CK, Tribble DR, Aliaga PA, Halvorson HA, Riddle MS. Infectious gastroenteritis and risk of developing inflammatory bowel disease. Gastroenterology. 2008;135(3):781–6.

Porter CK, Gormley R, Tribble DR, Cash BD, Riddle MS. The incidence and gastrointestinal infectious risk of functional gastrointestinal disorders in a healthy US adult population. Am J Gastroenterol. 2011;106(1):130–8.

Porter CK, Choi D, Cash B, Pimentel M, Murray J, May L, Riddle MS. Pathogen-specific risk of chronic gastrointestinal disorders following bacterial causes of foodborne illness. BMC Gastroenterol. 2013a;13:46.

Porter CK, Choi D, Riddle MS. Pathogen-specific risk of reactive arthritis from bacterial causes of foodborne illness. J Rheumatol. 2013b;40(5):712–4.

Prendergast AJ, Humphrey JH. The stunting syndrome in developing countries. Paediatr Int Child Health. 2014;34(4):250–65.

Prendergast AJ, Humphrey JH, Mutasa K, Majo FD, Rukobo S, Govha M, Mbuya MN, Moulton LH, Stoltzfus RJ. Sanitation hygiene infant nutrition efficacy (SHINE) trial team. Assessment of environmental enteric dysfunction in the SHINE trial: methods and challenges. Clin Infect Dis. 2015;61(Suppl 7):S726–32.

Prüss-Üstün A, Mathers C, Corvalán C, Woodward A. In: Prüss-Üstün A, Campbell-Lendrum D, Corvalán C, Woodward A, editors. Assessing the environmental burden of disease at national and local levels. Introduction and methods, Environmental burden of disease series. Geneva: World Health Organization; 2003.

Reddy KRN, Salleh B, Saad B, Abbas HK, Abel CA, Shier WT. An overview of mycotoxin contamination in foods and its implications for human health. Toxin Rev. 2010;29:3–26.

Richard SA, Black RE, Gilman RH, Guerrant RL, Kang G, Lanata CF, Mølbak K, Rasmussen ZA, Sack RB, Valentiner-Branth P, Checkley W, Childhood Malnutrition and Infection Network. Diarrhea in early childhood: short-term association with weight and long-term association with length. Am J Epidemiol. 2013;178(7):1129–38.

Richard SA, Black RE, Gilman RH, Guerrant RL, Kang G, Lanata CF, Mølbak K, Rasmussen ZA, Sack RB, Valentiner-Branth P, Checkley W, Childhood Malnutrition and Infection Network. Catch-up growth occurs after diarrhea in early childhood. J Nutr. 2014;144(6):965–71.

Riddle MS, Murray JA, Porter CK. The incidence and risk of celiac disease in a healthy US adult population. Am J Gastroenterol. 2012;107(8):1248–55.

Riddle MS, Murray JA, Cash BD, Pimentel M, Porter CK. Pathogen-specific risk of celiac disease following bacterial causes of foodborne illness: a retrospective cohort study. Dig Dis Sci. 2013;58(11):3242–5.

Roberts T, Kowalcyk B, Buck P, Blaser MJ, Frenkel JK, Lorber B, Smith J, Tarr PI. The long-term health outcomes of selected foodborne pathogens. The Center for Foodborne Illness Research & Prevention, November 12, 2009. www.foodborneillness.org.

Ruigómez A, Garcia Rodríguez L, Panes J. Risk of irritable bowel syndrome after an episode of bacterial gastroenteritis in general practice: influence of comorbidities. Clin Gastroenterol Hepatol. 2007;5(4):465–9.

Scallan E, Majowicz SE, Hall G, Banerjee A, Bowman CL, Daly L, Jones T, Kirk MD, Fitzgerald M, Angulo FJ. Prevalence of diarrhoea in the community in Australia, Canada, Ireland, and the United States. Int J Epidemiol. 2005;34(2):454–60. https://doi.org/10.1093/ije/dyh413.

Scallan E, Jones TF, Cronquist A, Thomas S, Frenzen P, Hoefer D, Medus C, Angulo FJ, FoodNet Working Group. Factors associated with seeking medical care and submitting a stool sample in estimating the burden of foodborne illness. Foodborne Pathog Dis. 2006;3(4):432–8. https://doi.org/10.1089/fpd.2006.3.432.

Scallan E, Hoekstra RM, Angulo FJ, Tauxe RV, Widdowson MA, Roy SL, Jones JL, Griffin PM. Foodborne illness acquired in the United States—major pathogens. Emerg Infect Dis. 2011a;17(1):7–15. https://doi.org/10.3201/eid1701.091101p1.

Scallan E, Griffin PM, Angulo FJ, Tauxe RV, Hoekstra RM. Foodborne illness acquired in the United States—unspecified agents. Emerg Infect Dis. 2011b;17(1):16–22.

Scharff R. Economic burden from health losses due to foodborne illness in the United States. J Food Prot. 2012;75(1):123–31.

Schlech WF 3rd. Foodborne listeriosis. Clin Infect Dis. 2000;31:770–5.

Shuaib FM, Jolly PE, Ehiri JE, Yatich N, Jiang Y, Funkhouser E, Person SD, Wilson C, Ellis WO, Wang JS, Williams JH. Association between birth outcomes and aflatoxin B1 biomarker blood levels in pregnant women in Kumasi, Ghana. Trop Med Int Health. 2010;15(2):160–7.

Siegler RL. The hemolytic uremic syndrome. Pediatr Clin North Am. 1995;42:1505–29.

Siegler R, Oakes R. Hemolytic uremic syndrome; pathogenesis, treatment, and outcome. Curr Opin Pediatr. 2005;17:200–4.

Siegler RL, Pavia AT, Christofferson RD, Milligan MK. A 20-year population-based study of post-diarrheal hemolytic uremic syndrome in Utah. Pediatrics. 1994;94(1):35–40.

Smit GSA, Padalko E, Van Acker J, Hens N, Dorny P, Speybroeck N, Devleesschauwer B. Public health impact of congenital toxoplasmosis and cytomegalovirus infection in Belgium, 2013: a systematic review and data synthesis. Clin Infect Dis. 2017;65:661–8.

Smith JL, Bayles D. Postinfectious irritable bowel syndrome: a long-term consequence of bacterial gastroenteritis. J Food Prot. 2007;70:1762–9.

Smith LE, Stoltzfus RJ, Prendergast A. Food chain mycotoxin exposure, gut health and impaired growth: a conceptual framework. Adv Nutr. 2012;3:526–31.

Suri RS, Clark WF, Barrowman N, Mahon JL, Thiessen-Philbrook HR, Rosas-Arellano MP, Zarnke K, Garland JS, Garg AX. Diabetes during diarrhea-associated hemolytic uremic syndrome: a systematic review and meta-analysis. Diabetes Care. 2005;28:2556–62.

Sutterland AL, Fond G, Kuin A, Koeter MW, Lutter R, van Gool T, Yolken R, Szoke A, Leboyer M, de Haan L. Beyond the association. Toxoplasma gondii in schizophrenia, bipolar disorder, and addiction: systematic review and meta-analysis. Acta Psychiatr Scand. 2015;132:161–79.

Swaminathan B, Gerner-Smidt P. The epidemiology of human listeriosis. Microbes Infect. 2007;9:1236–43.

Tam CC, Rodrigues LC, Petersen I, Islam A, Hayward A, O'Brien SJ. Incidence of Guillain-Barré syndrome among patients with Campylobacter infection: a general practice research database study. J Infect Dis. 2006;194(1):95–7.

Tam CC, O'Brien SJ, Petersen I, Islam A, Hayward A, Rodrigues LC. Guillain-Barré syndrome and preceding infection with Campylobacter, influenza and Epstein Barr virus in the general practice research database. PLoS One. 2007;2(4):e344.

Tam CC, Rodrigues LC, Viviani L, Dodds JP, Evans MR, Hunter PR, Gray JJ, Letley LH, Rait G, Tompkins DS, O'Brien SJ, IIDS Study Executive Committee. Longitudinal study of infectious intestinal disease in the UK (IID2 study): incidence in the community and presenting to general practice. Gut. 2012;61:69–77.

Ternhag A, Torner A, Svensson A, Ekdahl K, Giesecke J. Short- and long-term effects of bacterial gastrointestinal infections. Emerg Infect Dis. 2008;14:143–8.

Teunis PF, Havelaar AH. The Beta Poisson dose-response model is not a single-hit model. Risk Anal. 2000;20:513–20.

Teunis PFM, Chappell CL, Okhuysen PC. Cryptosporidium dose response studies: variation between isolates. Risk Anal. 2002;22(1):175–83.

Teunis PFM, Koningstein M, Takumi K, Van Der Giessen JW. Human beings are highly susceptible to low doses of Trichinella spp. Epidemiol Infect. 2012;140(2):210–8.

Thabane M, Kottachchi DT, Marshall JK. Systematic review and meta-analysis: the incidence and prognosis of post-infectious irritable bowel syndrome. Aliment Pharmacol Ther. 2007;26:535–44.

The MAL-ED Network Investigators. The MAL-ED study: a multinational and multidisciplinary approach to understand the relationship between enteric pathogens, malnutrition, gut physiology, physical growth, cognitive development, and immune responses in infants and children up to 2 years of age in resource-poor environments. Clin Infect Dis. 2014;59(Suppl 4):S193–206.

Torrey EF, Yolken RH. *Toxoplasma gondii* and schizophrenia. Emerg Infect Dis. 2003;9:1375–80.

Toval F, Schiller R, Meisen I, Putze J, Kouzel IU, Zhang W, Karch H, Bielaszewska M, Mormann M, Müthing J, Dobrindt U. Characterization of urinary tract infection-associated Shiga toxin-producing *Escherichia coli*. Infect Immun. 2014;82:4631–42.

Turner PC. The molecular epidemiology of chronic aflatoxin driven impaired child growth. Scientifica (Cario). 2013; https://doi.org/10.1155/2013/152879.

Turner PC, Collinson AC, Cheung YB, Gong Y, Hall AJ, Prentice AM, Wild CP. Aflatoxin exposure in utero causes growth faltering in Gambian infants. Int J Epidemiol. 2007;36:1119–25.

UNICEF, WHO, World Bank Group. Joint child malnutrition estimates 2017 edition. http://www.who.int/nutgrowthdb/jme_brochoure2017.pdf?ua=1. Accessed 20 Sept 2017.

Vally H, Glass K, Ford L, Hall G, Kirk MD, Shadbolt C, Veitch M, Fullerton KE, Musto J, Becker N. Proportion of illness acquired by foodborne transmission for nine enteric pathogens in Australia: an expert elicitation. Foodborne Pathog Dis. 2014;11(9):727–33. https://doi.org/10.1089/fpd.2014.1746. Epub 2014 Jul 29

Vincent C, Boerlin P, Daignault D, Dozois CM, Dutil L, Galanakis C, Reid-Smith RJ, Tellier PP, Tellis PA, Ziebell K, Manges AR. Food reservoir for *Escherichia coli* causing urinary tract infections. Emerg Infect Dis. 2010;16:88–95.

Walker SP, Wachs TD, Grantham-McGregor S, Black MM, Nelson CA, Huffman SL, Baker-Henningham H, Chang SM, Hamadani JD, Lozoff B, Gardner JM, Powell CA, Rahman A, Richter L. Inequality in early childhood: risk and protective factors for early child development. Lancet. 2011;378(9799):1325–38.

Werber D, Fruth A, Buchholz U, Prager R, Kramer MH, Ammon A, Tschape H. Strong association between Shiga toxin-producing *Escherichia coli* O157 and virulence genes stx2 and eae as possible explanation for predominance of serogroup O157 in patients with haemolytic uraemic syndrome. Eur J Clin Microbiol Infect Dis. 2003;22(12):726–30.

Wheeler JG, Sethi D, Cowden JM, Wall PG, Rodrigues LC, Tompkins DS, Hudson MJ, Roderick PJ. Study of infectious intestinal disease in England: rates in the community, presenting to general practice, and reported to national surveillance. The infectious intestinal disease study executive. BMJ. 1999;318(7190):1046–50.

Wild CP, Gong YY. Mycotoxins and human disease: a largely ignored global health issue. Carcinogenesis. 2010;31(1):71–82.

Winer JB. Guillain Barré syndrome. Mol Pathol. 2001;54(6):381–5.

Wong CS, Mooney JC, Brandt JR, Staples AO, Jelacic S, Boster DR, Watkins SL, Tarr PI. Risk factors for the hemolytic uremic syndrome in children infected with *Escherichia coli* O157:H7: a multivariable analysis. Clin Infect Dis. 2012;55:33–41.

Wu F. 2013. Aflatoxin exposure and chronic human diseases: estimates of burden of disease. In Aflatoxins: finding solutions for improved food safety. Unnevehr, LJ Grace D. 2020 vision focus 20(3). Washington, DC: International Food Policy Research Institute (IFPRI).

Part III
Case Studies in Applied Food Safety Economics

Chapter 10
Economic Incentives for Innovation: *E. coli* O157:H7 in US Beef

Tanya Roberts

Abbreviations

APC	Aerobic plate counts
ARS	USDA's Agricultural Research Service
BIFSCo	Beef Industry Food Safety Council
BPSTP	Bacterial Pathogen Sampling and Testing Program
BSPS	Beef steam pasteurization system
BSPS-SC	BSPS-Static Chamber unit
CCP	Critical control point
CDC	Centers for Disease Control and Prevention
DOJ	US Department of Justice
ERS	Economic Research Service
FDA	US Food and Drug Administration
FSIS	Food Safety and Inspection Service
HACCP	Hazard Analysis and Critical Control Point
HUS	Hemolytic uremic syndrome
IAFP	International Association for Food Protection
KSU	Kansas State University
MOU	Memorandum of understanding
NCBA	National Cattlemen's Beef Association
NSLP	National School Lunch Program
OIG	Office of Inspector General/USDA
PR	Pathogen reduction
STEC	Shiga toxin-producing *E. coli*
STOP	Safe Tables Our Priority
US	United States
USDA	US Department of Agriculture

T. Roberts (✉)
Economic Research Service, USDA (retired), Center for Foodborne Illness
Research and Prevention, Vashon, WA, USA
e-mail: tanyaroberts@centurytel.net

10.1   Introduction

This section of the book starts with three chapters on the three important foodborne pathogens: *Salmonella, E. coli* O157:H7, and *Campylobacter*. The route taken to control each pathogen is explored in three countries: the United States, Sweden, and New Zealand. The approaches taken emphasize different mixes of private economic incentives and public/regulatory approaches and incentives. Private marketplace incentives for food safety are relatively weak as discussed in this book's early chapters, because food safety is a "credence" good. Consumers cannot tell ahead of time whether food will make them sick because pathogens cannot be seen with the naked eye. Even after consumption, there is a lag of hours to days (and even weeks for *Listeria*) before illness occurs. In fact, in the United States, the Centers for Disease Control and Prevention (CDC) is only able to identify the causal pathogen with a food and a company in 1/1,000 case of foodborne illness, meaning that CDC can NOT identify the pathogen/food/company combination that is a prerequisite for legal action in 999/1,000 cases of foodborne illness (Chaps. 2 and 17).

To maintain a reputation or to meet contractual or regulatory requirements, companies choose different target levels of pathogen control and use different strategies to achieve their targets (Roberts 2005). In Table 10.1 seven strategies are listed that range from a minimal action (maintaining sanitation control) to investing in research and development to invent a method of preventing pathogens from appearing in the food product above a targeted level of control. Many companies use several of these methods, a multiple hurdle approach. In this chapter, private investment to innovate and create methods to control *E. coli* O157:H7 is examined most closely.

10.2   The Jack in the Box Outbreak

In 1993, the Jack in the Box outbreak with *E. coli* O157:H7 contamination in hamburgers set in motion many private and governmental actions seeking to control this pathogen. In total, 501 acute illnesses were reported, including 151 hospitalizations (31%), 45 cases of hemolytic uremic syndrome (HUS) (9%), and 4 deaths. Some cases developed long-term health outcomes, including neurological complications and kidney failure. Tests of hamburgers and patients found the same *E coli* O157:H7 strains using pulsed-field gel electrophoresis. Hamburgers had been cooked to internal temperatures below 60 °C (140 °F), temperatures insufficient to kill *E. coli* O157:H7. The stock of Foodmaker, Inc., the parent company of Jack in the Box, fell 30%, and Jack in the Box suspended sales of hamburgers in all 66 of its Washington state restaurants (Benedict 2011, p. 85). Sources of ingredients for the burgers could have come from 443 individual cattle from 6 different states and 3 different countries (Armstrong et al. 1996). To stop the outbreak and begin to salvage its reputation, management at Jack in the Box took immediate action (Benedict 2011; Theno 2001, Chap. 7):

Table 10.1 Company strategies to control foodborne pathogens

Strategy 1—Sanitation control. Cross-contamination of meat and poultry is minimized by regular sanitation of the conveyor belts and other equipment in the plant. Systematic cleaning of the plant's walls, drains, and air ventilation at regular intervals further reduces risk. Although HACCP requires certain sanitation practices, firms may choose to comply minimally (or do nothing) until receiving notice of a regulatory violation.
Strategy 2—Kill step for pathogens. A firm decontaminates food at the end of the production line, for example, pasteurizing milk, canning fruits, or irradiating hamburger patties in case-ready packages for sale in supermarkets.
Strategy 3—Pathogen prevention. A firm prevents pathogens from entering the plant at one or more locations, keeps pathogens from growing on food through control over temperature and shelf life, and minimizes cross-contamination between food products and between the plant environment and food products.
Strategy 4—Multiple hurdle approach. A firm improves control over all operations in the plant or at least at several prevention and decontamination steps. This is similar to the standard practice in food companies for designing new foods with several barriers or hurdles to keep pathogens from surviving or growing in foods.
Strategy 5—Identify key risk locations. A firm uses microbial testing at various locations in the plant to determine where the highest probability of pathogen contamination occurs. Pathogen data are used to identify key risk locations, where managers improve pathogen control using new processes and employee training. Or the data can be put into a risk model and various control scenarios evaluated to determine key risk locations and effective control strategies.
Strategy 6—Compare risk/cost trade-offs. A firm adds explicit consideration of the costs of alternative control options to strategy 5 and evaluates the risk/cost trade-offs of different control options.
Strategy 7—Invest in R&D. A firm adopts a long-run strategy to invest in research and development and invent new control options, either by adapting management systems or processes used in a related industry or by inventing a new management system or process (complete with new equipment) to control pathogens.

Source: Roberts (2005)

- Raised the cooking temperature for hamburgers in its retail stores to 160 °F to assure that *E. coli* O157 was killed and changed procedures to ensure that the minimum temperature was consistently reached.
- Hired a new consultant for food safety, Dr. David Theno, who had designed a HACCP system for Foster Farms' poultry operations. Later Dr. Theno was hired permanently and promoted to vice president of quality assurance, research and development, and product safety. He reported to the head of Jack in the Box.
- Canceled all existing contracts with hamburger patty suppliers and requested new proposals from suppliers to meet specific criteria to control *E. coli* O157.
- Developed a checklist for beef slaughterhouses: proper ways to skin animals, wash carcasses, and remove the internal organs without rupturing the intestines (Andrews 2013).
- Over time paid the medical bills of ill consumers and paid some compensation for illness (Chap. 17).

10.3   Texas American Foodservice's Response to Jack in the Box's Request for Proposals

Texas American Foodservice, the largest independent beef patty supplier in the United States, was one of the two beef patty suppliers that offered a proposal to meet Jack in the Box's criteria. Texas American was in a good position to compete for Jack in the Box's contract. In 1982, Texas American hired Timothy Biela who had a master's degree in engineering quality control to develop a systematic approach to control hazards in their hamburger patties. By 1992, Mr. Biela had conducted hazard analyses for bacterial, physical, and chemical hazards for the company's hamburger patties and had begun developing pathogen testing and management protocols (Golan et al. 2004).

With Dr. Theno's input, Mr. Biela developed a Bacterial Pathogen Sampling and Testing Program (BPSTP) for their hamburger patty production line (Table 10.2). The essential components in BPSTP are (1) a new sampling protocol/management

Table 10.2 Texas American Foodservice Corporation's Bacterial Pathogen Sampling and Testing Program (BPSTP) (and description of additional quality control procedures)

Temperature monitoring of incoming combo bins (2000 lbs.) of beef trim; reject if temperature is above 40 °F
Combo bins sampled based on type, supplier, and supplier performance record; sampled not less than every 100,000 lbs.; most raw material lots sampled daily
Test results given to supplier monthly for all lots tested; if lots test higher than standards, supplier is notified immediately, and testing is intensified, monthly review of supplier performance on microbiological criteria and in-plant audits to assess compliance with Texas American standards with performance compared to that of other suppliers
Temperature control (40 °F) and inventory management system for combo bins, first-in-first-out, use by 5th day after boning
Samples are taken at the final grind head for each 3000-lb batch of hamburger and tested for *E. coli* 0157
Samples of finished products are taken from each process line every half hour; half-hour samples are combined into "half shift" composites representing every 4 h of production and tested for complete microbial profile (APC, coliform, E. *coli*, *Staphylococcus aureus*, *Salmonella* sp., and *Listeria monocytogenes*); individual backup samples for each half hour are tested only if composites show spikes or high counts
Rework procedures in place; internal failures (e.g., the patty does not meet specifications) are continuously reworked during the day with quantity of rework recorded for each batch, end of day rework is only used during the last hour of production on the next day (segregated by product), and at the end of the week all remaining rework is destroyed
In-plant cleaning regime in continuous operation, monthly random pre-operational swab tests verify the efficacy of cleaning procedures and monitor the environment for pathogens
Temperature control (less than 10 °F) for frozen patties
Continuous review of procedures and results; adjustment of operating procedures to address problems and opportunities for improvement

Source: Golan et al. (2004)

system and (2) tests of inputs and product at different points in the production process for *E. coli* O157 and other pathogens (Biela 2001).

Sampling protocols are important for controlling pathogen risk because pathogens tend to appear sporadically, often occur at a low level, and sometimes are present at high levels (when bacteria grow in clumps within beef trim or when the cow/bull/ steer is a super-shedder of *E. coli* O157 and the hide/viscera contaminate the carcass at high levels). The BPSTP sampling protocol is designed to manage risk to an acceptably low level. At all testing points, action levels and actions to be taken if deviations occur are clearly defined. Beef is sampled at three locations:

1. *Raw beef ingredients entering the plant are sampled* based on type, supplier, and supplier performance but not less than every 100,000 pounds (daily tests for most suppliers). If lots test higher than BPSTP standards, the supplier is notified immediately, and testing is intensified. All raw materials are routinely screened for aerobic plate counts (APC), generic coliforms, generic *E. coli*, *Staphylococcus aureus*, *Salmonella*, and *Listeria monocytogenes*. These routine test results are reported to suppliers and reviewed with them monthly.
2. *Samples are next taken at the final grind head* where each lot of 3,000 pounds of hamburger is tested for *E. coli* O157. Note that many companies in the beef industry use larger volume grinder/mixers.
3. *Samples of the finished product* are taken from each process line every 15 min. Every hour, composites of the four samples are tested to detect *E. coli* O157:H7. These samples are also combined to make a "half-shift" composite, which is tested for an entire microbial profile (APC, coliform, *E. coli*, *Staphylococcus aureus*, *Salmonella* species, and *Listeria monocytogenes*). If the half-shift composites show spikes or high counts, more tests are run on the backup samples, also collected every 15 min.

The development of a good testing technology was as important as the sampling protocol to the success of the BPSTP. Both Dr. Theno and Mr. Biela believed that no one truly understood the incidence of contamination of beef with pathogens. When Dr. Theno tested the original suppliers of meat to Jack in the Box, he found that the amount of generic *E. coli* in the raw meat was "off the charts" (Benedict 2011, p. 79). Yet the meat had been inspected and approved by USDA.

Traditional microbiological testing methods were inadequate because they relied on culturing samples of meat that were not very sensitive, took time to run, and were not well defined for these organisms. Texas American started its quest for a new testing methodology by upgrading its own microbiology lab and investigating the availability of human clinical microbiological test technologies that could be adapted to use to monitor pathogens in the hamburger supply chain. Their investigation led them to DuPont Qualicon, which had developed the BAX® system, based on polymerase chain reaction (PCR) technology. The PCR technology allows users to target known DNA strands from specific organisms and is *capable of detecting the target organisms at levels much lower than the standard serological methods.* Serological tests take a sample and grow it out on selective media for the pathogen of interest, but a pathogen is only detected if the pathogen grows to relatively large numbers. This was the gold standard of FSIS regulatory testing.

Both Texas American and Qualicon wanted to apply the PCR technology to detect *E. coli* O157 in hamburger in an actual meat plant. To accomplish this, Texas American took hamburger samples and used the BAX® system equipment, selective media, reagents, and primers. Qualicon fine-tuned the protocols to achieve the best results. For validation, side-by-side test comparisons with a standard culture test for *E. coli* O157 were conducted. Texas American benefitted by having a more accurate, sensitive, and specific test it could use in ground beef. Qualicon benefitted from validating the BAX® system assay, equipment, and protocols for meat products.

To properly validate the technology, Texas American solicited the involvement of several other groups. There was significant speculation about the sensitivity of the PCR/DNA method and resistance to its use. It was also not well understood how organisms contained in food products (meat) reacted in typical grinding operations, for example, how they moved and the level of transfer from contaminated to non-contaminated meat. The validation collaboration involved parallel testing in samples, using different methods, by Texas American, by Silliker Laboratories (the largest independent commercial testing lab in the United States at that time), and by USDA's Food Safety and Inspection Service (FSIS), through its Office of Public Health and Science (Pruett et al. 2002). Texas American funded its technicians, the microbiological assays, and data analysis. The National Cattlemen's Beef Association funded the testing by Silliker Laboratories. FSIS funded testing at FSIS labs. Testing by three different laboratories was important to confirm that different technicians and laboratories would find the same *E. coli* O157:H7 results.

The successful collaboration of Texas American, Jack in the Box, Qualicon (DuPont), FSIS, and the National Cattlemen's Beef Association ultimately confirmed the test results of Texas American's BPSTP sampling and testing protocol to detect *E. coli* O157 in ground beef at different stages of the hamburger patty process. Mr. Biela believes that this innovation reduced risk of foodborne illness by 80% in raw ground beef (Biela 2001). Cooking to a high-enough temperature controls the remaining 20% of the risk of illness.

Texas American was able to reap several benefits from its food safety innovation. The major benefit was that Texas American shifted from being a commodity producer selling on a week-to-week basis to being a cost-plus contract supplier. The contract improved operational efficiency through better planning for capacity utilization, capital investment, spending plans, and other business activities. And the contract price covered the additional cost of providing additional food safety controls.

Another benefit of the innovation was Texas American's ability to use its superior knowledge and expertise in the area of pathogen control to attract new customers. Texas American has enhanced its reputation with quality control, superior knowledge, and risk management skills it has built over a period of almost a decade. The company's sales increased approximately 5% annually after it implemented the innovation. Over the 3 years up to 2001, Texas American estimates that about

25–30% of its new sales opportunities occurred because of the innovation. The increase in sales had the added important benefit of allowing Texas American to increase its utilization of fixed capital by 20% over the next 5 years.

Texas American also attributes significant savings and other financial benefits to adoption of the program. The superior knowledge about incidence rates and potential for product contamination that Texas American has gained through the program has enabled it to make better risk management decisions regarding suppliers of raw materials. Texas American's understanding of which raw material suppliers have higher incidence levels and at what times of the year to expect positive test readings in different types of raw materials allowed it to make better purchasing decisions.

The benefits of BPSTP outweighed the costs of the innovation. Texas American characterized the start-up expenses as very high. In addition, there were high costs related to destruction of product in the early stages of the implementation. To contain some of these costs, Texas American worked with USDA to identify sublots for purposes of testing and recall. Over time, costs have not increased, even though testing technology has become increasingly sensitive.

Texas American reports that costs were controlled due to several factors. First, the development of the sublot system has reduced the amount of product that needs to be removed by pinpointing product that is contaminated. Second, the raw material industry has reduced microbial contamination rates for incoming product under the Texas American program, since Texas American works with its suppliers to reduce contamination and the performance of the industry has generally been improving. Finally, Texas American has set a reasonable threshold level for the BAX® tests of its finished, frozen hamburger patties. Texas American set the threshold level for product rejection to eliminate the possibility of an outbreak and massive recall.

Texas American estimated that the cost per pound of the system is at between $0.001 and $0.01, without significant increases in labor, raw material consumption, or energy consumption (Biela 2001). To maintain a competitive edge, and its name as a food safety leader, Texas American continued to expend capital on research and development, with the bulk of these expenditures going to food safety improvements.

Two decades later, however, the original owners decided to sell the company. In 2009, Yucaipa, an investment company located in California, bought American Foodservice, Inc. and Texas American Foodservice. Under this new leadership, some bad investments were made, and American Food Service filed for (Chap. 11) Bankruptcy in April 2012 (Biela 2016). The Texas American and the American Foodservice beef plants were the only ones sold to new buyers; the other food plants were closed (Ibid.).

Texas American and Jack in the Box shared the new approach to pathogen control with other members of the hamburger patty supply chain. They believed that the reputation of the entire industry, including their own, is on the line anytime poor quality control results in illnesses and outbreaks associated with hamburger products.

Dr. Theno reported on pathogen test levels at many conferences on HACCP implementation in the 1990s and was often the only industry person in the room to give such detailed information (Roberts 2016). Texas American and Jack in the Box worked collaboratively over time to attain higher standards. Both companies were first motivated by the need for risk management to limit or eliminate the damage in reputation, sales, and liability stemming from inadequate quality control. Jack in the Box was successful in overcoming the negative publicity of the outbreak and won its hamburger-eating customers back (for details see Benedict 2011).

Both companies found that a reputation for quality has served as a foundation for growth. These companies helped develop a market for food safety—and through their reputations as safety leaders, both reaped benefits from supplying this market. In 2004, Jack in the Box earned the prestigious Black Pearl Award from the International Association for Food Protection for "Recognition for Corporate Excellence in Food Safety and Quality" (IAFP 2016).

In 2017 Dr. Theno died unexpectedly while swimming in the ocean. At IAFP, the organization Stop Foodborne Illness announced the Dave Theno Fellowship, an award now given annually to a new graduate from a food safety program or a public policy program (Beach 2017).

10.4   Invention of the Beef Steam Pasteurization System (BSPS)

The company closest to the consumer is the retailer and the easiest to associate with pathogen contamination. Hence, the retailer has the greatest incentive to produce safe food. Moving up the supply chain from the retailer to the beef patty supplier, Texas American had the next greatest incentive to produce safe food, especially with the inducement of a cost-plus contract. The next link in the supply chain is the slaughterhouse where the possibility of meat contamination with pathogens is most likely to occur during hide removal and evisceration processes (Kalchayanand et al. 2015). Repeat sales to Texas American were the incentive for slaughterhouses to practice careful hide removal. In 1996, Tanya Roberts (leader of the slaughterhouse module of USDA's *E. coli* O157:H7 risk assessment team) called sellers of hide pullers listed on the website of the American Meat Institute. Surprisingly, none of the sellers were marketing their systems for their food safety advantages. Each hide puller was produced to the specifications of the individual plant, and none of the sellers seemed to know that aerosols of *E. coli* O157 from the hide can occur during dehiding, leading to possible carcass contamination. Dehiding equipment rolls up the beef hide, and the machinery can be attached above or below the beef carcass. In general, a "down-puller" attached to the floor creates fewer aerosols than an "up-puller" near the ceiling, since gravity settles the aerosols on the carcass on the way down to the floor. If the beef carcass is contaminated during dehiding and evisceration, it is best to decontamination the carcass immediately, before pathogens attach firmly to the meat surface (Kalchayanand et al. 2015). This "decontamination" event in the slaughterhouse is the purpose of the next invention.

In 1993, Craig Wilson was working for Frigoscandia Equipment and designed and installed equipment in Excel/Cargill, Inc. beef slaughter plants (Golan et al. 2004). The Jack in the Box outbreak was discovered in his backyard, Seattle, and Brianne Kinner (who almost died in the outbreak) attended the same school as his children. Frigoscandia Equipment primarily invented and marketed equipment in cold storage and transportation to maintain product quality, control pathogens, and increase shelf life. The company's expertise and contacts in the beef world opened the door to try to invent a new kind of equipment to control pathogens in the slaughterhouse using steam. Since 1972, steam pasteurization had been studied on hog carcasses, meat surfaces, and chicken carcasses, with mixed results (Phebus et al. 1997). Craig Wilson was tasked with investigating steam pasteurization of the exterior of beef carcasses. Steam pasteurization was a new food safety technology that could be a complementary addition to the company's product line and a new marketing opportunity for Frigoscandia Equipment.

Given a positive initial assessment of the innovation, Frigoscandia funded an exploration of the technical feasibility of the project. Frigoscandia realized a substantial investment would be required to develop the equipment. Building the machinery and testing the efficacy of the procedure would require time and financial commitment. Whether the BSPS innovation would prove financially profitable would depend on how well the BSPS equipment reduced pathogens, the cost of the equipment, the enforcement of regulatory programs to control *E. coli* O157 and other pathogens, the requirements imposed by buyers, and the cost and benefits of alternative pathogen-control systems available to beef packing plants. Would the domestic beef industry consider the pathogen reduction benefits worth the purchase price of the equipment? Would the innovation succeed in global markets?

To reduce the technological risks and share the costs of creating the new BSPS technology, Frigoscandia contacted a business client, Excel, the second largest US beef packing company, which agreed to collaborate on the BSPS invention. Excel had the day-to-day knowledge of operating beef packing plants where the equipment was to be used. Though the two companies jointly developed the technology and applied for the patent, Frigoscandia Equipment holds the rights to the patent on this technology because the global beef industry was the target sales market. If Cargill/Excel co-held the rights to the patent, other beef companies might be reluctant to purchase equipment, thinking they would be supporting a competitor.

As a first test of the technology's efficacy, Frigoscandia Equipment built a prototype BSPS unit. Preliminary tests at Frigoscandia found that the BSPS prototype successfully killed the pathogen on small pieces of beef inoculated with *E. coli* O157. Next Frigoscandia and Cargill/Excel decided to add academic microbiologists to the team as outside, nonbiased evaluators of the performance of the BSPS prototype. Dr. Randall Phebus at Kansas State University (KSU) was chosen to head the academic team. Frigoscandia shipped the prototype steam pasteurization system to KSU. Cargill/Excel supplied six live market-weight steers. Both Frigoscandia and Cargill/Excel contributed the kits and other materials required for pathogen

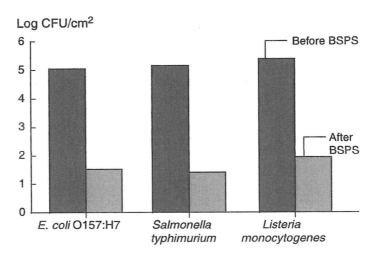

Source: Data from Phebus et al., 1997.

Fig. 10.1 Steam pasteurization reduces mean pathogen population on beef carcasses. Source: ERS/USDA, Golan et al. (2004)), https://www.ers.usda.gov/webdocs/publications/41634/18032_aer831.pdf?v=42265

tests of beef samples. After slaughter at KSU, meat samples were inoculated with 5 logs of a pathogen (100,000 organisms/cm^2) and then treated in the BSPS prototype. All three pathogens tested, *E. coli* O157, *Salmonella typhimurium*, and *Listeria monocytogenes*, were reduced by 4.65–5 logs at 15 s of steam treatment at 196–199 °F (Fig. 10.1). Dr. Phebus and his team concluded that "Steam pasteurization is an effective method for reducing pathogenic bacterial populations on surfaces of freshly slaughtered beef…" (Phebus et al. 1997, p. 476). The researchers found steam pasteurization provided numerically *greater pathogen reductions than any other single treatment studied.* One reason for this result is that steam vapor uniformly blankets irregularly shaped surfaces, in contrast to hot water coming from a nozzle aimed at carcasses. If there is any irregularity on the surface of the carcass, the back side of the irregularity will not receive the hot water treatment, and pathogens lurking there will not be killed. Properly applied steam can reach these problem areas. In addition, hot water quickly loses its temperature and any ability to kill pathogens, once it hits the carcass (Wilson 2016). BSPS is also superior to chemical rinses for carcasses because it does not entail treatment of potentially toxic wastewater.

In 1995, after the success of the prototype at KSU, Frigoscandia engineers designed, built, and installed a commercial-sized BSPS unit at an Excel plant in Sterling, Colorado. This stainless steel clamshell could hold four sides of beef at a time and moved along the slaughter line. It also used monitoring techniques for temperature and lot identification that Frigoscandia had developed for its food chilling and freezing equipment. After solving a number of technical issues related to the pressure, temperature, and application of the steam in the moving clamshell, BSPS, Wilson (Frigoscandia Equipment), Leising (Cargill/Excel), and other Frigoscandia Equipment inventors filed a patent application on November 6, 1995 (United States Patent 1995).

To test the efficacy of the commercial scale-up of the BSPS prototype, Frigoscandia and Cargill/Excel again invited the KSU team into the plant to conduct tests. The objective was to determine the effectiveness of the BSPS unit in reducing naturally occurring populations of indicator organisms on the surfaces of commercially slaughtered beef carcasses. Indicator microorganisms, not pathogens, were used because of the danger of introducing pathogens into a commercial facility. Over a 10-day testing period, 140 carcasses (70 cows and 70 fed cattle-steers/heifers) were tested with steam applied for 8 s at 195–201 °F. Twenty carcasses (9 cows and 11 fed cattle) were tested with steam applied for 6 s. An additional 20 control carcasses (10 cows and 10 fed cattle) received no steam treatment.

The KSU team found that steam treatment for 8 s was "very effective" in a commercial setting for reducing overall bacterial populations on beef carcass surfaces after 24 h in the chiller (Nutsch et al. 1997, p. 491). In most cases, the enteric bacteria were undetectable after pasteurization. Reductions in bacterial populations after a 6-s steam exposure time were very similar to those obtained with an 8-s exposure time. The equipment worked equally well with cows and steers/heifers, despite considerable variations in carcass size and shape.

For the third set of tests in 1996, Frigoscandia installed a moving clamshell BSPS in a larger commercial facility, Excel's plant in Fort Morgan, Colorado. Again, KSU conducted the testing (Nutsch et al. 1998). This time, the testing team made several changes to the testing protocol to more closely approximate an actual plant operation. Samples were randomly selected from 200 carcasses from two production shifts, rather than the known carcasses in the earlier test at the Sterling plant. Steam temperature was lowered to 180 °F for either 8 or 6.5 s. Instead of excising a small piece of meat to test, sponges were swabbed over the carcass, and the liquid was tested to see if microbes were detected. Twenty carcasses were sampled at five carcass locations to see if the steam treatment effectiveness differed at the five sites. The KSU team concluded that the BSPS-moving clamshell unit was effective in reducing natural bacterial populations on freshly slaughtered beef carcasses. Frigoscandia Equipment submitted the KSU's laboratory results on pathogen reduction to USDA.

USDA regulatory approval of the BSPS process was a necessary step for commercial acceptance. The KSU data was shared with regulators, industry members, and consumer groups. In December 1995, USDA certified that Frigoscandia Equipment's BSPS-moving clamshell can significantly reduce pathogens (Cargill 1995). The BSPS is equipped with recordkeeping capabilities: carcass identification, carcass surface temperature in the steam chamber, exposure time, and deviations are automatically logged into a computer for plant monitoring and regulatory review. The monitoring features make it feasible to use the BSPS as a critical control point under FSIS PR/HACCP regulations.

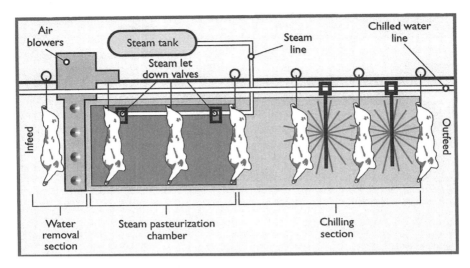

Fig. 10.2 Beef steam pasteurization system—static chamber. Source: ERS/USDA, Golan et al. (2004), https://www.ers.usda.gov/webdocs/publications/41634/18032_aer831.pdf?v=42265

Next, a BSPS-Static Chamber unit (BSPS-SC) was invented to perform the same three processes as the moving unit, except that with the static unit, the sides of beef travel through the enclosed chamber and sequentially receive (1) dewatering treatment, (2) steam treatment, (3) cold water shower (Fig. 10.2). With the BSPS-SC design, carcasses can travel through the chamber at any chosen line speed. The doors and the overhead rail (on which the carcasses hang) have seals to maintain the positive air pressure in the chamber. In January 1998, Wilson (Frigoscandia Equipment), Leising (Excel), and other Frigoscandia Equipment inventors filed a patent application for a static chamber system that uses steam to destroy surface pathogens on meat (United States Patent 1999a).

The BSPS-SC had several advantages over the moving clamshell BSPS. The unit did not break down as often or require as much maintenance as the moving unit, reducing the warranty costs to Frigoscandia Equipment (Brodziak 2001). This additional reliability is beneficial to customers as well: it facilitates the use of the BSPS-SC system as a control measure in a plant's PR/HACCP system. Control measures must be reliable because the whole slaughter line must stop production if any of the critical control points in the PR/HACCP system are not functioning correctly. The BSPS-SC units reliable enough to use as a critical control measure. Another benefit to beef packing plants with the BSPS-SC was a reduction in operating costs because the steam part of the tunnel can be kept at a constant high temperature (Leising 2002).

The three collaborators for the BSPS-SC invention, Frigoscandia Equipment, Excel, and KSU, contributed in different ways to the development of the technology—and benefitted differently. Frigoscandia Equipment, through Craig Wilson, initiated the innovation and contributed technical and administrative expertise. The costs to Frigoscandia Equipment of designing, building, and testing the BSPS prototype and the moving clamshell BSPS unit were $1.2 million spread over 3 years, mid-1994 to mid-1997 (Brodziak 2001). These costs were in-house labor and other variable costs, including contracting costs to the machine shop that produced the parts for the prototypes. The BSPS-SC modification took Frigoscandia Equipment 9 months and $100,000 to design and build. Frigoscandia Equipment's total investment was $1.3 million for the BSPS-SC innovation.

The two largest US beef packing companies, Cargill/Excel and IBP, bought the equipment for all of their slaughter plants. Frigoscandia Equipment earned a small profit on the BSPS-SC equipment sales and the installation (Brodziak 2001). From 1996 to 2001, Frigoscandia Equipment sold 28 BSPS-SC units: 20 large and 8 smaller units. Smaller units were sold at approximately $250,000 each, depending on site-specific requirements.

Cargill/Excel's contribution included paying for the beef used in the testing and all plant operation costs during the testing at the Cargill/Excel's Sterling and Fort Morgan plants. Cargill/Excel also invested a considerable amount of resources in adjustments and adaptations to the unit, including engineering maintenance. The company recouped some of these expenses because it was not charged for the first moving clamshell BSPS unit and adjustments were made in the purchase price of other BSPS units to compensate for Cargill/Excel's investment. Cargill/Excel also benefitted by taking advantage of its "first right of refusal" and being the first US company to install the BSPS and BSPS-SC in all its packing plants (Cargill 1997). This gave them an enhanced reputation as a leader in food safety research and development that led to an increase in beef sales and contracts.

KSU was brought into the development team to conduct a wide variety of microbiological tests on pathogens and indicator organisms using four different pieces of equipment, using different testing procedures, and using different combinations of steam temperature/time in the BSPS units. KSU contributed the time of two Ph.D. students and one professor to the project. Most of the testing equipment was purchased by Frigoscandia Equipment and Cargill/Excel, including about $40,000 to $50,000 worth of testing kits and other supplies.

All three collaborators boosted their food safety reputation through their involvement in the innovation. Frigoscandia Equipment strengthened its position in the food safety equipment industry. Cargill/Excel became known as a food safety leader and gained market share in the beef packing business (Leising 2002). KSU became known for its expertise in microbial food safety (Leising 2002). Two KSU students earned doctorates doing microbiological research on the BSPS technology. KSU now grants distance-learning degrees in food science, and this program has been recognized for its quality by the Institute of Food Technologists (Phebus 2002).

US government certification in 1995 that BSPS significantly reduces pathogens lessened the uncertainty facing industry purchasers regarding the efficacy of the technology and opened the door for the use of the BSPS as a critical control point in PR/HACCP (Cargill 1995; U.S. Department of Agriculture 2002). In addition, a number of government guidelines have explicitly endorsed the use of the technology. For example, in 2000, USDA's Agricultural Marketing Service specified that suppliers of beef trim and ground bison to agency-administered purchasing programs, such as the National School Lunch Program, must include an antimicrobial intervention as a critical control point (CCP) in the establishment's HACCP plan. "The CCP must be one of the following processes: steam pasteurization; an organic acid rinse; or 180 °F hot water wash" (U.S. Department of Agriculture 2000).

The BSPS-SC innovation enjoyed market success in the United States. Cargill/Excel, the second-largest US beef packing company, installed the technology in all seven of its beef packing plants by June 1997. IBP, Inc., the largest beef packing company, installed BSPS-SC equipment in all its beef slaughterhouses. (In 2001, IBP was purchased by Tyson Foods, Inc., and Tyson became the world's largest marketer of beef, pork, and chicken.) In 2018, Cargill states that steam-pasteurized cabinets are required for fed cattle beef harvest. Cow harvest facilities have either a steam pasteurization intervention or validated hot water treatment (Cargill 2018).

Costco requires that beef must come from plants that use steam pasteurization or an equally effective intervention (Andrews 2013). For ground beef, Costco has a "test and hold" program. Raw materials are tested for *E. coli* before and after grinding. Last, all beef is traceable from the production facility to the daily processing of beef at the warehouse (Talevich 2013).

The positive market response is also reflected in a beef-product recall insurance policy available through the American Meat Institute, the meat industry's largest trade association. This recall insurance, which is sold by MacDougall Risk Management, offers the possibility of reduced rates and higher likelihood of coverage for plants that have installed the BSPS-SC (MacDougall 2002). Other insurance programs covering product quality or safety are also sensitive to baseline plant risks and safety investments.

In 1996, Frigoscandia Equipment was purchased by FMC headquartered in Chicago, one of the world's largest manufacturers of food equipment. In 2008, FMC FoodTech became JBT FoodTech (2018). The BSPS-SC continues to be sold, and a Danish company is now developing a similar invention for hogs (Wilson 2016).

10.5   The Beef Industry Reaction to the Outbreak

The farm is the beginning of the beef supply chain, and economic incentives for food safety are the least here, since consumers have no ability to link a hamburger they consume in a restaurant or purchase in a supermarket to a particular farm. Yet, following the 1993 *E. coli* O157 outbreak in ground beef, the beef industry responded by founding the first-ever Blue Ribbon Task Force to focus on improving beef safety. From 1993 to 2001, the National Cattlemen's Beef Association (NCBA) spent $10 billion on research on how to reduce *E. coli* O157 in beef slaughter and processing, including Frigoscandia's steam pasteurization of carcasses (American Meat Institute Foundation 2001). Consumer research on how to inform consumers of best cooking practices for hamburgers was also a research concern, albeit minor. The focus on slaughter and processing may have been to search for a silver bullet, such as irradiation or steam pasteurization, that would have been a "kill" step analogous to milk pasteurization. If such a magic bullet could be found, then farmers would not have to change their practices in raising beef. Farmers, after all, were in the business of raising cattle to feed hungry people, and the notion of controlling pathogens that could cause human illness was a new concept.

In fact, human illnesses were not the focus of USDA when the Carter administration (1977 to 1981) asked USDA's Economic Research Service (ERS) to do a benefit/cost analysis of meat and poultry inspection. Inspectors told the new undersecretary for Food Safety and Quality Services, Carol Tucker Foreman, that infectious disease problems had been solved and that the inspectors were only identifying broken chicken wings and bruises on carcasses. In addition, the checklist that meat and poultry inspectors used to inspect and reject animals/carcasses only identified animal diseases. A link to human illnesses was not made, as I found out when I was the ERS employee assigned to this project. It was not until 1996 that USDA created the Office of *Public Health* and Science in the Food Safety and Inspection Service (FSIS).

In 1997, the Beef Industry Food Safety Council (BIFSCo) was formed to foster collaboration among all sectors of the beef industry from farmers to retailers. This led an annual meeting of the beef industry, starting in 2003, to discuss current knowledge on *E. coli* O157. Texas American's Tim Biela was asked to lead BIFSCo's processing sector from 2002 to 2010. The processing sector focused on how to produce safer ground beef and worked with suppliers of beef trim to get trim tested prior to shipping to the processing plants. Biela shared all his learnings with the group on input and finished product testing. In the beginning, there was resistance to testing, either trim or ground beef. Eventually companies such as McDonalds adopted the testing of finished products. In the words of Tim Biela:

"BIFSCo remains a fundamental part of the Beef Industry today. They continue to share research and information and have continued to discuss pertinent and timely topics to members on food safety issues of all types. BIFSCo was and is successful in that brings all portions of the beef supply chain together in one place to discuss and apply strategies to improve food safety in beef products. Texas American shared its strategy of auditing suppliers utilizing a comprehensive audit scheme, similar to what you see now in Global Food Safety Initiative type audits. At the end of each audit, we would sit down and discuss deficiencies and request actions to address them. We focused on sanitary dressing issues which contributed to contamination of meat from hides, hands and equipment. BIFSCo in some sense helped to hold suppliers accountable for producing safer trim by utilizing testing, and was pivotal in getting large companies to test, share data, and collaborate on best practices" (Biela 2001).

The American Meat Institute Foundation, the research arm of the American Meat Institute which is the trade association for the beef industry, commissioned a report on potential on-farm (preharvest) interventions to reduce *E. coli* O157 in cattle (2001). The report summarized what was known about on-farm contamination: "*E coli* O157:H7

(a) Is distributed throughout the U.S.
(b) Occurs in dairy, feedlot and range cattle.
(c) Does not cause clinical symptoms in carrier animals.
(d) Can be transmitted among cattle.
(e) Persists on farms for at least 2 years.
(f) Is shed variably among herds.
(g) Is shed intermittently by animals.
(h) Is shed more often in warm weather.
(i) Colonizes in deer, sheep, horses, dogs, flies and birds.
(j) Has no host or long-term reservoir that has been identified.
(k) Is present in multiple sources on the farm.
(l) Occurs as the same strain, on many sites, in many states; birds could be responsible for its widespread distribution.
(m) Incidence is not related to new-animal introduction into a herd.
(n) Incidence is not related to spreading manure on pastures or rangelands. (p. 81)"

In February 2000, USDA's Agricultural Research Service (ARS) reported that 28% of cattle presented for slaughter were infected with *E. coli* O157, higher than previously reported (USDA/FSIS 2002), and that an average of 43% of beef carcasses were contaminated with the pathogen. This research shows that both the farm and the slaughterhouse were important contributors to the contamination of beef with *E. coli* O157.

Transportation of live cattle has its own risks: cross-contamination of *E. coli* O157 from one animal to another, issues of feed and water withdrawal, distance and impact of travel stress on fecal shedding of *E. coli* O157 on hides, and the cleaning of trucks (Roberts et al. 1995). The cattle pens at feedlots and at the slaughterhouse can also amplify on-farm contamination.

10.6   Federal and State Actions Create Incentives to Control *E. coli* O157:H7

On March 11, 1992, the Washington State Board of Health adopted a new food service regulation that raised the minimum cooking temperature of ground beef to 155 °F. The notification and enforcement by counties in Washington were uneven, contributing to confusion about the change. At the Federal level, FDA's US Food Code recommended ground beef be cooked to 140 °F, further complicating compliance by Jack in the Box. Violating either a State or Federal law makes a company vulnerable to enforcement actions by regulators and legal liability suits by sickened consumers.

In January 1993, one of the first actions by President Bill Clinton was sending his Secretary of Agriculture, Michael Espy, to Washington state to address its legislature on the *E. coli* O157:H7 outbreak. On February 5, 1993, the US Senate convened a hearing on "Food Safety and Government Regulation of Coliform Bacteria," featuring the head of Jack in the Box, Robert Nugent. The TV news channels picked up the story, and the stock of Foodmaker, Inc., the parent company of Jack in the Box, fell 30% (Benedict 2011, p. 85). Adding to the news, some of the parents of children who were sickened, died, or experienced long-term health outcomes formed an advocacy organization, Safe Tables Our Priority (STOP) (Chap. 16).

In September 1994, Michael Taylor had been the USDA's new acting administrator for the Food Safety and Inspection Service for just 6 weeks when he took the podium at the annual meeting of the American Meat Institute and announced that raw ground beef contaminated with *E. coli* O157:H7 was adulterated under the Federal Meat Inspection Act. Until Taylor's announcement "adulterant" had been restricted to harmful chemicals or foreign objects. This was the first time a bacterium had been declared an "adulterant." In October 1994, USDA/FSIS began a sampling program to test for *E. coli* O157:H7 in federally inspected establishments and in retail stores (USDA/FSIS 2002). Raw ground beef was targeted because of its strong epidemiological link with *E. coli* O157:H7 infection. Although the beef industry sued, the courts upheld the action, and it was illegal to sell ground beef contaminated with *E. coli* O157:H7 in the United States.

In 1995, FSIS proposed that meat and poultry companies design and implement Hazard Analysis and Critical Control Point (HACCP) systems to control foodborne pathogens (Chap. 4). These regulations were finalized in 1996, per 9 CFR 417. FSIS, through various sampling programs, verifies the effectiveness of company control systems at preventing hazards from entering commerce. *Salmonella* was to be the performance standard for HACCP, but industry successfully fought this in the courts. When industry won, USDA did not appeal the Supreme Beef ruling (Chap. 16). As a consequence, not much progress has been made in controlling *Salmonella* (Chap. 11).

In January 1999, USDA/FSIS published a statement clarifying that the public health risk posed by *E. coli* O157:H7 includes intact raw beef products, such as trimmings that are often turned into ground beef (USDA/FSIS 2002). If beef trim test positive for *E. coli* O157, then they must be cooked and processed into ready-to-eat products, or the trim will be deemed adulterated. Since beef trim sells at a premium price in the fresh marketplace relative to trim destined for cooking, beef producers and slaughterhouses have an economic incentive to control *E. coli* O157 to get the higher price.

In 2010, the US Department of Justice (DOJ) initiated a new policy, the "Human Illness Standard." Whenever a food product becomes associated with an outbreak of foodborne illness, it is likely to trigger a federal criminal investigation of the company. Under this policy, "responsible corporate officials" can be found guilty and given prison sentences of up to a year *even if the food company has no knowledge it was producing contaminated products* (Flynn 2016a, b. This action ups the ante by making food safety a more personal concern to "responsible corporate officials" who could face jail time. From an economic perspective, this increases incentives for food safety and leads to more tests of raw materials/foods for pathogens, improves other food safety practices of the companies, and leads to fewer outbreaks.

On September 2015, USDA/FSIS added tests for six non-O157 Shiga toxin-producing *E. coli* (STEC) in raw ground beef products to increase protection of US consumers. Of interest is that the largest plants producing more than 600,000 pounds/day have a maximum of four samples that can be taken by FSIS per month. Not only is this a small number of samples, but the sampling rate per pound is higher for smaller plants compared to large plants (Table 10.3). For example, plants producing 50,001–250,000 pounds/per day have a maximum of three samples/month that can be tested by FSIS. On a per pound basis, *the larger the plant, the less likely its ground beef or trim will be tested for E. coli O157:H7 or non-O157 STEC.* This means that the large plants are less likely to have contamination detected by FSIS and less likely to have regulatory consequences. This favoritism for large plants may reflect their political clout.

Compared to the sampling rate of Texas American Foodservice where ground beef is sampled and tested in each lot of 3,000 pounds of hamburger, the FSIS sampling rate is very low, so low, in fact, that it provides a minimal economic incentive to improve control of STEC. Buchanan and Schaffner state that "…there are multiple forms of microbiological testing (e.g., process verification, lot release, investigational), each with its own protocols and underlying mathematics" (Buchanan and Schaffner 2015). This raises the question of what is the goal of the FSIS testing at such a low level.

10.7 Modeling Economic Costs Versus Risk Reduction of Pathogens in a Beef Slaughterhouse

In 2004, Malcolm et al. developed a probabilistic risk analysis model of beef slaughter plant practices to evaluate the cost/pathogen reduction trade-offs for generic *E. coli*. They identified typical slaughterhouse activities and put three pathogen reduction opportunities in the model:

- One prevention step: improved hide removal and evisceration to prevent carcass contamination from pathogens on the hide or in the viscera either by these parts touching the carcass or through cross-contamination via aerosols created or via knives/workers gloves, etc.
- Two decontamination procedures after the hide and viscera are removed and before the carcass goes into the chiller: steam pasteurization of the carcass as discussed above or carcass irradiation (Morrison 1989).

Combinations of three pathogen-reducing practices, careful hide and viscera removal from the carcass (dehiding = D), steam pasteurization of the carcass (S), and irradiation of the carcass (I) result in seven combinations for options of pathogen controls: D, DS, DI, DSI, S, and SI.

The model results are shown in Fig. 10.3. The most cost-effective options are joined by a dotted line, called the trade-off frontier, which indicates the least-cost

Table 10.3 Maximum FSIS monthly sampling frequencies for STEC in raw ground beef products, by plant size

Sampling	Plant size defined by production volume in lbs/day						
	Large		Medium	Small			Very small
	>600,000 lbs/day	250,001–600,00 lbs/day	50,001–250,000 lbs/day	6001–50,000 lbs/day	3001–6000 lbs/day	1001–3000 lbs/day	<1001 lbs/day
Max # of sampling tasks per plant/month	4	4	3	2	2	2	1

Source: USDA/FSIS (2015)

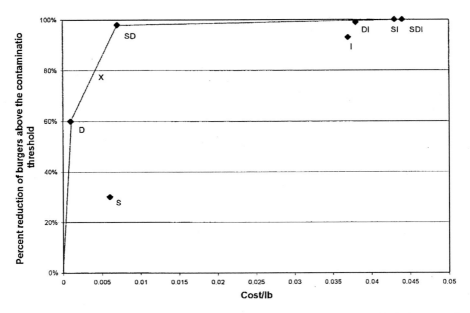

Fig. 10.3 Trade-off curve for combinations of three technology adoption strategies in large steer/heifer plants. Note: *D*, improved dehiding; *S*, steam pasteurization; *I*, irradiation. The risk threshold selected is 10,000 generic *E. coli* per hamburger patty. Source: Malcolm et al. (2004). Malcolm, SA, Narrod CA, Roberts T, Ollinger M (2004) Evaluating the economic effectiveness of pathogen reduction technologies in cattle slaughter plants, Agribusiness 20(1):109–23. https://naldc.nal.usda.gov/download/34673/PDF, https://docs.google.com/viewerng/viewer?url=eurekamag.com/ftext.php?pdf%3D004149491

pathogen reduction strategies. Note that improved dehiding lies on the frontier and note the synergy in combining steam pasteurization with improved dehiding procedures. This result supports the multiple hurdle approach used by the food industry for pathogen control. While irradiation provides marginal improvement over careful dehiding plus steam pasteurization, this strategy comes at a significant cost increase.

What was not modeled is Texas American Foodservice's Bacterial Pathogen Sampling and Testing Program (BPSTP) with a risk reduction of 80% for *E. coli* O157 at a cost of $0.001 to $0.01. This risk/cost trade-off is similar to that of careful dehiding. BPSTP comes at a slightly greater cost, but yields a greater reduction in *E. coli* O157, as indicated by the × on the trade-off frontier in Fig. 10.3 at the 80% reduction point.

Also, the model was constructed for large slaughter plants. Yet, small plants may have a comparative advantage in careful hide removal, if they have lower worker turnover and higher morale. Anecdotal evidence suggests that large plants do have high turnover rates and thus a workforce with, on average, less experience than smaller plants. In addition, line speeds at the largest plants have increased to the point where 400 cattle/h is common in the United States. The faster line speeds and greater crowding of carcasses in a plant can increase the probability of the air becoming contaminated during hide removal and increase the chances of carcass-to-carcass cross-contamination (Hauge et al. 2012). Both the less experienced workforce and faster line speeds in the largest plants suggest a greater chance for errors and increased odds of carcass contamination with *E. coli* O157 and other pathogens (Malcolm et al. 2004).

10.8 Company Response to Outbreak Risk

Food companies have a choice of strategies for dealing with outbreaks of *E. coli* O157 and other pathogens and the government (Federal, State, or local) regulations designed to control pathogens. The first option is *to innovate and solve the problem of pathogen contamination*, as Jack in the Box/Texas American Foodservice and the Frigoscandia/Cargill/Excel cases have shown. These companies had the strong economic incentive of protecting their company from bankruptcy (Jack in the Box), the incentive of expanding their business (Texas American, Frigoscandia), or the incentive to protect their beef market (Cargill/Excel). Another case where there is a strong economic incentive to exert pathogen control is international trade and access to markets. In 2009, Gill reports that "…the microbiological conditions of frozen trimmings from Australia and New Zealand continue to be superior to that of U.S. product, although in those countries most carcasses are not pasteurized or treated with antimicrobial solutions" (p. 1797) (Gill 2009).

A second option is to create doubt that your company caused the outbreak (Oreskes and Conway 2010). There are many ways to create doubt: (1) create confusion (we have no pathogen test data indicating any pathogen contamination in our products); (2) blame someone else for causing the illness (a competitor's hamburger caused the human illness, the consumer did not cook the product sufficiently and caused their own illness, or a supplier sold us contaminated beef trim); (3) criticize the science, the test method, the epidemiology, or the regulators' ability to enter your plant and examine your data; (4) falsify/destroy any records the company does have; and/or (5) challenge the authority of the regulatory body to hold the company accountable for the illnesses. An example of a company taking legal action against a regulatory authority is the Supreme Beef Co. case that challenged the authority of USDA's Food Safety and Inspection Service's to set performance standards for the very common pathogen, *Salmonella* (Chap. 16).

A third option is to use political pressure or persuasion to gain exemption from the regulations. ConAgra employed this strategy to seek exemption from FSIS *E. coli* O157 testing. ConAgra made a presentation on their pathogen intervention system to a few FSIS personnel who also invited Mike Ollinger and Tanya Roberts from the Economic Research Service, USDA. Apparently, this presentation was part of the exemption strategy. After an outbreak was detected and the USDA's Office of Inspector General (OIG) was charged with investigating, OIG found:

ConAgra was generally exempted from the testing program because it used its own validated pathogen reduction interventions on beef carcasses....This exemption was provided
by FSIS Directive 10,010.1 (Microbiological Testing Program for *E. coli* O157:H7 in Raw
Ground Beef), dated February 1, 1998. (USDA 2003, p. 5)

In May 2002, FSIS tests found *E. coli* O157 in ground beef at a meat grinder that
used beef inputs purchased from ConAgra in Greeley, Colorado. Beginning in mid-
June 2002, 46 people in 26 states became ill from contaminated beef. Test by epidemiologists in Colorado and the Centers for Disease Control and Prevention (CDC)
confirmed that 23 of the illnesses were from the "same genetic strain of *E. coli*"
found in the FSIS previous tests. In all, FSIS recalled 18 million pounds of beef
product or 18 days of production at the Greeley plant. In September 2002, the
ConAgra Beef Company was sold to Swift Foods Company, and the Greeley plant
became known as Swift and Company. It would be interesting to know the sale price
of the plant: was it a bargain or sold at market value? Certainly, the goal of deflect-
ing bad publicity away from ConAgra was achieved by selling the Greeley plant. In
June 2003, ConAgra's annual report stated that it had sold all its fresh beef and pork
plants and announced plans to sell its chicken plants: "the company will no longer
have any current Meat Processing segment activities" (ConAgra 2003, p. 2).

A fourth option is to be a "free rider" or cheater and have no interest in complying with any regulations and see if the company gets caught. Actually, ConAgra
used this strategy too. ConAgra did not verify that their control processes were
"working as planned" or test their outgoing product for contamination with *E. coli*
O157:H7. They took their chances that they would not be caught. As discussed earlier in this chapter and at length in Chap. 2, the chances that a foodborne illness
would be traced back to a food company were found to be roughly 1 case in 1,000
cases of foodborne illness—a very weak incentive to comply with food safety regulations. Furthermore, the lack of US animal identification (RFID or ear tags) is in
marked contrast to our trading partners and hinders tracing pathogen contamination
back to the farm. The history of the US beef industry challenging USDA/FSIS regulatory changes and enforcement actions is another indication of mixed willingness
to provide food safety (Chap. 17).

10.9   Conclusion

Today, US-reported illnesses caused by *E. coli* O157:H7 are 47% lower than in 1996–
1998 (Fig. 10.4). Children under five, with their immature immune systems, are at
greatest risk of illness (Fig. 10.5). Progress has been made by US beef companies from
farm to fork, but progress has been minimal in recent years. There are several ways that
control of *E. coli* O157:H7 and other Shiga toxin *E. coli* (STEC) could be improved.

FSIS could follow the lead of the USDA's purchasing programs for US schools.
The National School Lunch Program (NSLP) has lowered the level of *Salmonella* in
its ground beef to 0.1%, significantly below the FSIS performance standard of 9.5%
and below the average contamination rate of commercial sales at 2% (Fig. 10.6)

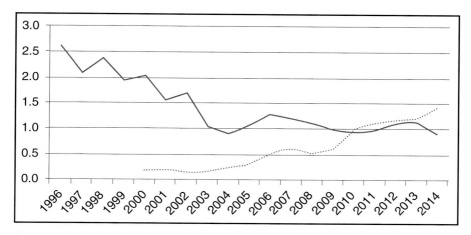

Fig. 10.4 Today, *E. coli* O157:H7 reported illnesses are 47% lower than in 1996–1998, United States. Note: Solid line is *E. coli* O157:H7 infections. Dotted line is other STEC infections. Source: CDC (2016). CDC. Foodborne Diseases Active Surveillance Network (FoodNet): FoodNet Surveillance Report for 2014 (Final Report). Atlanta, Georgia: U.S. Department of Health and Human Services, CDC (2014), http://www.cdc.gov/foodnet/reports/annual-reports-2014.html

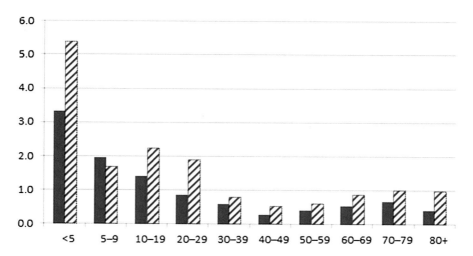

Fig. 10.5 Age and sex distribution of US Foodborne Illnesses with *E. coli* O157:H7. Note: solid bars are males, striped bars are females. Source: CDC (2014). CDC. Foodborne Diseases Active Surveillance Network (FoodNet): FoodNet Surveillance Report for 2014 (Final Report). Atlanta, Georgia: U.S. Department of Health and Human Services, CDC (2014). http://www.cdc.gov/food-net/reports/annual-reports-2014.html

(Ollinger and Rhodes 2017). Since some companies have found ways to reduce the contamination of their ground beef, other companies can follow their lead. The NSLP requires frequent testing for pathogens and removal of tendons and other high-risk animal parts. If FSIS required more testing of ground beef for *E. coli* O157 and other STECs, companies would be incentivized to perform more careful dehiding, use beef carcass steam pasteurization (which costs less than 1 cent/pound), and copy Texas America's procedures to reduce contamination of trim and burgers (which raises costs less than 1 cent/pound). If all companies in the United States were required to meet these strict requirements, the small increase in costs would be passed onto their customers and to some extent US consumers. This would be a win-win situation. Beef companies would no longer face outbreaks associated with ground beef, and consumers would be spared the acute illnesses, deaths, and long-term health outcomes caused by STECs.

Another approach FSIS could pursue, perhaps in tandem with the NSLP approach, is to seek authority to regulate on-farm production of cattle and be delegated authority over transport of cattle to the slaughterhouse. Recently, FDA and USDA signed a memorandum of understanding (MOU) to collaborate and cooperate in food safety activities (FDA 2018). An MOU about on-farm production of cattle and transport could be next. The justification for such action is USDA's Agricultural Research Service agency's finding that 28% of cattle presented for slaughter are infected with *E. coli* O157:H7 (USDA/FSIS 2002) and the need for knowledge of pathogen risks on individual farms shown in Table 10.4.

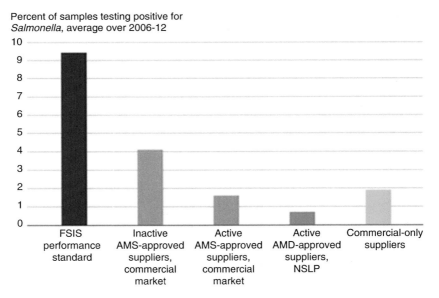

Fig. 10.6 Plants that supply AMS with ground beef for the NSLP had fewer samples test positive for *Salmonella* than did other plants. *AMS*, Agricultural Marketing Service; *FSIS*, Food Safety and Inspection Service; *NSLP*, National School Lunch Program. Source: USDA, Economic Research Service using data from the USDA, Agricultural Marketing Service and USDA, Food Safety and Inspection Service's Public Health Information System

Table 10.4 Variables of potential concern in estimating foodborne disease risks in beef production, handling, and consumption

Farm input use	Farm production practices	Animal transportation
Production animal: Animal breeds (e.g., Holstein, Hereford), animal purpose (dairy, beef, veal), gender, age. **Type of housing:** Open range, feeding shed, group pens, individual calf housing, enclosed barn for all animals (concrete/wood/dirt floor), etc. **Feed inputs:** • Use of colostrum (fresh or frozen) fed to newborns and protective effect against pathogens (amount fed, timing of feeding). • Calf feed type (udder or pail milk, formula, milk replacer). • Other types of feed (pelleted feed, roughage, additives, silage, etc.) and treatment of feed (irradiated, steam sterilization, medicated). • Use of pasture, rotation, and manure management on pasture. **Water sources and access:** Well water, municipal water supply (chlorinated, filtered, etc.), on-farm pond, irrigation water, manure lagoon. **Wildlife access to farm:** Rodents, birds or other animal access to farm ponds, food animals, and pathways to contamination (aerosols, urine, feces; ingestion of vermin). **Geographic factors:** Local climate and pathogen survival, local wildlife vectors, and trade patterns/impact on replacement stock and pathogen probability. **Pathogen testing:** Testing farm inputs and the environment to identify pathogens (test sensitivity, sampling design, and frequency and breadth of pathogen testing).	**Type of operation:** Product (e.g., dairy, veal calf, grow-out, finishing, feedlot, range-fed beef), vertical integration (owned, contract, independent), and purchased and/or farm-produced feed. **Herd management systems:** Calving management, calf rearing management (e.g., weaning practices), breeding practices, replacement strategies, barn cleaning practices. **Feed handling:** Delivery system (bulk feeding at-will, computer-programmed rations), types of rations/roughage, additives (rumensin, vitamins, antibiotics, idophones), and cleaning of system. **Animal health practices:** Herd health monitoring, use of veterinarian services, source of drugs and drug-use patterns. **Pathogen testing:** Use of pathogen test information from the individual/herd in farm management and to support probability modeling; quality of test information (sensitivity/specificity, sampling design, frequency/breadth.) **Water delivery:** Delivery system (pipes, troughs, etc.), testing for pathogens, and cleaning of system. **Manure handling practices:** Type of system (open pit, other liquid, dry, free range), disposal, and cleaning of system; animal exposure to manure. **Wildlife control:** Pest surveillance and control (e.g., traps, poison); presence or absence of cats. **Control of visitors/trucks:** Restrictions on truck/human entry. **Animal identification system:** Maintenance of animal identification.	**Transport type:** Independent trucker, company truck, and/or railroad car, ship, plane. **Travel:** Length (local/regional/national/international), timing (season). **Feeding system:** Feed and water practices during transit. **Manure handling:** Loading procedures, stanchions in transportation, number of layers of animals. **System cleaning:** Type of cleaning, location, and timing of cleaning of transportation vehicles. **Identification:** Maintenance of animal identification.

(continued)

Table 10.4 (continued)

Animal slaughter system	Beef processing system	Product transportation system
Type of operation: Integrated with processing or farm operation, single/multiple types of animals slaughtered, single/multiple slaughter lines.	**Type of operation:** Integrated with slaughter or retail operation, single/multiple types of products/animal species processed, single/multiple processing lines.	**Transport type:** Independent trucker, company truck, and/or railroad car, ship, plane.
Antemortem treatment: Live animal inspection, hide wash, dry manure removal from animal.	**Cross-contamination control:** Physical separation of incoming trucks/personnel that may be contaminated from plant workers and product; equipment cleaning program, knife sterilization, worker glove use/ hand-washing, and refraining from handling food when ill; air ventilation system; separation of raw from cooked products; special control procedures for ground (comminuted) products.	**Length of trip:** Local, regional, national, or international movement of raw or cooked product.
Carcass preparation: Hide removal, opening of abdominal cavity, tying off digestive tract, and other procedures to minimize manure contamination of meat.		**Temperature control:** Refrigeration practices during transit, temperature monitoring, use of temperature indicators.
Cross-contamination control: Plant air ventilation system; physical separation of product/ workers from beginning to end of line; equipment cleaning and/or sterilization between carcasses (e.g., knives); worker glove use/ handwashing; and refraining from handling food when ill.	**Meat cutting/trimming:** Removal of fecal contamination; minimization of cross-contamination along the processing line and from workers to product.	**Cross-contamination:** Requirements for driver's hygiene; sanitary handling practices.
Digestive tract removal: Minimizing spillage on meat, organoleptic examination of organs.	**Meat temperature control:** Control of meat temperature during fabrication, cooling of processed products, and temperature maintenance after processing.	**Cleaning system:** Type of cleaning, location, and timing of vehicle cleaning.
Carcass treatment: Removal of fecal contamination and minimization of cross-contamination among carcasses and from workers. Carcass cooling and temperature maintenance.	**Inventory control:** Special date control programs for ground (comminuted) products and for cooked products, and coordination with lot identification system for products.	**Pathogen testing:** Testing vehicles, raw product, processed product, and environment for pathogens.
Plant sanitation program: Weekly/daily/shift/lot schedule for cleaning building, drains, and equipment.	**Plant sanitation program:** Weekly/daily/shift/lot schedule for cleaning building, drains, conveyor belts, grinders, and the like.	**Identification:** Maintaining lot and company identification.
Pathogen testing: Testing animals, equipment, the environment, workers, and meat for pathogen occurrence.	**Pathogen testing:** Testing raw meat, finished product, equipment, workers, and the environment for pathogens.	
Identification: Maintaining farm/carcass/lot identification linkages.	**Identification:** Maintaining slaughterhouse/lot identification linkages.	

Meat wholesale/retail system	Kitchen handling/consumption	Food link to human health
Type of operation: Degree of integration with processing and food preparation activities. **Cross-contamination:** Physical separation of plant workers and product from incoming trucks/personnel that may be contaminated; equipment cleaning program; knife sterilization; worker glove use/handwashing; air ventilation system; separation of raw from cooked products. **Product fabrication:** Minimization of cross-contamination from raw to cooked product, among products, and from workers to product. Risk level may vary depending upon fabrication practices (origins of meat used in ground product, age of pieces, location of grinding, number of regrindings), and reworking practices. **Temperature control:** Control of meat temperature during fabrication, cooling of fabricated products, and temperature maintenance after processing. **Inventory control:** Special date control programs for ground (comminuted) products and for cooked products, and coordination with lot identification system for products. **Sanitation program:** Weekly/daily/shift/lot schedule for cleaning knives, grinders, display cases, drains, and the like. **Pathogen testing:** Testing raw meat, finished product, equipment, workers, and the environment for pathogens. **Identification:** Maintaining lot/company identification linkages.	**Type of kitchen:** Home, restaurant, fast food chain, schools, group homes/elder care, military, detention facilities, and the like. **Cross-contamination:** Kitchen sanitation (washing utensils, counters); glove use/handwashing; separation of raw from cooked products; refraining from handling food when ill. **Temperature control:** Rapid refrigeration after purchase and maintenance at low temperature. Cook raw animal protein products well-done. **Inventory control:** Use of raw product within a few days, especially ground meats. **Pathogen testing:** Testing raw meat, finished product, equipment, workers, and the environment for pathogens. **Consumption:** Avoid ordering undercooked beef, sending restaurant food back for more cooking if not well-done. **Identification:** Source of product.	**Surveillance and monitoring:** Foodborne association is variable depending on the pathogen/food combination and whether the illness is caused by cells or their toxins. Improved monitoring and surveillance combined with epidemiological investigations using new tests can link disease outbreaks to causative pathogens/food. **Acute illness:** Variable incidence for various chronic illnesses and variable severity distribution, depending on the pathogen and food combination. **Chronic illness:** Human-to-human transmission is variable, depending on the pathogen and food combination. **Secondary cases:** Variable, depending on the pathogen and food combination. **High-risk individuals:** Increased risk of acute illness, more severe illness, and chronic illness in immunocompromised and other individuals. Variable, depending on the pathogen and food combination. **Outrage factors:** Variable, depending on who is at high risk, the pathogen/food combination, the ability to control the pathogen using a variety of risk-reducing techniques, and the like. **Anticipating the future:** Improved knowledge about evolving foodborne pathogens and human illness, especially linkages to chronic disease.

source: Roberts, Ahl, McDowell 1995

Last, there are two other actions FSIS could take today that would increase the economic incentives for pathogen control: (1) testing on farm or in transit, perhaps by requiring that all trucks delivering animals to the slaughterhouse be swabbed down and the samples tested to see if the floor of the trucks show evidence of STECs or *Salmonella* or (2) requiring that before animals could enter the slaughterhouse, all incoming herds and flocks be tested on farm for the presence of STECs, *Salmonella*, and *Campylobacter*. For poultry, on-farm boot sock samples are reliable (Chap 11). Once in the slaughterhouse, composite samples of fecal material or swabs of cattle hides could be collected before evisceration for cattle. For poultry, neck skin samples have been shown to be an excellent indicator of pathogen contamination (Chap. 11). These tests of trucks and animals would give FSIS excellent information on the pathogen load on the animals entering the slaughterhouse. In addition, these tests would be useful in tracing back contamination to the farm of origin, in case of an outbreak. The data could be entered into CDC's databases and be used in analyzing patterns of pathogens moving geographically from farm to farm and from region to region in the United States.

References

American Meat Institute Foundation. White paper—minimizing microbiological food safety risks: potential for preslaughter (preharvest) interventions. 2001, p. 111.

Andrews J. Jack in the box and the decline of *E coli*. Food Safety News. 2013.

Armstrong GL, Hollingsworth J, Morris JG. Emerging foodborne pathogens: *Escherichia coli* O157:H7 as a model of the entry of a new pathogen into the food supply of the developed world. Epidem. 1996;Rev 18:29–51.

Beach C. Dave Theno fellowship unveiled at IAFP food safety conference. Food Safety News. 2017.

Benedict J. Poisoned: the true story of the deadly *E. coli* outbreak that changed the way Americans eat. February Books. 2011.

Biela T. Vice President of Food Safety and Quality Assurance, Texas American Foodservice Corporation. Interviews in 2001, 2002, 2003, and 2016.

Brodziak M. Director of Customer Services, Frigoscandia Equipment, FMC FoodTech. Interviews in 2001 and 2002.

Buchanan RL, Schaffner D. FSMA: testing as a tool for verifying preventive controls. Food Prot Trends. 2015;35(3):228–37.

Cargill. Frigoscandia, Cargill and Excel receive USDA approval for steam pasteurization of fresh beef. Press Release. 1995. http://www.cargill.com/today/releases/12061995.htm.

Cargill. Cargill to install steam pasteurization system in its north American beef plants by June 1 Press Release. 1997. http://www.cargill.com/today/releases/0517b1997.htm.

Cargill. Beef harvest HACCP letter. 2018. https://www.cargill.com/doc/1432077212954/mfs-haccp-ltrs-beef-pdf.pdf.

CDC. Foodborne diseases active surveillance Network (FoodNet): FoodNet Surveillance Report for 2014 (Final Report). U.S. Department of Health and Human Services, CDC. 2014.

Centers for Disease Control and Prevention (CDC). FoodNet 2014 Annual Foodborne Illness Surveillance Report. 2016. http://www.cdc.gov/foodnet/reports/annual-reports-2014.html.

ConAgra News Release, ConAgra Foods Reports Solid Fourth Quarter & Fiscal 2003 Results; Portfolio Progress on Track, June 2003, available @ http://www.conagrabrands.com/newsroom/news-conagrafoods- reports-solid-fourth-quarter-fiscal-2003-results-portfolio-progress-on-track-426023

FDA. FDA, USDA announce formal agreement to bolster coordination and collaboration. FDA news release. 2018. https://www.fda.gov/NewsEvents/Newsroom/PressAnnouncements/ucm594424.htm

Flynn D. 150-plus Peter Pan restitution claims possible in ConAgra case. Food Safety News. 2016a.

Flynn D. Letter from the editor: the new normal. Food Safety News. 2016b.

Frenzen PD, et al. Economic cost of illness due to Escherichia coli O157 infections in the United States. J Food Prot. 2005;68:2623–30.

Gill CO. Effects on the microbiological condition of product of decontaminating treatments routinely applied to carcasses at beef packing plants. J Food Prot. 2009;72(8):1790–1801. https://doi.org/10.4315/0362-028X-72.8.1790.

Gill CO. Visible contamination on animals and carcasses and the microbiological condition of meat. J Food Prot. 2003;67(2):413–9.

Golan E, Roberts T, Salay E, Caswell J, Ollinger M, Moore D. Food safety innovation in the United States: evidence from the meat industry. AER 831, ERS/USDA. 2004. http://www.ers.usda.gov/media/494174/aer831.pdf.

Hauge SJ, Nafstad O, Rotterud OJ, Nesbakken T. The hygienic impact of categorization of cattle by hide cleanliness in the abattoir. Food Control. 2012;27:100–7.

International Association for Food Protection (IAFP). The black pearl award, Food Protection Trends. 2016:158.

JBT FoodTech website. Our legacy. http://www.jbtfoodtech.com/en/Our-Company/Our%20Legacy. Accessed 29 Jan 2018

Kalchayanand N, Arthur TM, Bosilevac JM, Schmidt JW, Wang R, Shackelford S, et al. Efficacy of antimicrobial compounds on surface decontamination of seven Shiga toxin–producing *Escherichia coli* and *Salmonella* inoculated onto fresh beef. J Food Prot. 2015;78(3):503–10. https://doi.org/10.4315/0362-028X.JFP-14-268.

Leising J. SPS Co-inventor, Vice President at Excel and Director of Research and Development at Cargill/Excel. Interview in November 2002 (now Vice President of Meat Sector R&D, Technology Center, General Mills).

MacDougall E. Head of MacDougall risk management. Interview in November 2002.

Malcolm SA, Narrod CA, Roberts T, Ollinger M. Evaluating the economic effectiveness of pathogen reduction technologies in cattle slaughter plants. Agribusiness. 2004;20(1):109–23. https://naldc.nal.usda.gov/download/34673/PDF; https://docs.google.com/viewerng/viewer?url=eurekamag.com/ftext.php?pdf%3D004149491

Morrison RM. An economic analysis of electron accelerators and cobalt-60 for irradiating food+ USDA Economic Research Service. Technical bulletin 1762. 1989

Morrison R, Buzby J, Lin C-TJ. Irradiating ground beef to enhance food safety. FoodReview. 1997;20:33–7. http://www.ers.usda.gov/publications/foodreview/jan97e.pdf

Nutsch AL, Phebus RK, Riemann MJ, Schafer DE, Boyer JE Jr, Wilson RC, et al. Evaluation of steam pasteurization process in a commercial beef processing facility. J Food Prot. 1997;60(5):485–92.

Nutsch AL, Phebus RK, Riemann MJ, Kotrola JS, Wilson RC, Brown TL. Steam pasteurization of commercially slaughtered beef carcasses: evaluation of bacterial populations at five anatomical locations. J Food Prot. 1998;61(5):571–7.

Ollinger M, Rhodes MT. Regulation, market signals and the provision of food safety in meat and poultry. Amber Waves, ERS/USDA. 2017. https://www.ers.usda.gov/amber-waves/2017/may/regulation-market-signals-and-the-provision-of-food-safety-in-meat-and-poultry/.

Oreskes N, Conway EM. Merchants of doubt: how a handful of scientists obscured the truth on issues from tobacco smoke to global warming. Bloomsbury Press. 2010.

Phebus RK. Associate Professor, Animal Sciences & Industry, Kansas State University. Interview in November 2002. A description of the Food Science Institute's BS and MS programs can be found at http://foodsci.k-state.edu/academics/distance/grad-courses.html.

Phebus RK, Nutsch AL, Schafer DE, Wilson RC, Riemann MJ, Leising JD, et al. Comparison of steam pasteurization and other methods for reduction of pathogens on surfaces of freshly slaughtered beef. J Food Prot. 1997;60(5):476–84.

Pihkala N, Bauer N, Eblen D, Evans P, Johnson R, et al. Risk profile for pathogenic non-O157 Shiga toxin-producing Escherichia coli (non-O157 STEC). FSIS/USDA. 2012.; https://www. fsis.usda.gov/shared/PDF/Non_O157_STEC_Risk_Profile_May2012.pdf

Pruett W, Payton T Jr, Biela CP, Lattuada PM, Mrozinski W, Barbour M, et al. Incidence of Escherichia coli O157:H7 in frozen beef patties produced over an 8-hour shift. J Food Prot. 2002;65(9):1363–70.

Roberts T. Economics of private strategies to control foodborne pathogens, choices, 2nd quarter. 2005;20(2). http://www.choicesmagazine.org/2005-2/safety/2005-2-05.htm.

Roberts T. Personal observation at several food safety conferences. 2016.

Roberts T, Ahl A, McDowell R. Risk assessment for foodborne microbial hazards. Tracking food-borne pathogens from farm to table: data needs to evaluate control options. In: Roberts T, Jensen H, Unnevehr L, editors. U.S. Department of Agriculture, Economic Research Service. 1995;MP-1532, p. 95–115. http://www.ers.usda.gov/publications/MP1532/mp1532.pdf.

Roberts T, Morales RA, Lin C-TJ, Caswell JA, Hooker NH. Worldwide opportunities to market food safety. 1997;161–78. Government and the food industry: economic and political effects of conflict and cooperation, editors. Tim Wallace and William Schroder, (Dordrecht, The Netherlands: Kluwer Academic Publisher).

Roberts T, Malcolm SA, Narrod CA. Probabilistic risk assessment and slaughterhouse prac-tices: modeling contamination process control in beef destined for hamburger. pp. 809–15, Probabilistic Safety Assessment PSA '99: Risk-Informed Performance-Based Regulation in the New Millennium, ed. Prof. Mohammad Modarres, American Nuclear Society. 1999. http:// www.ers.usda.gov/briefing/IndustryFoodSafety/pdfs/psa9.pdf.

Roberts T, Narrod CA, Malcolm SA, Modarres M. An interdisciplinary approach to developing a probabilistic risk analysis model: applications to a Beef Slaughterhouse. 2001; pp. 1–23, Interdisciplinary Food Safety Research, editors. Neal H. Hooker and Elsa A. Murano (Boca Raton, FL: CRC Series in Contemporary Food Science).

Talevich T. Steps to safer meat: Costco focuses on all points along the supply chain. The Costco Connection. 2013, pp. 66 and 67. http://www.costcoconnection.ca/connectioncaeng/2013030 4?pg=68#pg68.

Theno D. Senior Vice President for Quality and Logistics. Jack in the box. Interviews in 2001 and 2002.

U.S. Department of Agriculture, Agric Mark Service (AMS). Technical Data Supplement (TDS) for the Procurement of Beef Special Trim, Boneless, Frozen for Further Processing, TDS-139. 2000. http://www.ams.usda.gov/lsg/cp/beef/TDS-139%20DEC2000.pdf.

U.S. Department of Agriculture, Food Safety and Inspection Service. 1996. Pathogen reduction: Hazard Analysis and Critical Control Point (HACCP) Systems: final rule. Federal Register, July, includes page 38965. U.S. Department of Agriculture, Food Safety and Inspection Service. 2001.

U.S. Department of Agriculture, Food Safety and Inspection Service. Guidance for Minimizing the Risk of Escherichia coli O157:H7 and Salmonella in Beef Slaughter Operations in Docket No. 00-022N, E. coli O157:H7 Contamination of Beef Products. 2002. http://www.fsis.usda.gov/ oppde/rdad/frpubs/00-022N/BeefSlauterGuide.pdf.

United States Patent #5,711,981. 1998. Method for Steam Pasteurization of Meat. Inventors: Wilson; Robert C. (Redmond, WA); Leising; Jerome D. (Shorewood, MN); Strong; John (Kirkland, WA); Hocker; Jon (Bothell, WA); O'Connor; Jerry (Issaquah, WA) Assignee: Frigoscandia Inc. (Redmond, WA) Appl. No: 553852 Filed: 1995 Nov 6.

United States Patent #5,976,005. 1999a. Apparatus for Steam Pasteurization of Meat. Inventors: Wilson; Robert C. (Redmond, WA); Strong; John (Kirkland, WA); Hocker; Jon (Bothell, WA); O'Connor; Jerry (Issaquah, WA); Leising; Jerome D. (Shorewood, MN) Assignee: Frigoscandia Equipment Inc. (Redmond, WA) Appl. No: 014358 Filed: 1998 Jan 23.

United States Patent #6,019,033. 2000. Apparatus for Steam Pasteurization of Food. Inventors: Wilson; Robert C. (Redmond, WA); Leising; Jerome D. (Shorewood, MN); Strong; John

(Kirkland, WA); Hocker; Jon (Bothell, WA); O'Connor; Jerry (Issaquah, WA) Assignee: Frigoscandia, Inc. (Bellevue, WA) Appl. No: 259036 Filed: 1999b Feb 26.

USDA, Office of Inspector General. Food safety and inspection service oversight of production process and recall at ConAgra plant (Establishment 969). 2003; Report No. 24601-2-KC, September, 149 p.

USDA, FSIS. Draft risk assessment of the public health impact of *Escherichia coli* O157:H7 in ground beef. 2001. https://www.fsis.usda.gov/OPPDE/rdad/FRPubs/00-023nreport.pdf.

USDA/FSIS. Backgrounder: new measures to address *E. coli* O157:H7 contamination. 2002. http://www.fsis.usda.gov/Oa/background/ec0902.pdf?redirecthttp=true.

USDA/FSIS. Directive 10,010 Rev. 4: sampling verification activities for Shiga toxin-producing *Escherichia coli* (STEC) in raw beef products. 2015. http://www.fsis.usda.gov/.

Wilson C. SPS co-inventor, Director of Special Projects, Frigoscandia Equipment, interviews in 11/02 and 2001 (now Assistant Vice President for Food Safety and Technology, Costco).

Wilson C. Interview with Tanya Roberts. 2016.

Wilson RC, Leising JD. Method and apparatus for steam pasteurization of meats. U.S. Pat. Appl. 08/335, 437. 1994

Chapter 11
Benefits and Costs of Reducing Human Campylobacteriosis Attributed to Consumption of Chicken Meat in New Zealand

Peter van der Logt, Sharon Wagener, Gail Duncan, Judi Lee, Donald Campbell, Roger Cook, and Steve Hathaway

Abbreviations

ASC	Acidified sodium chlorite
BCR	Benefit-cost ratio
CBA	Cost-benefit analysis
COI	Cost of illness
COP	Poultry processing code of practice
DALY	Disability-adjusted life years
ESR	Environmental Services and Research Ltd.
GHP	Good hygienic practice
HACCP	Hazard Analysis and Critical Control Point
IRR	Internal Rate of Return
MAF	Ministry of Agriculture and Forestry
MoH	Ministry of Health
MPI	Ministry for Primary Industries
NMD	National Microbiological Database
NPV	Net Present Value
NZFSA	New Zealand Food Safety Authority

P. van der Logt (✉) · D. Campbell · R. Cook · S. Hathaway
Science and Risk Assessment Directorate, Regulation and Assurance Branch,
Ministry for Primary Industries, Wellington, New Zealand
e-mail: Peter.vanderLogt@mpi.govt.nz

S. Wagener · G. Duncan
Assurance Directorate, Regulation and Assurance Branch, Ministry for Primary Industries,
Wellington, New Zealand

J. Lee
Market Access Directorate, Regulation and Assurance Branch, Ministry for Primary
Industries, Wellington, New Zealand

© Springer International Publishing AG, part of Springer Nature 2018
T. Roberts (ed.), *Food Safety Economics*, Food Microbiology and Food Safety,
https://doi.org/10.1007/978-3-319-92138-9_11

OECD   Organisation for Economic Co-operation and Development
PIANZ   Poultry Industry Association New Zealand
POAA   Peroxyacetic acid
RMP   Risk Management Programme
VLT   Very low throughput
VS   Verification services
WTP   Willingness to pay

11.1 Introduction

Campylobacteriosis has been the most commonly reported gastrointestinal disease in New Zealand for many years with the notification rate rising progressively since the disease first became notifiable in 1980.

This chapter describes the poultry-associated human campylobacteriosis burden in New Zealand that came to a head in 2006 and the regulatory measures that have since been put in place to reduce human infection.

A specific *Campylobacter Risk Management Strategy* (*Campylobacter* Strategy) was implemented by the New Zealand Food Safety Authority (NZFSA) in 2006 to reduce this food safety risk. It has included improvements in hygienic dressing, comprehensive monitoring of broiler carcasses for *Campylobacter*, setting of national and premises level performance targets and regulatory-driven corrective actions when performance targets have not been met.

This chapter discusses the costs of illness and costs of regulatory activities and discusses the net benefit of implementing the *Campylobacter* Strategy.

The *Campylobacter* Strategy initially rapidly achieved more than 50% reduction in the rate of notified human cases of campylobacteriosis, and there has been improvement in subsequent years. The rate in 2016 still appears to be higher than the rates reported by other countries. The *Campylobacter* Strategy is under annual review. In seeking new risk management options to further reduce the number of foodborne cases, transmission pathways other than poultry meat are also being investigated.

11.2 *Campylobacter* and Campylobacteriosis

There are presently 34 species and 14 subspecies assigned to the genus *Campylobacter*, with *C. jejuni* (subspecies *jejuni*) and *C. coli* being the most frequently reported in human infections. The other species rarely cause illness.

Campylobacter species are widely distributed in warm-blooded animals. Poultry and ruminants are important reservoirs. The modes of transmission to humans include the handling and consumption of contaminated meat, consumption of raw drinking milk and untreated water and contact with infected pets and farm animals.

Person-to-person transmission of *Campylobacter* appears infrequent. The relative contribution of each of the recognised sources to the overall burden of human disease varies from country to country.

Attribution and epidemiological studies have clearly identified handling and consumption of contaminated chicken meat as a major contributor to human campylobacteriosis in a number of countries. In Iceland, the introduction of various control measures to reduce the prevalence of *Campylobacter* on chickens resulted in a significant decline in human disease (Stern et al. 2003).

Campylobacteriosis causes gastrointestinal symptoms of varying intensity (diarrhoea, fever and abdominal pain) in humans. The probability of illness depends on the ingested dose. According to a commonly used dose-response model, one *Campylobacter* organism can cause disease (FAO/WHO 2009). However, the probability of disease due to product contaminated with small numbers of *Campylobacter* appears to be low.

The high incidence of campylobacteriosis, its duration (typically 3–6 days) and its possible sequelae (reactive arthritis and Guillain-Barré syndrome), makes it highly significant from a socio-economic perspective worldwide. Typically, no treatment is required as infections are self-limiting. Death is rare and confined usually to very young or elderly patients or to those suffering from a serious concurrent disease.

11.3 Campylobacteriosis in New Zealand

Campylobacteriosis has been the most commonly reported gastrointestinal disease in New Zealand for many years. It has been a statutorily notifiable disease under the Health Act 1956 since 1980. Medical practitioners and medical laboratories are required to report confirmed or suspected cases to their local public health service for case investigation, performed under contract to the Ministry of Health (MoH). Food-associated investigations are led by the regulator, the Ministry for Primary Industries (MPI). National data on case diagnosis, demographics, risk factors and information on outbreaks are compiled by Environmental Services and Research Ltd. (ESR) on behalf of the MoH.

Campylobacteriosis notifications rose progressively after the disease first became notifiable, peaking at 379 per 100,000 population in 2006 (Fig. 11.1). A simultaneous increase in campylobacteriosis hospitalisations also occurred (Baker et al. 2006). As with all communicable disease reporting systems within countries, there is a considerable degree of under-reporting, and there may be significant regional differences in reporting rates. However, as long as any such differences stay constant over the period studied, temporal variations can still be distinguished. In recent times, apart from minor changes in laboratory methods and data collection systems, the notification system in New Zealand has remained stable and would not have biased the reported notification rate to any degree.

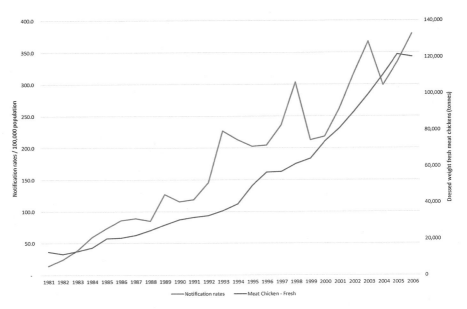

Fig. 11.1 Annual production of poultry meat (tonnes) and notification rate (cases /100,000 population) of human campylobacteriosis

From 1994 to 2005 cross-government collaboration was focused on identifying the probable contributors to the dramatic rise in reported rates of campylobacteriosis. Initiatives included improvement in detection, isolation and molecular typing methods to better collate and interpret molecular typing data.

11.4   Attribution of Campylobacteriosis to Consumption of Chicken Meat

Prior to 2006, a number of scientific initiatives were undertaken under the auspices of the Enteric Disease Research Steering Committee, a collaboration between the NZFSA and the MoH, to verify the long-held assertion that chicken meat handling and consumption were major sources of human campylobacteriosis.

Case-control studies in New Zealand also implicated chicken meat as the significant source of foodborne sporadic campylobacteriosis. A relatively small case-control study in Christchurch reported several chicken-associated risk factors, including consumption of undercooked chicken meat (Ikram et al. 1994). A larger national case-control study conveyed similar findings with a combined population-attributable risk of chicken meat-related exposures greater than 50% (Eberhart-Phillips et al. 1997).

These findings were supported by the results of a comprehensive microbiological survey of *Campylobacter* (and other foodborne pathogens) in retail meats that showed that chicken meat was by far the predominant contaminant (Wong et al. 2007).

In 2004, NZFSA commissioned a systematic review that concluded that poultry consumption was a leading risk factor for sporadic campylobacteriosis in New Zealand (Wilson 2005). The rise in campylobacteriosis was associated closely with the increase in national production of chicken meat intended for sale as chilled rather than frozen or cooked (Fig. 11.1).

In 2006, scientific publications from public health researchers, with associated media interest, dramatically publicised fresh chicken meat as the source of the high rates of campylobacteriosis (Baker et al. 2006; Wilson et al. 2006). Similarly, NZFSA issued a press release expressing its concern about the human burden of campylobacteriosis (NZFSA 2006a).

The heightened political, academic and media attention resulted in a concerted effort by NZFSA and the poultry industry in late 2006 to begin reducing the degree of *Campylobacter* contamination on broiler chicken carcasses (NZFSA 2006b).

In 2005, NZFSA commissioned a molecular source attribution study by Massey University that provided robust estimates of the contribution of the various sources of *Campylobacter* to human campylobacteriosis (French and Molecular Epidemiology and Veterinary Public Health Group 2009). Faecal samples were collected from humans with clinical campylobacteriosis and from food and environmental sources (poultry, ruminants, water, etc.) that were most likely to be possible sources of *Campylobacter* within the same region.

Campylobacter isolates from the samples were typed using multilocus sequence typing (MLST). Three statistical models were used to estimate the probability of human cases being acquired from the various animals and environmental sources (Fig. 11.2). The predominant source was poultry.

11.5 New Zealand Regulatory Framework

Under New Zealand's food safety legislation, primary responsibility for the provision of safe, suitable and properly labelled food to the New Zealand domestic and overseas marketplace lies with the food processor.

Regulation of poultry meat production in New Zealand, i.e. from growing broiler chickens to retail sale of chicken meat and export, has evolved over time. Prior to 1999, food safety regulations for processing chicken meat for domestic consumption in New Zealand were administered by the MoH under the Food Act 1981. Processing for export, albeit a small amount, was regulated by the Ministry of Agriculture and Forestry (MAF) under the Meat Act 1981.

Replacement of the Meat Act 1981 with the risk-based Animal Products Act 1999 enabled the shift of food safety regulation for both domestic and export markets to a single legislative base administered by MAF, subsequently NZFSA and, most recently, MPI.

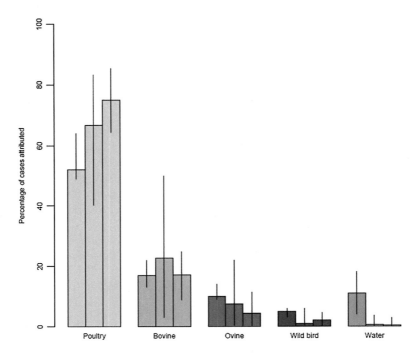

Fig. 11.2 Molecular source attribution estimates of human campylobacteriosis using three models over the period of March 2005–February 2008

Under the Animal Products Act 1999, broiler chicken producers (i.e. growers) and primary processors (i.e. those who slaughter and dress chickens) are required to implement a number of regulatory programmes (Fig. 11.3).

A whole flock health scheme is a programme designed to identify and manage hazards associated with the birds that are likely to affect animal and human health. Control measures include disease control or eradication, control of agricultural compounds and veterinary medicines and feed management.

A supplier guarantee programme, or a supplier statement that accompanies each delivery of birds to the processor, confirms that the delivered birds are healthy and compliant with any withdrawal period for medications.

Ante- and postmortem examination of all birds is required by regulation.

From 1 July 2004, all primary processors that slaughter and dress chickens were required under the Animal Products Act 1999 to have developed, registered and implemented a Risk Management Programme (RMP). A company's RMP describes its specific procedures under good hygienic practice and hazard control measures that ensure that its products are fit for their intended purpose. MPI-recognised verifiers check that the food business is operating in accordance with its RMP and any other regulatory requirements.

Fig. 11.3 Regulatory programmes for chicken growers and processors

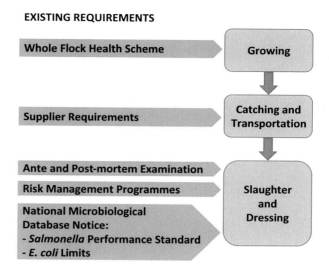

Guidance, including a generic model RMP for broiler chicken processors developed by MPI and the Poultry Industry Association New Zealand (PIANZ), is available to assist businesses to write their own programmes (NZFSA 2002). Hazard control measures described in model RMPs are established using the internationally recognised Hazard Analysis and Critical Control Point (HACCP) system.

Since 2001, chicken broiler processors have participated in MPI's National Microbiological Database (NMD) nationwide monitoring programme (current version: MPI 2016). The NMD is a systems assurance component of the food safety system that provides an objective view of the performance of industry and government hazard reduction measures. Whole carcass rinse samples for microbiological analysis are collected daily at the end of primary processing (post-spin chill).

In the NMD, samples were initially tested only for *Salmonella* and *E. coli*, with microbiological limits for *Salmonella* equivalent to those in the USA's 1996 "Pathogen Reduction and HACCP rule". *Campylobacter* testing of carcasses (three per day or five per week depending on throughput) was included in the NMD programme in April 2007. In addition, caecal samples were collected and tested for *Campylobacter* to provide a measure of *Campylobacter* carriage in the gut of chickens presented for slaughter. The intent was to establish a baseline against which the effectiveness of on-farm biosecurity control measures could be evaluated. MPI subsequently removed the requirement to test for *Campylobacter* in the caeca as it was not adding value to the risk management programme. Regulatory *E. coli* testing was also disestablished as it was not used as a regulatory tool for risk management.

11.6   Managing the Risk of Foodborne Campylobacteriosis

In response to the NZFSA and poultry industry agreement in 2006 that the level of campylobacteriosis in New Zealand attributable to chicken meat was unacceptably high, NZFSA and PIANZ worked collaboratively to develop a formal *Campylobacter* Risk Management Strategy to minimise the extent of *Campylobacter* contamination on broiler chicken carcasses and hence reduce human campylobacteriosis.

A biosecurity manual for broiler growers (current version, PIANZ 2015) and a Poultry Processing Code of Practice (COP) (NZFSA 2007) were developed. Companies on a voluntary basis reviewed, improved and formally documented their good hygienic practice (GHP) and HACCP-based procedures.

Despite these initiatives, microbiological monitoring under the NMD showed that *Campylobacter* prevalence and counts remained at high levels. The NMD also showed what the better-performing processors could actually achieve under commercial conditions.

It was evident from the early years of monitoring via the NMD that stringent action under the *Campylobacter* Strategy was required to bring about a reduction in *Campylobacter* levels on chicken meat.

11.6.1   Regulatory Options

Several options were considered to drive a reduction in *Campylobacter* contamination of chicken meat. These included:

- Microbiological limits for *Campylobacter*
- Implementation of on-farm measures to reduce carriage of *Campylobacter* by chickens
- Regulatory process interventions over and above good hygienic practice
- Public disclosure of NMD data and company performance

11.6.2   Microbiological Limits for Chicken Meat

In 2007, the poultry industry suggested a national target for *Campylobacter* on chicken carcasses of 1 $\log_{10}$ less than the current national mean $\log_{10}$ count per carcass as measured under the NMD programme, a reduction from an average of 3.07 to 2.07 $\log_{10}$ CFU/carcass rinse. This target was agreed by NZFSA for chicken carcasses at the end of primary processing, i.e. post-spin chill.

Process limits encourage premises that do not consistently meet the specified level of *Campylobacter*, i.e. are performing poorly, to improve while not penalising processors that are performing well. Process limits rather than prescriptive hygiene controls also allow processors to implement changes that are optimal for their specific processing conditions.

Three types of limits were developed based on what the better processors could achieve for both standard throughput (>1 million chickens processed per year) and very low throughput (VLT) premises (<1 million chickens processed per year).

For standard throughput processors, the *Campylobacter* limit was based on a "processing period" of five consecutive processing days and hence 15 carcass rinse samples in total. Any of the following failures required corrective action by the processor:

- *High count limit*: Four or more samples in one processing period exceeding 5.88 $\log_{10}$ CFU/carcass rinse.
- *Moving window limit*: Seven or more samples in three consecutive processing periods (45 samples in total) exceeding 3.78 $\log_{10}$ CFU/carcass rinse (90th percentile of the NMD distribution of counts). The latest processing period's five samples displace the earliest processing period's five samples in this moving window.
- *Quarterly limit*: The company's quarterly (calendar year) median value exceeding 4.16 $\log_{10}$ CFU/carcass rinse.

VLT premises were required to take five *Campylobacter* samples per processing week allowing a 3 week, 15 sample, moving window. The design of the limits for VLT premises was similar to those of standard throughput premises, but the number of samples allowed to be non-compliant was adjusted proportionately to reflect the smaller sample number.

Failure of a company to meet any of the limits resulted in an escalating response over consecutive failures. Initially, companies were expected to review their processing equipment set-up and control measures. If non-compliance was not resolved within a specified time, a response team consisting of NZFSA/MPI technical and verification experts, a compliance auditor and, on occasion, industry experts would visit the company to review the actions taken. Additional actions might then be specified by the audit team.

Sanctions such as freezing product or a regulatory direction to cease operation were available to the regulator when corrective actions were not effective. To date, direction to cease processing has not been required.

11.6.3 Implementation of On-Farm Measures to Reduce Carriage of Campylobacter by Chickens

On-farm measures that have been reported internationally to reduce carriage of *Campylobacter* in chickens, e.g. avoiding partial depopulation and installing fly screens, were extensively evaluated in the New Zealand context. Company on-farm practices were correlated against 2 years of caecal sampling under the NMD programme as part of these investigations. However, specific biosecurity measures for *Campylobacter* that would make a significant difference to flock contamination levels if mandated on a national basis were not identified. While individual companies still monitor *Campylobacter* in the caeca of each flock's first group of birds, and are encouraged to occasionally review whether or not on-farm controls might be effective for them, compulsory caecal sampling and evaluation of biosecurity measures were discontinued in July 2009.

11.6.4 Regulatory Interventions

Much effort had been put into good hygienic practice (GHP) and HACCP, and all premises have in place a validated Risk Management Programme (RMP). Nevertheless, implementation of RMPs and routine microbiological testing as specified by the regulator to assess effectiveness of these measures had not impacted upon *Campylobacter* levels in process. This indicated that new process interventions should be considered.

Specific process interventions including antimicrobial washes that have been reported internationally to reduce contamination of *Campylobacter* in chicken carcasses were extensively evaluated as part of the *Campylobacter* Strategy. Some companies have implemented antimicrobial washes on their own initiative, e.g. acidified sodium chlorite, where they consider them practical and effective.

While it is recognised that freezing can substantially reduce the level of *Campylobacter* on chicken meat, this is not considered to be a workable option in New Zealand. There is a strong consumer preference for fresh chilled chicken meat; further, freezing of large quantities of product would require large infrastructural changes. Importantly, contamination of the kitchen environment during thawing and undercooking of frozen chicken meat would remain significant risk factors.

11.6.5 Public Disclosure of NMD Data and Company Performance

Some public advocates suggest that publishing a company's microbiological data and processing performance can provide the consumer with the choice to purchase chicken meat from a better-performing company and, hence, by inference lessen their risk of foodborne campylobacteriosis. They also submit that such a requirement would encourage poorly performing companies to improve.

This was not deemed a useful risk management option in New Zealand. The processing companies that might be named as poor performers have little control on the level of infection of chickens entering their premises due to the ineffectiveness of biosecurity measures on-farm. Further, the performance of individual premises varies over short time periods due to a number of factors. Another consideration is the increasing market demand for free-range birds that roam free without biosecurity protection.

It was considered imperative that the poultry industry worked together in a cooperative rather than competitive manner. Companies that performed poorly invited experts from competitors to review procedures at their plants to identify means to bring them back into compliance. The industry's rationale was that a food safety problem at one plant will affect all companies through the ensuing negative publicity about the safety of chicken meat. Scientific research through NZFSA and others was supported and shared openly, as were engineering designs and intervention validation trials. Publicly publishing the microbiological data or premises ranking would have undermined the cooperation that existed between the processors and their willingness to share information with NZFSA.

Thus, NZFSA chose not to disclose the individual processor's *Campylobacter* results to the public. Instead, it provided confidential information to each premises to show how they were performing compared to other anonymised premises so as to encourage improvement.

11.7   Success of the *Campylobacter* Strategy

Implementation of the *Campylobacter* Strategy rapidly achieved the one log reduction of *Campylobacter* on chicken carcasses that had been agreed with industry. In fact, the greatest decrease was observed as industry pre-emptively reacted to upcoming promulgation of the mandatory limits in 2008.

Concurrently, a 58% reduction in the rate of notified human cases of campylobacteriosis was observed (Fig. 11.4). MPI believes that the principal factor that contributed to this was the reduction in *Campylobacter* contamination on chicken meat due to implementation of the *Campylobacter* Strategy. This was supported by Sears et al. (2011).

11.8   The Costs of Human Campylobacteriosis

11.8.1   Cost Framework

Handling and consumption of chicken meat was, and likely still is, the predominant pathway for human campylobacteriosis in New Zealand. The cost framework for New Zealand's campylobacteriosis epidemic consists of the escalating costs of human illness; the cost of mitigation measures, including regulatory compliance; and the subsequent reduction in the numbers of cases hence cost of illness.

This section discusses application of the cost framework (Fig. 11.5) to determine the cost-benefit of implementation of the *Campylobacter* Strategy, in particular the implementation of measures to reduce the level of *Campylobacter* on chicken meat.

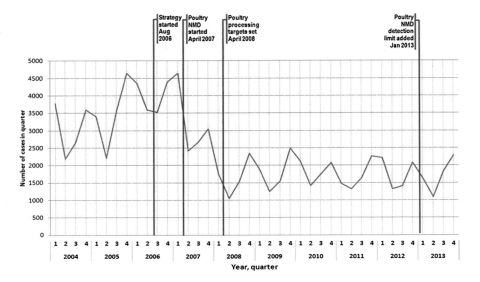

Fig. 11.4 Relationship between the implementation of the *Campylobacter* Strategy and NMD *Campylobacter* limits and the incidence of notified cases of human campylobacteriosis in New Zealand

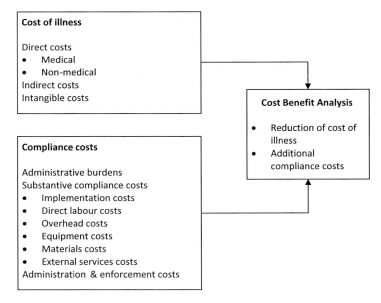

Fig. 11.5 Overview of the framework for evaluating the cost-benefits attributable to the *Campylobacter* Strategy

11.8.2   Cost of Illness

Campylobacteriosis generally manifests itself as a severe gastrointestinal illness, with occasional more serious sequelae such as Guillain-Barré syndrome.

The costs of human campylobacteriosis can be categorised in direct medical and non-medical costs, indirect costs and intangible costs.

- Direct medical costs include self-treatment, consultation with a physician or medical specialist, hospitalisation, medication, medical and laboratory tests. Direct non-medical costs are those associated with transport to physicians, medical specialists, hospitals and pharmacies.
- Indirect costs result from absenteeism and cause lost productivity when people are in paid employment. These costs do not only apply to the patients but also to their caregivers. There is debate whether indirect costs should be applied to people in paid employment only, or also to those who are not, such as homemakers.
- The intangible costs are the costs of suffering. There are methods to express human suffering in monetary terms. In recent times, the concept of disability-adjusted life years (DALY) has become widely used. Broadly speaking, it measures the severity of suffering and the time over which this is experienced.

Cost of foodborne illness and DALY studies were carried out in New Zealand from 2000 to 2014 (Cressey and Lake 2008, 2014; Gadiel 2010; Scott et al. 2000). The studies differed to some degree in their methodology and scope. Indirect costs associated with persons who were not in paid employment were estimated by Cressey and Lake (2008) using DALYs, whereas Gadiel (2010) accounted for such costs in their willingness to pay (WTP) model. Notwithstanding which method was used, the study results were generally comparable, albeit with some uncertainty.

Estimates of the number of cases were based on notification data from EpiSurv (National Notification System database) and public hospital discharge data, the latter identifying the number of hospitalised cases. Minor changes to New Zealand reporting practices have occurred in recent times but are unlikely to have had a major effect on the under-reporting rates.

While the degree of under-reporting of campylobacteriosis is unknown, a value between 7.6 and 10 based on overseas studies has been applied in New Zealand. This range obviously introduced a degree of uncertainty into the estimated costs.

The studies also estimated the foodborne proportion of campylobacteriosis to be ~57% at the time (Cressey and Lake 2007). This estimate was based on expert elicitation and the wide range of opinion expressed during that process also introduces uncertainty into the cost estimates.

The studies took a societal perspective on costs. The cost of treatment for campylobacteriosis is generally funded by the government (Scott et al. 2000), but the cost of productivity losses is mainly incurred by employers (Gadiel 2010). Self-employed people carry their own productivity losses.

Estimating the period of illness also adds uncertainty to the estimates. Most gastroenteritis is of a short-term nature. In the case of campylobacteriosis, 97.3% of illness is short-term gastroenteritis lasting for 3–7 days. The economic burden is influenced by the large number of persons absent from work for short periods. However, campylobacteriosis can have long-lasting effects such as Guillain-Barré syndrome or, at worst, death.

Long-term effects add to future costs, albeit at a discounted rate, i.e. a year of healthy life gained 10 years from now is worth less than a year gained now. Scott et al. (2000) acknowledged the issue of long-term illness but include it for 1 year only due to a lack of data. Cressey and Lake (2008) and Gadiel used a discount rate of 3.5% in their cost estimates.

Notwithstanding uncertainty, the cost per illness and total cost for New Zealand per annum based on these studies are summarised as follows. All estimates henceforth are expressed in 2015 New Zealand dollars, adjusted for inflation using the Reserve Bank of New Zealand inflation calculator.

The direct medical cost estimates to a patient across the three cost of illness (COI) studies varied from ~$NZ 23 to ~$NZ 58. Scott et al. (2000) and Cressey and Lake (2008) estimated the direct non-medical costs to be ~ $NZ 2.50 and the indirect costs ~ $NZ 670.

Importantly, the total of the direct and indirect costs per annum of foodborne campylobacteriosis to New Zealand was $NZ 116.1 million when determined at the peak of the epidemic. Of this, ~94% was indirect costs, ~6% direct medical costs and the remaining minute proportion direct non-medical costs.

11.8.3   Compliance Costs

The purpose of the *Campylobacter* Strategy and *Campylobacter* limits under the NMD programme was to drive improvement in company performance, which comes at a cost. The consequences of not complying are significant, with the ultimate possible sanction for continued non-compliance being processing plant closure. Chicken meat processors spent substantial budget on measures to remain below the limits.

Industry and government were subject to costs that resulted from implementing or meeting regulatory requirements. The OECD compliance cost assessment guidance (OECD 2014) calls such costs "regulatory" and describes five cost categories: compliance, financial, indirect, opportunity and macroeconomic costs. This section discusses only the "compliance cost" category because of its particular relevance to the *Campylobacter* Strategy.

Compliance costs are discussed below. They consist of administrative burdens, substantive compliance costs and administration and enforcement costs. Government and the poultry processing industry incur the costs of developing and implementing the *Campylobacter* Strategy. It is beyond the scope of this analysis to establish whether such costs are ultimately recoverable through changes to the price of chicken meat products at retail.

11.8.3.1 Administrative Burden Costs

Administrative burdens are the costs that processors incur when new information is required by the regulator. These were minimal as a result of the *Campylobacter* strategy. Processors (or their contracted laboratory) were required to submit *Campylobacter* test results to the NMD but were already reporting *E. coli* and *Salmonella* results to the NMD from the same samples.

Processors that exceeded the *Campylobacter* limits were also required to inform their NZFSA-recognised verifier of the corrective actions that had been taken or that were planned. This information was usually emailed. The associated costs were for the time taken to write a brief report and compile any supporting information/evidence. The administrative burden costs are considered to be minor and are likely to have been absorbed in the day-to-day running of the operations.

11.8.3.2 Substantive Compliance Costs

Substantive compliance costs consist of implementation, direct labour, overhead, equipment, materials and external services costs. While processing tasks can generally be categorised accordingly, separating the costs of tasks into each category is difficult. The following summarises reports by Gadiel (2010) and Duncan (2014) in this regard.

Implementation Costs

Processors incurred implementation costs during familiarisation with the new regulatory requirements implicit in the *Campylobacter* Strategy.

By regulating *Campylobacter* targets that required action if exceeded, companies had to review their processes to identify areas where improvements could ensure compliance with the new regulatory requirements. Not prescribing specific interventions provided companies with the flexibility to identify the most cost-effective measures to take in order not to exceed these limits. In general, this involved review and optimisation of existing processes and processing equipment. Duncan (2011) reported that estimating costs of this review activity was problematic.

Direct Labour Costs

There was extra work involved as a result of the implementation of the *Campylobacter* strategy. Quality control and laboratory staff were required to take and analyse extra samples. Samplers and NMD controllers were already required prior to *Campylobacter* Strategy, but time commitments increased.

When processors exceeded the *Campylobacter* limits, additional personnel resource was required to audit processes and procedures and implement corrective actions. These increased labour costs remain unquantified due to lack of data.

Overhead Costs

Any increases in the overhead costs such as rent, office equipment, utilities and other inputs used by the staff engaged in regulatory activities are likely to have been minor. The cost of routine corporate overheads such as management resource would have been low but increased proportionately to the consequences when processors exceeded the *Campylobacter* limits.

Equipment Costs

Commercial chicken slaughter and dressing is heavily mechanised, and consequently equipment costs are important. Equipment is designed to meet commercial quality specifications, e.g. minimise carcass damage, as well as minimise contamination by spoilage organisms and pathogens. Only the specific costs required for pathogen control should be classified as food safety costs, not those to achieve commercial specifications or shelf life.

In response to the *Campylobacter* limits, individual companies optimised or replaced evisceration equipment, sprays on carcasses and equipment, post-chilling decontamination equipment and washing machines for live bird crates. The improvement of evisceration was critical to all, and some companies replaced obsolete equipment, while others improved its performance (Biggs 2012).

The cost to each company differed depending on their equipment optimisation or replacement programmes. Duncan (2014) reported capital expenditure of the entire poultry processing industry to be $NZ 2.4 million. Similarly, Gadiel (2010) reported capital upgrades to be $NZ 1.9 million from 2007 to 2009. The proportion of capital costs specifically attributable to the requirement to comply with the *Campylobacter* limits is unknown.

Materials Costs

Materials costs increased substantially. Processors use large amounts of water to wash contamination from poultry carcasses and to clean dressing equipment. The regulatory *Campylobacter* limits have led processors to re-evaluate their water use. Installation of new sprays, optimisations of the position of the spray heads and water pressure were identified as essential to reduce the level of *Campylobacter* on carcasses. Consequently, the volume of water used may have changed, but no information is available as to what extent.

Materials costs include the use of chemicals for decontamination and are incurred on an ongoing basis. Most processors optimised, or installed new, antimicrobial wash systems to further reduce residual contamination in addition to improvements in dressing hygiene. Control of pH in chlorine-containing spin chillers was identified as a priority for the control of *Campylobacter*. Industry carried out comprehensive and costly validation trials to determine the effectiveness of new antimicrobial wash regimes.

Specifically, industry trialled, and some companies have now implemented, the use of acidified sodium chlorite (ASC) as a decontamination wash and the use of peroxyacetic acid (POAA) is currently under evaluation. Duncan (2014) reported that the estimated total cost to industry of citric acid (for pH correction) and ASC was $NZ 586,000 per annum.

The application of new chemicals not only reduced the *Campylobacter* concentration but also the spoilage organism concentration, resulting in an improved shelf life. The new hygiene measures therefore had a direct commercial as well as a food safety benefit.

External Services Costs

Additional external services costs were incurred on several fronts. External contractors were required to help correctly install and calibrate equipment, e.g. antimicrobial wash systems, and to identify areas where equipment performance could be improved.

Independent poultry experts and veterinarians were contracted across industry to evaluate processing for good hygienic practices and to provide options for improvement to processes. Costs incurred included travel, accommodation and time but have not been quantified in the overall cost analysis.

Microbiological Monitoring Costs

Infrastructure costs associated with the NMD, i.e. sample collection, courier delivery, ISO 17025 accreditation of laboratories, cost of laboratory analysis and reporting of results to NZFSA, were put in place prior to implementation of the *Campylobacter* Strategy. The *Campylobacter* Strategy required companies to carry out additional testing of carcasses for *Campylobacter* and the collection and testing of caecal samples (Fig. 11.6). NZFSA incurred a small cost to amend the NMD database to receive and analyse the industry data.

Laboratory costs were not separated into the categories described above. Duncan (2014) estimated the total industry components costs (including the laboratory costs) of the poultry *Campylobacter* NMD programme for the period April 2008– end of March 2009 to be ~ $NZ 438,000. This was a considerably greater cost than that of the sampling and testing programme prior to implementation of the *Campylobacter* Strategy.

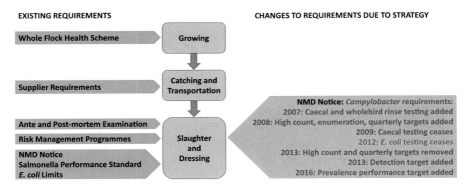

Fig. 11.6 Summary of regulatory requirements for chicken processors associated with implementation of the *Campylobacter* Strategy

11.8.3.3   Administration and Enforcement Costs

While the substantive compliance costs fall on the poultry processors, the administration and enforcement costs fall on government in its duty to develop, implement, administer and enforce the new requirements. Further, these regulatory actions must be supported by a robust scientific evidence base and an operational research programme.

Development of the *Campylobacter* Strategy was informed by an extensive working group consisting of officials with expertise in the areas of database management, epidemiology, food safety, microbiology, poultry processing, public health, risk assessment, risk management, standard setting and verification. This working group guided the development of standards and guidance materials (e.g. poultry processing code of practice and biosecurity manual) and legal notices. The working group continues to advise on new initiatives to further reduce the food-borne burden of campylobacteriosis in New Zealand.

MPI's Verification Services (VS) and/or the *Campylobacter* response team followed up on non-compliances. The VS costs are recovered from the industry and hence, although directly attributable to the *Campylobacter* Strategy, could be classified as external services costs.

The annual ongoing cost to NZFSA of the *Campylobacter* Risk Management Strategy was estimated to be $NZ 1.0 million in 2009 (NZFSA 2009).

11.8.4   Cost-Benefit Analysis

The *Campylobacter* Strategy has been highly successful as illustrated in Figs. 11.4 and 11.7.

Duncan (2014) applied various evaluation techniques for the period 2007–2017 to express the effectiveness of the *Campylobacter* Strategy in monetary terms.

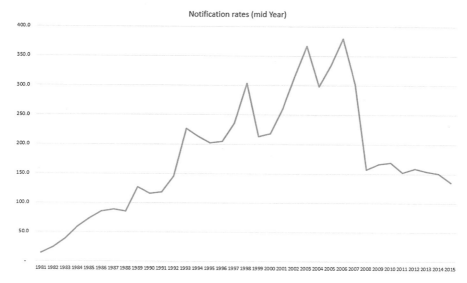

Fig. 11.7 The impact of the *Campylobacter* Strategy on human campylobacteriosis notifications

The benefit was measured as a reduction in number of illnesses and hence the cost of illness. In the absence of other measures that may have substantially reduced consumer exposure to *Campylobacter,* the total reduction in human campylobacteriosis was attributed to the *Campylobacter* Strategy. There was a 58% reduction in campylobacteriosis notifications (15,873 cases in 2006 compared to 6694 cases in 2008).

The cost-benefit analysis (CBA) was applied to the COI estimate of $NZ 116.1 million at the beginning of 2007 which Duncan (2014) calculated from the three New Zealand COI studies. The COI had been reduced to $NZ 48.8 million over 2007/2008. Consequently there was a benefit of $NZ 67.3 million. This gain was ongoing, i.e. it has been occurring every year since the start of the *Campylobacter* Strategy.

This benefit was combined with the compliance and investment costs provided by the industry and the regulator. The various cost categories have been explained above. In brief, the total industry compliance costs including those of PIANZ were estimated to be $NZ 438,000. The ongoing annual costs of the chemicals were $NZ 586,000. The cost to the regulator was $NZ 1.0 million per year. The capital investments that were made to stay below the regulatory limits were $NZ 2.4 million. For the purpose of these evaluations, these capital investments were deemed to have been amortised over the full period of the CBA calculations.

It was assumed that the industry compliance levels would be maintained, the actual number of human cases was ten times the number of reported cases, and there was a linear relationship between notifications and the health effects. The number of human notified cases over the period 2007–2010 was the actually notified cases, while the number of cases over the period 2011–2017 was estimated to be 7000.

The Net Present Value (NPV) as estimated for the period 2007–2017 was $NZ 398.5 million. A large proportion of the benefits are due to indirect costs. The NPV based on the same COI but on direct and non-direct health benefits was $NZ 9.9 million.

The benefit-cost ratio (BCR) similarly was calculated with and without indirect costs. The BCR with indirect costs was 25.7, while the BCR without these costs was 1.6.

A discount rate of 10% was used as recommended by the New Zealand Treasury (2005). Duncan evaluated the sensitivity to this percentage by varying it between 0 and 10%. The BCR varied between 28.0 and 25.7 when the indirect costs were included in the calculations and between 1.8 and 1.6 when they were not included.

Finally, the Internal Rate of Return (IRR) was calculated. It was 1925% if the indirect costs were included and 63% if they were not included.

This cost-benefit analysis clearly demonstrates that from an economic perspective, regardless of the evaluation method and accompanying assumptions, there is a significant gain from the *Campylobacter* Strategy. Nonmonetary considerations, especially consumer trust and confidence that an active risk management strategy is in place, considerably add to this economic gain.

11.9   Ongoing Developments

Setting *Campylobacter* limits under the NMD programme proved to be an effective measure in the *Campylobacter* Strategy to reduce *Campylobacter* levels on chicken meat.

In January 2013, the NMD *Campylobacter* limit was tightened through the addition of a detection limit to reduce the number of positive carcasses:

- A maximum of 29 samples out of a moving window of 45 samples (i.e. three processing periods) are permitted to be *Campylobacter*-positive for standard throughput premises.
- Very Low Throughput (VLT) premises are permitted to have a maximum of five out of nine carcass samples taken from a three successive processing period moving window.

In 2015 the poultry industry and MPI agreed to concentrate on poor performing processors and in 2016 added a further *Campylobacter* limit:

- Standard throughput processing premises (>1 million carcasses per year) should not have more than 30% *Campylobacter*-positive carcass rinsates on a quarterly basis.

While reducing consumer exposure to *Campylobacter* through broiler chickens has to date been the priority of the *Campylobacter* Strategy, surveys of meat from turkeys, ducks, end-of-lay hens and breeder birds have been carried out. Recently, microbiological monitoring of ducks, turkeys, end-of-lay hens and breeders has been added to the NMD programme to improve MPI's understanding of their contribution to consumer exposure. *Campylobacter* performance limits have not been set for these species as yet.

Previous retail surveys show that a large proportion of samples of fresh chicken meat at retail are positive for *Campylobacter* (Marshall et al. 2016; Moorhead et al. 2015). Continued tightening of *Campylobacter* limits until low human disease burdens attributable to poultry are achieved is a priority for MPI.

The cut-off value of 3.78 $\log_{10}$ CFU/rinsate may seem rather high given that low *Campylobacter* doses can cause illness. Reducing this limit will require the development of new enumeration methods, most probably molecular, with increased sensitivity and hence lower limits of detection (LOD).

Industry continues to evaluate processing improvements and decontamination procedures that will further reduce the level of *Campylobacter* on chicken meat and result in a further decrease in human notification rates due to this exposure pathway. Industry is undertaking research for measures that will be practical in a commercial context.

References

Baker M, Wilson N, Ikram R, Chambers S, Shoemack P, Cook G. Regulation of chicken contamination is urgently needed to control New Zealand's serious campylobacteriosis epidemic. N Z Med J. 2006;119(1243):1–8.

Biggs R. Beating the bacteria. Food Technol N Z. 2012;12(1):17–9.

Cressey P, Lake R (Institute of Environmental Science & Research Limited, Christchurch, New Zealand). Risk ranking: Estimates of burden of foodborne disease for New Zealand. Final report. New Zealand Food Safety Authority. 2007. Report FW0724.

Cressey P, Lake R (Institute of Environmental Science & Research Limited, Christchurch, New Zealand). Risk ranking: Estimates of the cost of foodborne disease for New Zealand. Final report. New Zealand Food Safety Authority. 2008. Report FW07102.

Cressey P, Lake R (Institute of Environmental Science & Research Limited, Christchurch, New Zealand). Risk ranking: Updated estimates of the burden of foodborne disease for New Zealand in 2013. Final report. New Zealand Ministry for Primary Industries. 2014. MPI Technical paper No: 2016/59.

Duncan GE. The economic benefits of food safety regulation. Thesis submitted to Department of Public Health, University of Otago, Wellington for Master of Public Health; 2011

Duncan GE. Determining the health benefits of poultry industry compliance measures: the case of campylobacteriosis regulation in New Zealand. N Z Ned J. 2014;127(1391):22–37.

Eberhart-Phillips J, Walker N, Garrett N, Bell D, Sinclair D, Rainger W, Bates M. Campylobacteriosis in New Zealand: results of a case-control study. J Epidemiol Community Health. 1997;51(6):686–91.

FAO/WHO [Food and Agriculture Organization of the United Nations/World Health Organization]. Risk assessment of Campylobacter spp. in broiler chickens: technical report. Microbiological risk assessment series no 12. Geneva; 2009. pp. 132

French N Molecular Epidemiology and Veterinary Public Health Group (Massey University, Palmerston North, New Zealand). Enhancing Surveillance of Potentially Foodborne Enteric Diseases in New Zealand: Human campylobacteriosis in the Manawatu: Project extension incorporating additional poultry sources. 2009. Final report: FDI/236/2005 Report prepared for the New Zealand Food Safety Authority. http://www.foodsafety.govt.nz/elibrary/industry/enhancing-surveillance-potentially-research-projects/finalreportducketc2009.pdf. Accessed 14 Dec 2016.

Gadiel D (Applied Economics Pty Ltd, Sydney, Australia). The economic cost of foodborne disease in New Zealand. 2010. Report prepared for the New Zealand food safety authority. http://www.foodsafety.govt.nz/elibrary/industry/economic-cost-foodborne-disease/foodborne-disease.pdf. Accessed 18 Feb 2016.

Ikram R, Chambers S, Mitchell P, Brieseman MA, Ikam OH. A case control study to determine risk factors for *Campylobacter* infection in Christchurch in the summer of 1992–3. N Z Med J. 1994;107(988):430–2.

Marshall J, Wilkinson J, French N (Massey University, Palmerston North, New Zealand). Source attribution January to December 2015 of human *Campylobacter jejuni* cases from the Manawatu. Completion of sequence typing of human and poultry isolates and source attribution modelling. Final Report. New Zealand Ministry for Primary Industries. 2016. MPI Agreements 17433 and 17509.

Moorhead S, Horn B, Hudson JA (Institute of Environmental Science & Research Limited, Christchurch, New Zealand). Prevalence and enumeration of *Campylobacter* and *E. coli* on chicken carcasses and portions at retail sale. Final report. New Zealand Ministry for Primary Industries. 2015. MPI Technical paper No. 2015/32.

MPI. National Microbiological Database (NMD). 2016. http://www.foodsafety.govt.nz/industry/general/nmd/. Accessed 14 Dec 2016.

New Zealand Treasury. Cost benefit analysis primer. Version 1.12. Departmental CFISNET Release; 2005.

NZFSA. Guidance and generic risk management programme for slaughter and dressing of broilers. 2002. http://www.foodsafety.govt.nz/elibrary/industry/Generic_Model-Been_Produced.pdf. Accessed 14 Dec 2016.

NZFSA. NZFSA concerned at increase in human *Campylobacter* infection. 2006a. http://www.foodsafety.govt.nz/elibrary/industry/Nzfsa_Concerned-Zealand_Food.htm. Accessed 14 Dec 2016.

NZFSA. NZFSA moves to curb *Campylobacter* rates. 2006b. http://www.foodsafety.govt.nz/elibrary/industry/Nzfsa_Moves-Along_With.htm. Accessed 14 Dec 2016.

NZFSA. Poultry processing – code. Poultry processing – code of practice. 2007. http://www.foodsafety.govt.nz/elibrary/industry/processing-code-practice-poultry/index.htm. Accessed 14 Dec 2016.

NZFSA. New Zealand Food Safety Authority Annual Report 2008/2009; 2009.

OECD. OECD regulatory compliance cost assessment guidance. OECD Publishing; 2014. https://doi.org/10.1787/9789264209657-en.

PIANZ. 2015. http://pianz.org.nz/farming-systems/management/biosecurity-manual-for-nz-meat-chicken-growers. Accessed 14 Dec 2016.

Scott WG, Scott HM, Lake RJ, Baker MG. Economic cost to New Zealand of foodborne infectious disease. N Z Med J. 2000;113(1113):281–4.

Sears A, Baker MG, Wilson N, Marshall J, Muellner P, Campbell DM, Lake RJ, French NP. Marked Campylobacteriosis decline after interventions aimed at poultry, New Zealand. Emerg Infect Dis. 2011;17(6):1007–15.

Stern NJ, Hiett KL, Alfredsson GA, Kristinsson KG, Reiersen J, Hardardottir H, Briem H, et al. *Campylobacter* spp. in Icelandic poultry operations and human disease. Epidemiol Infect. 2003;130(1):23–32.

Wilson N. (2005) Report to the food safety Authority of New Zealand. A systematic review of the aetiology of human campylobacteriosis in New Zealand. http://www.foodsafety.govt.nz/elibrary/industry/Systematic_Review-Literature_Evidence.pdf

Wilson N, Baker M, Simmons G, Shoemack P. New Zealand should control Campylobacter in fresh poultry before worrying about flies. N Z Med J. 2006;119:U2242.

Wong TL, Hollis L, Cornelius A, Nicol C, Cook R, Hudson JA. Prevalence, numbers, and subtypes of *Campylobacter jejuni* and *Campylobacter coli* in uncooked retail meat samples. J Food Prot. 2007;70:566–73.

Chapter 12
Sweden Led *Salmonella* Control in Broilers: Which Countries Are Following?

Tanya Roberts and Johan Lindblad

Abbreviations

BCA	Benefit/cost analysis
CDC	Centers for Disease Control and Prevention
CSPI	Center for Science in the Public Interest
DVM	Doctorate of Veterinary Medicine
EPA	Environmental Protection Agency
ERS	Economic Research Service, USDA
EU	European Union
FSCP	Finnish *Salmonella* control program
FSIS	Food Safety and Inspection Service, USDA
GIPSA	Grain Inspection, Packers and Stockyards Administration, USDA
HACCP	Hazard Analysis and Critical Control Points
Kg.	kilogram
Lb.	pound
NCC	National Chicken Council
PEW	PEW Research Center
SVA	Swedish Board of Agriculture
USDA	United States Department of Agriculture
WHO	World Health Organization

T. Roberts (✉)
Economic Research Service, USDA (retired), Center for Foodborne Illness
Research and Prevention, Vashon, WA, USA
e-mail: tanyaroberts@centurytel.net

J. Lindblad
The Swedish Poultry Meat Association (Retired), Stockholm, Sweden

© Springer International Publishing AG, part of Springer Nature 2018
T. Roberts (ed.), *Food Safety Economics*, Food Microbiology and Food Safety,
https://doi.org/10.1007/978-3-319-92138-9_12

12.1   Sweden's *Salmonella* Control Program for Broilers

Sweden was the first country to achieve control of *Salmonella* in the production of broilers. In 1994, the World Health Organization's Veterinary Public Health Unit published a report summarizing the steps Sweden took, such as monitoring critical control points in production with *Salmonella* tests and depopulating flocks when tests were positive. This chapter explores the evolution of Sweden's successful control strategy and the role of economic incentives embodied in strong regulations, in private insurance policies, and in consumer demand. Spread of the control in other Nordic countries is discussed. Swedish researchers estimate that *Salmonella* control costs 2.6 US cents per broiler or less than 1 cent per pound of meat. Comparisons are made to the US poultry industry, US *Salmonella* regulations, and the demand for Salmonella control by retailers. The economic externalities imposed on the US public by the current low level of *Salmonella* control in broilers are also explored. One externality, the societal cost of foodborne salmonellosis, is estimated at $5–$16 billion annually (Chap. 8). The Centers for Disease Control and Prevention estimates that poultry, at 29%, is the largest cause of US *Salmonella* illnesses.

Sweden's *Salmonella* regulations were a response to public concern over a large outbreak of *Salmonella typhimurium* due to red meat that caused 8,845 illnesses (people who tested positive for *Salmonella*) and more than 90 deaths in 1953. In 1961, the Swedish Salmonellosis Order regulation required *Salmonella* testing in food-producing animals and reporting of positive test results. If *Salmonella* is detected, the lot is unfit for human consumption (§16 of the Food Act). In the late 1960s, several outbreaks of human illness were traced to *Salmonella*-contaminated broilers.

In 1970 the broiler industry initiated a "voluntary" *Salmonella* control program that was approved by the government's Swedish Board of Agriculture (SVA). In addition to basic requirements concerning day-old chicks, housing, and feed, all links in the chain of broiler production were monitored via *Salmonella* tests. Farmers wanting to participate in the program applied to the SVA, and a veterinarian examined the facilities before approval was granted. The government paid 90% of all costs due to *Salmonella* infection in poultry flocks affiliated with the voluntary program. This Swedish *Salmonella* control program is an example of a public/private partnership discussed in Chap. 14.

In 1984, Sweden made *Salmonella* testing compulsory 10–14 days before slaughter for broiler flocks. By walking through the enclosed broiler house, boot socks on farmworkers feet collect samples of what is in the litter of the house (Fig. 12.1). The boot socks are taken to the laboratory and tested for *Salmonella*. If the test is positive for *Salmonella*, the flock could not enter the slaughter facility and was depopulated. The sensitivity of detecting *Salmonella* using boot socks was confirmed by researchers at the United States Department of Agriculture (USDA), Agricultural Research Service (Buhr et al. 2007). Sock samples had the highest rates of detecting *Salmonella* of the four methods tested.

Fig. 12.1 Boot sock sampling in broiler house. Source from internet: Technical Service Consultants Ltd. http://www.tscswabs.co.uk/full-product-range/hygiene-enviroscreen-range/poultry-boot-swabs

At the same time, the Swedish government stopped compensating the broiler sector for any expenses or absence of income caused by the *Salmonella* testing and control program. Now commercial broiler producers had the possibility to be covered by private insurance, but only if they participated fully in the *Salmonella* control program. Then 90% of losses due to *Salmonella* infections were covered by the private insurance plan. Fewer than 20 broiler flocks a year were infected with *Salmonella* and depopulated after testing was made mandatory (Fig. 12.2). As in any private insurance plan, the premiums go up whenever there is a claim. Consequently, farms not able to control *Salmonella* in their broilers went out of business. The economic lesson here is that private insurance works for *Salmonella* when a successful *Salmonella* control program is implemented and enforced from farm to fork. Below are the details of the Swedish program that Johan Lindblad, DVM, former chief veterinarian of the Swedish Poultry Meat Association, presented at the 2007 USDA Agricultural Outlook Forum in Washington, D.C. (Lindblad 2007).

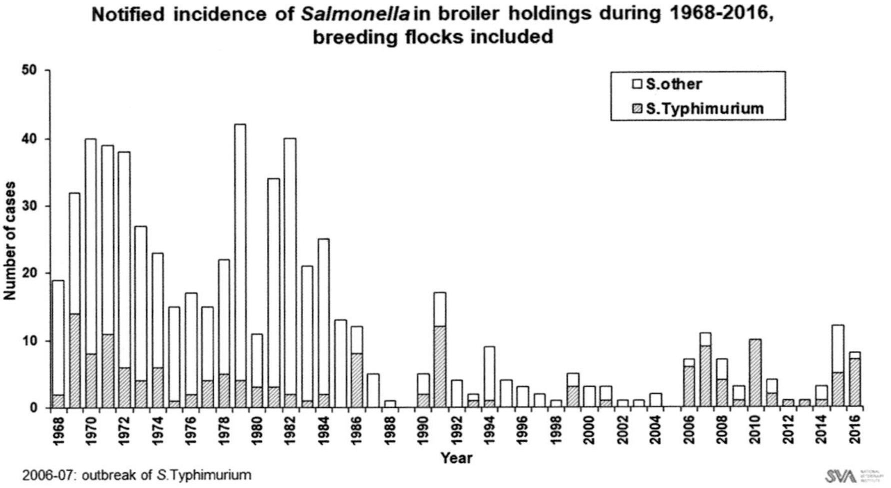

Notified incidence of *Salmonella* in broiler holdings during 1968-2016, breeding flocks included

2006-07: outbreak of S.Typhimurium

Fig. 12.2 Decline in infected flocks after the 1984 mandatory control of *Salmonella* in Sweden

12.2 The Five Principles of the Swedish Program of *Salmonella*-Free Broiler Production Are:

1. Start with *Salmonella*-free day-old chickens.
2. Rear chicks in a *Salmonella*-free environment.
3. Provide feed and water free from *Salmonella*.
4. Regularly monitor and test for *Salmonella* in the whole production chain.
5. Take immediate action whenever *Salmonella* is detected.

How each of the five principles is implemented in the Swedish control program is discussed in the following paragraphs (Lindblad 2007).

Prevention in day-old chicks: All live poultry (layers, broilers, turkeys, geese, and ducks) imported to Sweden are quarantined. Commercial poultry are imported as day-old GrandParents (GP) from the international breeder Ross and Cobb in the United Kingdom. During quarantine, the GP flock is regularly tested for diseases not present in the Swedish commercial poultry population as well as for *Salmonella*. If *Salmonella* is isolated, regardless of serotype, the flock is immediately destroyed. Since 1970, *Salmonella* has only been detected in one GrandParent flock after release from quarantine. (In 2016 a GP flock turned positive after release from quarantine; the GP flock was immediately killed, as well as two newly placed Parent (P)-flocks originating from these GP-hens as well as a few broiler flocks—by-products placed as broilers.)

Prevention in feed: Initially feed factories voluntarily analyzed samples of imported feed ingredients that were considered high-risk raw materials. In 1993, Swedish legislation made testing of all feed mandatory. Samples for testing are usually taken when feed ingredients are loaded onto trucks or vessels in the shipping country. Results of the *Salmonella* tests are usually known when feed ingredients arrive at the feed mill. The other option is to sample and test after arrival; the ingredients are not unloaded until *Salmonella*-negative test results are confirmed. If *Salmonella* is found, the whole consignment must be decontaminated before arrival and unloading at the feed mill.

In the feed mill, heat treatment to 75 °C (168 °F) is required and only lines, silos, and trucks designated for heat-treated feed are acceptable. The one exception to the heat treatment is whole grain, and no *Salmonella* contamination in broilers has been traced back to the use of whole grain. Many Swedish farmers raise whole wheat on their farms to use as a major ingredient in feed to broilers. There are strict legal regulations regarding fertilizing, harvesting, transportation, and storage for whole wheat and other whole grains used in broiler feed. In 2015, however, *Salmonella* was detected in a commercial broiler flock due to contamination of soy used in the flock's feed.

Prevention in the environment: All broiler houses in the voluntary program are enclosed with *solid* floors, walls, and ceilings that are easily cleanable. The houses are required to be rodent and wild bird proof. Due to these requirements, organic production cannot be accepted into the voluntary program.

All breeder and broiler houses are furnished with an "anteroom" at the entrance. The "anteroom" is divided into "dirty" and "clean" areas with a "hygiene barrier." When crossing the "hygiene barrier," footwear and preferably coveralls should be changed. In many GrandParent operations, the "hygiene barrier" consists of a shower. Broilers are raised in a *Salmonella*-free house.

After all the broilers have been taken to the slaughter house in one batch, all the litter must be removed within 24 h. The broiler house then must to be cleaned and sanitized (often by contract cleaners) and sit idle for a minimum of 2 weeks. Today, thinning of the flock is allowed, if well performed. It can be economically beneficial with very little or no increase in risk of *Salmonella* introduction. If *Salmonella* were isolated in the flock, the flock is destroyed, and all litter is composted for at least 6 months to prevent contamination of the surrounding environment.

Monitoring and testing for Salmonella: A rigorous testing program is in place throughout the commercial broiler supply chain. Imported day-old chicks are quarantined and tested to assure that they remain S- before placement in breeding. When chicks arrive on the farm as Grandparent breeding flocks, they are tested immediately, at 2 weeks, at 4 weeks, at 10 weeks, at 17 weeks, at 24 weeks and thereafter every 2 weeks, and within 2 weeks of slaughter. Every second week meconium samples from newly hatched Parent chicks representing every GP-flock is analyzed. Parent breeder flocks that supply the eggs for the broilers are also tested with the same frequency except the meconium samples. Twice a year, a state-appointed veterinarian visits each facility in the supply chain (breeding flocks, hatchery, and on-farm broiler flocks) and takes samples that are tested for *Salmonella*.

Every broiler slaughterhouse takes neck skin samples two times a day to test for *Salmonella*, to verify the control program. Any S+ earlier in the supply chain triggers eradication of the flock, as well as an investigation and testing back to the GrandParent breeding flock that produced the initial eggs, and the feed mill delivering the feed.

Actions taken when Salmonella is suspected: Complete sampling and testing procedures are performed whenever *Salmonella* is suspected in broilers or their

Table 12.1 Swedish industry *Salmonella* control costs per broiler, 1993

Cost category		Swedish ore
Grandparent rearing extra cost		2
Production of parents		13
Hatching of broilers		4
Growing broilers		52
Testing for *Salmonella*	5	
Improved hygiene	20	
Higher feed costs, etc.	27	
Private insurance for S+		8
Buildings		7
Slaughterhouse (vet, admin.)		8
Total		94 ore (16 US cents/broiler)
2003 costs 2016 costs estimation		10 US cents/broiler, or 2 cents/lb. 5 US cents/broiler, or 1 cent/lb.

Source: Engvall et al. (1994), Lindblad (2017)

feed. In all suspected or verified index cases, a veterinarian is appointed by the NBA to carry out an investigation to try to find the source of introduction. Whenever *Salmonella* is verified, regardless of prevalence or serotype, the flock is destroyed. All manure is composted for at least 6 months.

After thorough cleaning and disinfection of the broiler house and the feed mill if involved, they are visually inspected, and environmental swabs are taken for further *Salmonella* testing. These actions continue until the tests come back negative. Only after all *Salmonella* tests are negative can the feed mill start operating again or the broiler house be repopulated with chicks.

12.3   Economic Costs of Sweden's *Salmonella* Control Program in Broilers

In 1993 at a World Health Organization (WHO) conference, Engvall et al. (1994) reported that the marginal cost of producing *Salmonella*-free broilers in Sweden was 16 US cents per broiler. Most of the costs (55%) were in the direct farm grow-out of the broilers, The major cost components were S- feed costs and improved hygiene, while *Salmonella* tests were a minor cost (Table 12.1). The second highest cost category was the production of *Salmonella*-free, day-old chicks to be the parents of the broilers followed by the hatching of broilers and rearing of grandparents. The third category of costs included the minor costs of private insurance for *Salmonella*-positive flocks, enclosed housing, and the veterinarian and testing in the slaughterhouse. Today, the Swedish producers have found more cost-effective production methods, and the marginal cost of *Salmonella*-free broilers is estimated at 2.6 US cents per broiler or less than 1 cent/pound of meat (Table 12.1, Lindblad 2017).

12.4   Finland, Norway, and Denmark Follow Sweden

Finland and Norway followed Sweden's lead and their governments instituted similar control programs for broilers from farm to fork (Hopp et al. 1999). In 1992, the European Community passed *Salmonella* regulations (EU 1993). In 1995, Finland joined the European Union and negotiated its own program, a stricter Finnish *Salmonella* control program (FSCP) (Kangas et al. 2007). The objective of the FSCP is to maintain the annual prevalence of *Salmonella* below 1% at the national level. The program covers broilers for all *Salmonella*, whereas the EU regulations only cover *Salmonella enteritidis* and *Salmonella typhimurium*. Feedstuffs have been regulated in Finland for 40 years to prevent *Salmonella* contamination. Also, if *Salmonella* is found in the production chain from farm to fork for broilers, the FSCP investigates broiler primary and secondary production as well as the interventions taken after detection of any *Salmonella* serovar that can include freezing chicken meat or fully cooking it before sale to consumers. Based on the monitoring results and economic evaluation studies, the FSCP achieved its targets with minimal cost to industry and the government (Kangas et al. 2007).

In 1993 COOP Denmark, the largest Danish retail chain with 40% of the market, announced that suppliers must have no positive *Salmonella* tests (Terry 2014). If on-farm tests were positive, the broiler flock would be destroyed. COOP Denmark also required companies to test a sample of their butchered meat. Broiler meat was rejected from any positive flock. These were powerful economic incentives, yet the farmers found such strict measures hard to take seriously, until COOP Denmark started selling Swedish boilers. Next COOP Denmark introduced a "salmonella-free" label to entice consumers. To entice poultry farmers, a higher price was offered. Danpo, Denmark's largest poultry producer, decided to follow COOP Denmark's requirements and sell to them to gain a larger share of the market (Terry 2014).

Some of the smaller farmers thought *Salmonella* control was not possible and sought Danish government support. In 1996, Denmark began its National *Salmonella* Control Program that compensated farmers for slaughtered flocks and other losses. The European Union (EU) had passed a *Salmonella* testing and indemnification program for positive on-farm breeding stock (EU 1993). Denmark took advantage of the program and provided 50% of the indemnification that was matched by the EU's 50% (Wegener et al. 2003). The total amount spent for *Salmonella* testing and indemnification of contaminated eggs and broilers was 103 million Danish kroner. The EU spent 55 million Danish kroner on the shared costs of indemnification from 1996 to 2003. The private sector contributed 30 million Danish kroner to the *Salmonella* control program as well as investing in new equipment and houses and implementing new control procedures (Wegener et al. 2003). In the words of the authors:

> *Salmonella* can be effectively reduced (nearly eliminated) from broiler chickens by intensive flock-level testing and top-down eradication. Essential to success is a sufficiently sensitive testing program in the breeding and rearing flocks as well as in the hatcheries, i.e., one

that involves intensive sampling and a combination of serologic and bacteriologic testing methods. Bacteriologic testing alone is not sufficiently sensitive to achieve control, especially if *S*. Enteritidis infections are present. Removal of all organic material, thorough cleaning and disinfection of the poultry house, and an empty resting period of 10–14 days between flocks can effectively eliminate residual infections. In Denmark, most infections appear to be vertically transmitted (nearly always traceable to an infected hatchery or parent flock), whereas horizontal transmission from the environment and wild fauna appear to play a minor role. Competitive exclusion cultures, vaccines, or antibiotics have not been used in the Danish control program.
—Wegener et al. (2003).

In response to outbreaks with unusual *Salmonella* serotypes at two Danish hatcheries in 1997, researchers studied what on-farm actions were most effective in eradicating *Salmonella* contamination from Danish broiler houses. In all, 44 broiler houses were contaminated with *Salmonella enteritidis* 8; and 40 broiler houses were contaminated with *Salmonella typhimurium* 66. These houses were studied for practices that were most successful in eradicating the *Salmonella* contamination in 11 subsequent broiler flocks. Gradel and Rattenborg (2003) found that five control practices were the most significant:

- Combined surface-and-pulse fog disinfection of the house, after all manure is removed, must be done properly. When the humidity is 100% and the temperature is 60 °C (140 °F), all five disinfectants tested were effective in killing all *Salmonella*.
- Gravel strip along broiler house, about 4′ deep and 4′ wide, for rodent control.
- Antiseptic soap/water in anteroom of the broiler house for hand washing.
- Assuring that equipment does *not* cross the hygiene barrier, for example, when dead broilers are removed from the flock.
- Good system is in place to check indoor rodent-bait stations on a regular schedule.

In 2001, the Danish society saved US $25.5 million by controlling *Salmonella* and preventing human illnesses. The total annual *Salmonella* control costs in year 2001 were US $0.02/kg for broilers or US $0.01/lb. for broilers. "The control principles described are applicable to most industrialized countries with modern intensive farming systems" (Wegener et al. 2003).

12.5   The US Broiler Industry

US broiler production is characterized by major brand companies (called integrators) contracting with local farmers to raise broilers. Ninety-eight percent of all US broilers are produced under contract (Ahearn et al. 2005). Through the contract growing system, not only is the genetic stock of the birds controlled, but the feed and on-farm production practices are also standardized to produce a homogeneous commodity. The average length of a contract was 13 months in 2001, and more than one-third of contracts were flock-to-flock. Yet renewals are common and growers

Table 12.2 Largest US broiler integrators, 2012

Rank and company	Slaughter plants (#)	Aver. weekly slaughter (million head)	Aver. bird size (live weight lbs.)	Comments
1. Tyson Foods	33	35.40	5.53	$37 billion, sells in 115 countries Revenue: beef 38%, chicken 39%, prepared foods 20%, pork 11%; Arkansas
2. Pilgrim's Corp.	26	33.10	5.46	Second largest chicken producer in the world; purchased by JBS, Brazil
3. Perdue Farms, Inc.	12	12.01	5.67	Chicken 100% antibiotic free (no antibiotics ever label); veggie feed; controlled atmosphere stunning Family owned, Maryland
4. Koch Foods, Inc.	8	12.00	5.10	Small birds because of higher quality; innovative processing; products sold in United States and overseas; privately held, Illinois
5. Sanderson Farms	9	8.62	7.53	$2.8 billion sales of poultry (fresh, frozen, and prepared foods) Mississippi
6. Foster Farms	5	5.84	6.07	$2.3 billion; owns and operates most of its grow-out; outbreaks of multidrug-resistant *Salmonella* and refused to recall product; California
All USA production		162.1	5.85	

Source: Adapted from MacDonald (2014a) p. 4; comments added from company websites, etc.

report that 9 years is the average length of contracting with the same firm. While contract terms vary from company to company, the contracts basically outline the responsibilities of both farmers and companies. The contracting company generally provides chicks, feed, veterinary supplies and services, management services or field personnel, and transportation for the birds to and from the farm. The grower provides land and housing facilities, utilities, labor and management skills, and other operating expenses, such as repairs and maintenance, to raise broilers over a 6–7-week period. In summary, the integrator provides some of the variable inputs (chicks and feed), while the grower provides other variable inputs, labor, and capital (land and housing).

While the continuation of contracting and the bonus payment reflect the grower's management skills, there are increasing concerns about market power of the few US poultry companies located near a local slaughter plant (McKenna 2017; USDA/GIPSA 2016a, b). In Table 12.2 note the few slaughter plants that are owned relative to the number of broilers raised and slaughtered. Around 1,000,000 broilers are killed weekly at each slaughter plant. USDA's Grain Inspection. Packers and Stockyards Administration (GIPSA) proposed regulations to clarify the conduct or action by integrators that are "unfair, unjustly discriminatory, or deceptive" (USDA/

Contract grower fees cover a wide range

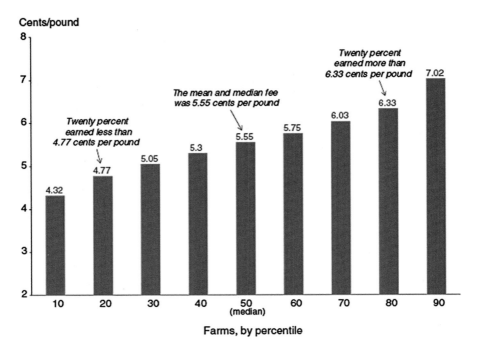

Fig. 12.3 US broiler contract fees vary widely, 2011 (fees are based on live weight of broilers produced). (In MacDonald (2014a, b)) Source: USDA agricultural resource management survey (2011), https://www.ers.usda.gov/webdocs/charts/61223/august14_feature_macdonald_fig03.png?v=41843

GIPSA 2016a) and to examine the poultry grower ranking systems (USDA/GIPSA 2016b). "Commenters noted that variations in chicks, feed, and medications have a significant influence on the poultry grower's performance, but the grower has no control or influence over the quality of those inputs." (USDA/GIPSA 2016b, p. 97274). The Trump administration, on October 18, 2017, declined to take any further action on the GIPSA proposed rule, meaning that the proposed rulemaking is cancelled (2017). One of the reasons given for the cancellation is the increase in litigation that could be expected *if* the proposed rules were enacted. Note that this reason does not address the need for protection of growers from the market power of integrators. For further information on the ambiguity in marketing contracts, see Chap. 3 for a discussion of economic incentives in contracts and the economics of principal agency theory.

In MacDonald's (2014a) publication, the average payment per live weight is 5.55 cents/lb. to growers (Fig. 12.3). But 10% of growers earn less than 4.32 cents/lb. and 10% earn more than 7.02 cents/lb., a difference of 63% for a relatively homogenous product, a live broiler. Does this reflect the different market segments US broilers are raised for organic chicken vs. processed meat used in frozen meals or fast food items? Or is this an indicator of market power gone awry? USDA's farm-level data "suggest that fees received by growers tend to be lower for growers in markets with few integrators" (Wu and MacDonald 2015). Another indicator is that in concentrated markets, integrators "appear to be making firmer commitments on duration, quantity or flock placements, and pay to new growers but *not* to existing growers" (emphasis added, Ibid.).

The National Chicken Council (NCC) website (NCC 2017a, b) states that US broilers today reach an average market weight of 6.24 pounds in 47 days and that broilers have an average feed conversion ratio of 1.89 pounds of feed/pound of live weight. On June 15, 2017, President Mike Brown of the NCC declared:

> The United States is the most efficient producer of poultry products in the world. Our comparative advantage in producing and marketing these products is both our access to America's abundant production of high quality feed grain and soybean products which are used to feed our flocks; and from America's technological leadership in poultry genetics and breeding, precision feed formulations, and animal health practices.

Broilers used to be sold whole, but by the early 1980s, US consumers preferred cut-up and further-processed broilers that required less home preparation, were in more convenient portions, or were in new processed products (Eales and Unnevehr 1988). In 1995 chicken consumption overtook beef consumption in the United States due to its cheap price, "healthy" attributes of chicken meat (beef implicated in colon cancer), and new convenience products. By 2000 whole broilers had fallen to 10%, while cut-up parts were 42%, and further-processed broiler meat led at 48% of broiler meat produced (MacDonald 2014a, p. 10). In 2016 consumers paid $1.42/pound in their supermarkets for a whole carcass broiler (NCC 2017a, b) and higher prices for more highly processed choices.

In 2011, US broilers houses averaged 18,618 sq. ft. ($42' \times 440'$), while the largest houses were 40.00 ft^2 ($66' \times 600'$) (MacDonald 2014a, p.19). An average contract grower produced 504,180 broilers in 2011 in just over four houses. Over two-thirds of growers had one to four houses, and these farms accounted for almost half of all US production (MacDonald 2014a, p. 15). The earthen floor of the house is covered with bedding material consisting of organic matter such as wood chips, rice hulls, or peanut shells. Because dry bedding helps maintain flock health, most grow-out houses have nipple drinkers for water rather than open troughs. Most new US broiler houses are fully enclosed with tunnel ventilation, evaporative cooling, and improved lighting operated with automatic controls.

Most broiler houses are in the so-called "Broiler Belt" stretching from Delaware to Texas. While the South has the advantage of mild winters, it can have extremely hot summers. If the environment in a grow-out house is not properly maintained, this could lead to extensive mortality of the broilers who lack sweat glands and are unable to regulate their own temperature. The solution has been the installation of large fans in conjunction with tunnel ventilation that keeps air moving throughout the house.

12.6   Discussion of Sweden's *Salmonella*-Free Production Versus US Broiler Production

The different broiler farming practices in the United States vs. Sweden are outlined in Table 12.3. When comparing Sweden's broiler production to the United States, we focus on the grow-out house, starting with size and other features, and discuss what is different and what is the same. We do this because from a risk assessment perspective, the flock is the unit of interest. And both the grow-out house size and number of flocks per grower are the economic variables of most interest, since grower household income depends on the number of flocks harvested, the contract price, and any bonus payment. Interestingly, the size of the average broiler grow-out house is similar in both countries, as is the number of houses per farm, so the impact on household investments in grow-out houses are similar as would be grower income. In conclusion, the most important issue of the technical ability of growers to raise broilers is similar in Sweden and the United States. This means that the economic cost of raising broilers is similar.

Genetics are also not different, since both countries use fast-growing birds. For example, today 99% of all broilers slaughtered in Sweden are standard Ross and Cobb broilers. The maturity of the birds at time of slaughter differs somewhat, and that alters the number of flocks per year. Sweden harvests 7 flocks, while the United States has 5.5 flocks per year. The United States raises birds a little longer to get heftier breasts, legs, thighs, and wings sold by the piece and in processed products. The pounds of boiler meat grown per year similar in a grow-out house in the United States or Sweden.

What technical changes in US broiler production would need to be adjusted to produce *Salmonella*-negative flocks, as Sweden does? And how expensive are these changes thought to be? The main changes would be higher cost for *Salmonella*-free feed; construction, sanitation, and management changes of the grow-out house to prevent *Salmonella* from infecting flocks; increased cost of providing *Salmonella*-free chicks; changes to clean-out of houses and litter disposal; and cost of on-farm testing for *Salmonella*.

- *Salmonella*-free feed: Heat treatment of feed, and transportation and storage to assure that it remains *Salmonella*-free, adds the greatest cost (Table 12.1). Daily attention to the feed storage and distribution inside the grow-out house are also required.
- Grow-out house construction, management, and hygiene: While the house construction costs can be amortized over the life of the house, many of the hygiene and management features require daily or weekly attention to assure that the sanitation barriers are in place. For example, mice entering the house for the feed are a constant issue that the rock perimeter and other barriers in house construction are designed to prevent. The changing of clothing and footwear every time the house is entered also requires diligence.
- *Salmonella*-free chicks: The most important cost here is the production of *Salmonella*-free parents with minor costs associated with the hatching of broilers and the extra costs of rearing *Salmonella*-free Grandparents.

Table 12.3 Comparison of broiler production in Sweden and the United States

Feature	Sweden	United States[a]	Comments
Genetics SIMILAR	Ross and Cobb 99% (fast) 4.3 lbs. in 33 days Feed conversion = 1.6 7 grow outs/year	Fast growing birds 6.13 lbs. in 49.5 days Feed conversion = 1.9 5.5 grow outs/year	Sweden harvests birds earlier At higher weights, US broilers can develop "wooden breast"[b]
Grow out house size SIMILAR	Aver. size 15, 780 ft^2 Aver. 4.4 houses/farm 100 broiler growers	Aver. size 18,618 ft^2 Aver. 4 houses/farm The United States has more growers	**US slaughters 162 million broilers/week vs. Sweden 90 million/ year**
Housing and cleanout between flocks DIFFERS	Enclosed with sanitary perimeters, as hygiene barrier, with change of foot wear and coveralls 100% total cleanout and sanitation between flocks	Enclosed, fan ventilation 77% not cleaned out and sanitized between flocks, 56% no change of clothing, 17 days between flocks	**The US has NO *Salmonella* control regulations in housing design or cleanout regulations between flocks**
Litter disposal DIFFERS	All litter removed between flocks and sold as fertilizer to farmers. But if test *S*+ on-farm composting of litter	33% of litter sold for use by other farms in 2011	**The US has NO *Salmonella* requirements for litter** US contamination of soil and water with excess nutrients and pathogens[c]
Feed DIFFERS	Farm-grown wheat Purchased feed must be *S*- Fish meal NOT allowed in feed	Provided by contractor— mostly corn and soybeans No regulations prohibiting *Salmonella* in feed	**The US has NO regulations on *Salmonella* in feed Sweden NO fishmeal, because of sustainability concerns**
Antibiotic use in feed DIFFERS	Not allowed	Antibiotics important in human health being phased out of animal feed by FDA	**US antibiotics used on-farm add to antibiotic resistance of human pathogens- Chap. 16**
On-farm *Salmonella* tests DIFFERS	Negative *Salmonella* test in GP stock, hatchery, breeders, feed. Flock tested before slaughter: must be *S*- to get into slaughterhouse	No on-farm regulations for *Salmonella* control	**The US has NO on-farm regulations for *Salmonella* control and no on-farm tests** On-farm control highest likelihood of controlling *Salmonella*[d]
Slaughter house *Salmonella* tests DIFFERS	Neck skin tests daily for *Salmonella*, gas kill, blast air chill, no chlorine, or other chemicals allowed. If test is *S*+, flock is depopulated	Performance standard allows 9.8% *Salmonella* rate in whole carcass rinse test	**The US does NOT test daily and allows 9.8% S+ tests** Chlorine, water absorption and US kill method reduce flavor and tenderness[e]

(continued)

Table 12.3 (continued)

Feature	Sweden	United States[a]	Comments
Farm contract DIFFERS	Slaughter plant allots the broiler farmer to hatchery Contract with the slaughter plant	Bullying of farmers: short contracts[f], blackballing, tournament payment fraud (in states other than Iowa)	Proposed US GIPSA regulations[g] withdrawn by President Trump

Sources: [a]MacDonald (2014) (USDA/GIPSA 2017), [b]Halley (PEW and CSPI 2014), [c]EPA (Bourassa et al. 2015), [d]PEW (Gamble et al. 2016), [e]Smart Chicken (CSPI 2015), [f]GIPSA (USDA/GIPSA 2016a, b), [g]GIPSA (CDC 2011)

- Clean-out of house and litter disposal: This activity requires increased attention to detail to make sure that a thorough clean-out after each flock is done and that the house is thoroughly disinfected to assure no *Salmonella*. Disposal of the litter is also a critical activity to assure that the litter does not contaminate the environment on the farm, the stream, and fields nearby or be present in the litter sold to others as fertilizer.
- *Salmonella* testing of flock: This new activity is a minor, but important, cost. If all the above items are done well and are effective, then the *Salmonella* tests will be negative and only the costs of the test itself will be involved. If not, the flock will be depopulated and not enter the human supply of food.

The first three changes are the most expensive in the analysis shown in Table 12.1. While there will be a steep learning curve for some growers and integrators (who specify the types of houses they want to build), the good news is that the international broiler industry has a lot of experience to share about these improvements.

How expensive would these changes be? Today, researchers in both Sweden and Denmark estimate that the increased cost of producing *Salmonella*-free broilers is about 1 US cent per pound of broiler meat sold to the consumer (Lindblad 2017; Wegener et al. 2003). Which US consumers would willingly pay, in our opinion, to escape the probability of salmonellosis acute illness, premature death, and long-term health outcomes such as arthritis. This answer is based on a supermarket survey of consumer's willingness to pay for pork with a reduced level of *Salmonella*. Sundström et al. (2014) found consumers would be willing to pay $4.92 per pound for pork chops. Since the average price for pork chops in this market was $3.00 per pound, consumers would be willing to pay a price premium of up to $1.92 per pound for a safer pork chop. The large willingness to pay answers in this survey indicate a strong public preference for stronger US food safety regulations to eliminate/reduce *Salmonella* contamination in food.

Sweden requires that the supply chain from breeder to fork be *Salmonella*-free for broilers. The United States has no such requirement, although the PEW foundation report found that the farm is the most effective location for *Salmonella* control in the poultry supply chain (PEW and CSPI 2014).

Another essential feature of *Salmonella* control in Sweden is pathogen testing. *Salmonella* tests are required in the production of chicks, the feed the chicks are fed that must be verified to be *Salmonella*-free, and the flock is tested with boot socks before the flock can be sent to the slaughterhouse. In the broiler slaughterhouse, Sweden takes neck skin samples twice a day each and every day. These samples are taken following air chilling of the carcasses. In 2015, USDA Agricultural Research Service researchers compared the Swedish method to the US sampling method (Bourassa et al. 2015). In the United States, sampling is not done daily. The test follows a chlorine carcass rinse thought to reduce *Salmonella*-positive results, and the test is the whole carcass rinse method. Bourasse et al. concluded: "When chlorine was present during chilling, Neck Skin sampling was more effective than Whole Carcass Rinse sampling." More *Salmonella* serogroups were detected with the neck skin samples; and the neck skin excision method samples the carcass without removal from the processing line.

12.7 The Political Economy of *Salmonella* Control in Sweden Versus the United States

In Sweden a large *Salmonella* outbreak galvanized the public, regulators, and the industry to work together to eradicate *Salmonella* in poultry. Sweden used national legislation and collaboration between government and industry to provide incentives for control. Together they developed, identified, and implemented production processes from breeding through grow-out to control *Salmonella*. The government, along with the cooperatives, provided research funds to discover how to eradicate *Salmonella* in the broiler supply chain. Initially, the Swedish government paid indemnities for destroyed flocks. After the control programs were implemented and *Salmonella* levels were below 1% in broilers, private insurance replaced government indemnification. Swedish regulations require private insurance before slaughter can occur. This collaboration between industry and government is an example of a public/private partnership discussed in Chap. 14.

At a 1993 WHO conference, Swedish researchers shared the details of how they achieved *Salmonella* control in their broilers (Brockotter 2017a). In summary, the broiler supply chain controls are as follows: purchase of *Salmonella*-negative (S-) chicks, all-in/all-out movement of flocks, requirements for enclosed building (easy to clean, disinfection after each flock, hygiene barrier, ventilation, rodent control measures), use of S- feed and emptying and cleaning bins after flock, regular cleaning and disinfection of the drinking water system, use of S- floor litter, and regular *Salmonella* tests of each flock (destroy if test S+; broiler flock is tested and must be S- on-farm before it can be sent to the slaughterhouse). Similar *Salmonella* control regulations apply to the hatchery and feed mill. The regulatory requirement for *Salmonella* control, or else the flock is eradicated, is a strong incentive for control since the government enforces these regulations.

In the United States, the meat and poultry industry has fought *Salmonella* control, as shown by these actions:

- The *Salmonella* performance standard in the 1996 HACCP regulations has been weakened by challenges in court (see Chap. 16).
- FSIS's *Salmonella* test for broilers is a whole carcass rinse fraught with problems, such as industry use of chemicals that reduce the probability that FSIS will detect *Salmonella* (Brockotter 2017b). However, the new testing of parts might provide a more sensitive test (Brown 2017).
- In 2011, the Center for Science in the Public Interest (CSPI) petitioned USDA to declare multidrug resistant *Salmonella* an adulterant, but USDA has declined to act (Cabazan 2004).
- In 2013 and 2014, two outbreaks of multidrug resistant *Salmonella* sickened 22,574 people in 27 states and Puerto Rico and were linked to Foster Farms, but the company refused to recall the products. USDA does not have the authority to impose a mandatory recall (Gieraltowski et al. 2016) and refuses to ask Congress to give it that authority.
- CDC reports that poultry is responsible for 29% of *Salmonella* outbreaks in the United States and that control of *Salmonella* has not improved in the last decade (Commission of the European Communities 2001). *Salmonella* and *Campylobacter* (also linked to poultry) are two of the most costly foodborne illnesses in the United States (see Chap. 8).

12.8   Benefit/Cost Analysis (BCA)

In Sweden, a BCA compared current *Salmonella* regulations to the less-rigorous controls in Denmark and the Netherlands. Sundstrom et al (2014) found the expected increase in human salmonellosis cases and the associated increase in reactive arthritis and irritable bowel syndrome (IBS) to be more costly than any reduction in *Salmonella*-control costs.

Turning to the United States, evaluating the costs and benefits of *Salmonella*-free broilers involves tallying up the externalities that broiler production now imposes on the US public. Externalities are defined as costs imposed on the society that the sellers of a product do not compensate society for. In other words, the integrators producing and selling broilers get a "free ride" on these costs that they avoid paying. Most important are the acute and long-term health costs, productivity losses, and pain and suffering due to illness from eating or working with *Salmonella*-contaminated broilers. In Chap. 8, Scharff estimates that salmonellosis costs the US $5–16 billion annually. CDC estimates that poultry is responsible for 29% of salmonellosis cases or $1.5–4.6 billion annually (Commission of the European Communities 2001). These billions of US dollars of medical costs and productivity losses due to salmonellosis are a huge externality that the sickened consumers and workers endure without being compensated by the US integrators. If the cost of preventing *Salmonella* in chickens is $US 0.01/lb., as Sweden and Denmark economists estimate, then the BCA becomes \$1.5–4.8 billion/\$0.33 billion = 4.5–14.5. In other words, for each dollar spent in controlling *Salmonella*, 4.5–14.5 times this is saved in human illness costs (based on ERS estimates of 33 billion pounds of chicken produced in the United States in 2016 (PEW and CSPI 2014).

Related to human illness costs, people often get salmonellosis that is resistant to treatment with antibiotics. Not only is the initial illness more difficult to treat, but it involves another externality. This externality is the spread of antibiotic resistance to other pathogens and the possibility of losing the effectiveness of antibiotics in treating human disease. While it is impossible to estimate this probability, it is important to note that the international broiler industry is contributing to antibiotic resistance and to the loss of antibiotics to treat human diseases.

Another important externality that the broiler companies impose on society is the environmental costs of *Salmonella*-contaminated water runoff into streams and rivers. Especially costly to society is the multidrug resistant *Salmonella* contamination that can cause illness in consumers and workers as well as contaminate the soil and water near grow-out facilities. For example, only 20% of the tetracycline in poultry feed is used by the bird, so 80% ends up in the litter (see Chap. 15).

The nutrient-rich runoff from broiler grow-out farms has caused algae blooms and "dead zones" in US bays and oceans and has been litigated by the Environmental Protection Agency (EPA). The Chesapeake Bay contamination was extremely notable in causing losses to the fishing and crabbing industries. A collaborative program initiated by EPA has reduced nitrogen by 57% and phosphorus by 75% and reduced toxic cyanobacteria (EPA 2016). A related issue is poultry litter placed on fields that can contaminate fields with *Salmonella* and become a problem, especially where fruits and vegetables are grown and could become contaminated with *Salmonella*.

One final externality is the costs of bullying growers by some of the US broiler companies. Bullying alters the profitability and debt of the growers as broiler companies require new investments in housing, electronics, etc. and manipulate the stocking of new flocks. In 2016, GIPSA proposals were published to address this bullying (USDA/GIPSA 2016a, b). But these proposed regulations were withdrawn in 2017 under President Trump (USDA/GIPSA 2017). The National Chicken Council praised the withdrawal (NCC 2017a).

12.9 Conclusion

In response to a large outbreak of salmonellosis, Sweden initiated a farm-to-fork control program of the broiler industry jointly with the government. The mandated *Salmonella* tests throughout the supply chain, in conjunction with flock destruction when positive tests were found, provided a strong economic incentive for growers to follow the control procedures. The costs of this control program in both Sweden and Denmark are estimated at 1 US cent/pound at retail.

In contrast, the United States does not have farm-level regulations for *Salmonella*. A comparison of US human illness costs of poultry-caused salmonellosis cases with the control costs of one cent per pound of retail weight results in benefit/cost ratio of 4.5–14.5. This means that for every dollar spent to control *Salmonella*, 4.5–14.5 dollars are saved in human illness costs. Furthermore, other externalities that are not put in dollars would be added benefits, making this ratio even higher.

References

Ahearn MC, Korb P, Banker D. Industrialization and contracting in U.S. agriculture. J Agric Appl Econ. 2005;37(2):347–64.

Bourassa DV, Holmes JM, Cason JS, Cox NA, Rigsby LL, Buhr RJ. Prevalence and serogroup diversity of *Salmonella* for broiler neck skin, whole carcass rinse, and whole carcass enrichment sampling methodologies following air or immersion chilling. J Food Prot. 2015;78(11):1938–44. https://doi.org/10.4315/0362-028X.JFP-15-189.

Brockotter F. Big in niche markets. Poult World. 2017a;23(7):18–9.

Brockotter F. Focus on cost price and quality. Poult World. 2017b;23(7):11–3.

Brown M. NCC Statement on USDA proposed rule to allow importation of cooked chicken from China, National Chicken Council website June 15, 2017. www.nationalchickencouncil.org.

Buhr RJ, Richardson LJ, Cox NA, Fairchild BD. Comparison of four sampling methods for the detection of *Salmonella* in broiler litter. Poult Sci. 2007;86(1):21–5.

Cabazan C. Key factors influencing day old chick quality. Int Hatchery Pract. 2004;18(5):11. 13, 14

Center for Science in the Public Interest (CSPI). Petition to the USDA re: antibiotic-resistant *Salmonella*. 2015. https://cspinet.org/resource/petition-usda-re-antibiotic-resistant-salmonella.

Centers for Disease Control and Prevention (CDC). Foodborne illness from *Salmonella* and *Campylobacter* associated with poultry, United States. 2011. https://www.fsis.usda.gov/wps/wcm/connect/00023142-2971-40b2-bc17-1a26f12693d3/Salmonella_Campylobacter_011811_190.pdf?MOD=AJPERES.

Commission of the European Communities. Proposal for a Regulation of the European Parliament and of the Council on the control of salmonella and other food-borne zoonotic agents and amending Council Directives 64/432/EEC, 72/462/EEC and 90/539/EEC. Annex II, Section E, Council Directive 92/117/EEC. 2001. http://europa.eu.int/eur-lex/en/com/pdf/2001/en_501PC0452_01.pdf13.

Eales JS, Unnevehr L. Demand for beef and chicken products: separability and structural change. Am J Agric Econ. 1988;70:521–32.

Engvall A, Andersson Y, Cerenius F. The economics of Swedish *Salmonella* control: a cost/benefit analysis. Control of foodborne diseases in humans and animals: the Swedish *Salmonella* control program. 1994. pp. 16–32, WHO/Zoon./94.171.

Environmental Protection Agency (EPA). Chesapeake Bay progress: wastewater pollution reduction leads the way. 2016. https://www.epa.gov/sites/production/files/2016-06/documents/wastewater_progress_report_06142016.pdf. Accessed 10 May 2017.

European Union (EU). Council Directive 92/117/EEC of 17 December 1992 concerning measures for protection against specified zoonoses and specified zoonotic agents in animals and products of animal origin in order to prevent outbreaks of food-borne infections and intoxications. Off J Eur Commun. 1993; L062: 15/03/1993, pp. 38–48. http://eur-lex.europa.eu/legal-content/EN/TXT/HTML/?uri=CELEX:31992L0117&from=EN

Gamble GR, et al. Effect of simulated sanitizer carryover on recovery of *Salmonella* from broiler carcass rinsates. J Food Prot. 2016;79:710–4.

Gieraltowski L, Higa J, Peralta V, Green A, Schwensohn C, Rosen H, et al. National outbreak of multidrug resistant Salmonella Heidelberg infections linked to a single poultry company. PLoS One. 2016;11(9):e0162369. https://doi.org/10.1371/journal.pone.0162369.

Gradel KO, Rattenborg E. A questionnaire-based, retrospective field study of persistence of *Salmonella* Enteritidis and *Salmonella typhimurium* in Danish broiler houses. Prev Vet Med. 2003;56(4):267–84. https://doi.org/10.1016/S0167-5877(02)00211-8.

Hopp P, Wahlström H, Hirn J. A common *Salmonella* control programme in Finland, Norway and Sweden. Acta Vet Scand Suppl. 1999;91:45–9.

Kangas S, Lyytikäinen T, Peltola J, Ranta J, Maijala R. Costs of two alternative *Salmonella* control policies in Finnish broiler production. Acta Vet Scand. 2007;49(1):35. Published online 2007 Dec 4. https://doi.org/10.1186/1751-0147-49-35.

Lindblad J. Lessons from Sweden's control of *Salmonella* and *Campylobacter* in broilers, USDA Agricultural Outlook Forum. 2007. https://ageconsearch.umn.edu/record/8109/files/fo07li01.pdf.

Lindblad J Updated estimate of costs of producing *Salmonella* negative broilers. 2017.

MacDonald JM, Technology, organization and financial performance in US broiler production, EIB 126, ERS/USDA, June 2014.

MacDonald JM. Technology, organization, and financial performance in U.S. broiler production, EIB-126, U.S. Department of Agriculture, Economic Research Service. 2014a. https://www.ers.usda.gov/webdocs/publications/43869/48159_eib126.pdf?v=41809

MacDonald JM. Technology financial risks and incomes in contract broiler production. Amber Waves, U.S. Department of Agriculture, Economic Research Service. 2014b. https://www.ers.usda.gov/amber-waves/2014/august/financial-risks-and-incomes-in-contract-broiler-production/

McKenna M. Big chicken. Washington, DC: National Geographic; 2017.

National Chicken Council (NCC) website, worried about *Salmonella*? 2017a. http://www.nationalchickencouncil.org/. Accessed 3 Dec

National Chicken Council (NCC) website, NCC praises USDA's withdrawal of controversial 'GIPSA Rules' on competitive injury, unfair practices, and undue preferences. 2017b. http://www.nationalchickencouncil.org/. Accessed 3 Dec

PEW and CSPI, Meat and Poultry Inspection 2.0: How the United States can learn from the practices and innovations in other countries. 2014. http://www.pewtrusts.org/en/research-and-analysis/reports/2014/10/meat-and-poultry-inspection-20.

Sundström K, Wahlström H, Ivarsson S, Lewerin SS. Economic effects of introducing alternative Salmonella control strategies in Sweden. PLoS One. 2014:e96446. https://doi.org/10.1371/journal.pone.0096446.

Terry L, Contaminated chicken: illnesses surge, Denmark attacks *Salmonella* in program that proves a success. The Oregonian. 2014. http://www.oregonlive.com. Accessed 16 Sep 2017.

USDA/GIPSA. Unfair practices and undue preferences in violation of the packers and Stockyards Act. Fed Reg. 2016a;81(244):92703–23.

USDA/GIPSA. Poultry grower ranking systems. Fed Reg. 2016b;81(244):92723–40.

USDA/GIPSA. Unfair practices and undue preferences in violation of the packers and Stockyards Act. Fed Reg. 2017;82(200):48603–4.

Wegener HC, Hald T, Wong DLF, Madsen M, Korsgaard H, Bager F, Gerner-Smidt P, Mølbak K. *Salmonella* control programs in Denmark. Emerg Infect Dis. 2003;9:774–80. https://wwwnc.cdc.gov/eid/article/9/7/03-0024_article

Wu SJ, MacDonald J. Economics of agricultural contract grower protection legislation, Choices, 2015; 3rd quarter, 30(3):1–5. https://www.aaea.org/publications/choices-magazine.

Chapter 13
The Role of Surveillance in Promoting Food Safety

Robert L. Scharff and Craig Hedberg

Abbreviations

CDC	Centers for Disease Control Research and Prevention/US
CIFOR	Council to Improve Foodborne Outbreak Response
FDA	Food and Drug Administration/US
FDOSS	Food Disease Outbreak Surveillance System
FOOD Tool	Foodborne Outbreak Online Database
HUS	Hemolytic uremic syndrome
NNDSS	National Notifiable Diseases Surveillance System
NORS	National Outbreak Reporting System
PFGE	Pulsed-field gel electrophoresis
PulseNet	National Molecular Subtyping Network for Foodborne Disease Surveillance
STEC	Shiga-toxin *E. coli*
WGS	Whole genome sequencing
USDA	United States Department of Agriculture

13.1 Introduction

Government agencies expend substantial resources to obtain and document information about foodborne illnesses in the United States and elsewhere. Foodborne illness surveillance systems have been designed to improve public health efforts by collecting, analyzing, and disseminating data on incidence of illness. The information

R. L. Scharff (✉)
Department of Human Sciences, The Ohio State University, Columbus, OH, USA
e-mail: scharff.8@osu.edu

C. Hedberg
Division of Environmental Health Sciences, University of Minnesota,
Minneapolis, MN, USA
e-mail: hedbe005@umn.edu

© Springer International Publishing AG, part of Springer Nature 2018
T. Roberts (ed.), *Food Safety Economics*, Food Microbiology and Food Safety,
https://doi.org/10.1007/978-3-319-92138-9_13

obtained through these systems helps to better characterize the burden of illness and to identify the sources of the illnesses. Though these surveillance efforts do not typically have a direct effect on incidence of illness, the information created by these efforts influences the behaviors of consumers, industry, and public health officials in ways that can lead to significant reductions in illnesses and associated costs.

In this chapter we examine the effects of foodborne illness surveillance systems on illness incidence using an economic approach. First, we describe the surveillance systems used in the United States today. Next, we examine how problems with consumer's lack of information that contribute to the occurrence of illnesses can be mitigated through the introduction of surveillance systems and examine how the information created by this surveillance aids in overcoming market failures caused by information problems. Finally, we report results of a study building on a recent paper that demonstrates the effectiveness of PulseNet.

13.2   Foodborne Illness Surveillance Systems

Food safety efforts in the United States are supported by several national and numerous state and local surveillance systems. Although the Food and Drug Administration and the USDA have jurisdiction over the interstate shipment of foods, public health surveillance of foodborne diseases is under the jurisdiction of state law and in most states is under the jurisdiction of local health departments. The Centers for Disease Control and Prevention (CDC) compiles data on a national level and coordinates multistate outbreaks and surveillance networks. The Council to Improve Foodborne Outbreak Response (CIFOR) has identified three general categories of surveillance systems that can be used to identify illnesses and detect outbreaks. These are (1) pathogen-specific surveillance, (2) complaint systems, and (3) syndromic surveillance (CIFOR 2014). Each of these systems provides information to public health officials, though there are also limitations associated with each.

Pathogen-specific systems produce information about illnesses from laboratory-confirmed agents (e.g., *Listeria monocytogenes* or *Salmonella*) or other clinically identifiable syndromes linked to specific pathogens (e.g., hemolytic-uremic syndrome) (CIFOR 2014). These systems are used to identify outbreaks from seemingly unrelated illnesses. One drawback of pathogen-specific surveillance, however, is that it takes a long time for outbreaks to be identified in this way. Identification of illnesses associated with specific *pathogens* is also used to inform surveillance systems that measure burden of illness. PulseNet, the primary federal outbreak detection system, informs the National Notifiable Diseases Surveillance System (NNDSS), a system designed to estimate burden of illness.

Complaint surveillance systems rely on complaints from the public to identify potential outbreaks (CIFOR 2014). This can take the form of an individual notifying the health department about his/her observation that a large number of people at a particular event became ill shortly afterward (e.g., after a wedding reception or church picnic) or could be based on multiple individual illness reports that are linked by the health department due to location or consumption commonalities.

Complaint systems are primarily used at the local or state level. They have the potential to very rapidly identify outbreaks that occur from contamination of foods consumed by groups at restaurants or at large events and are necessary for identifying outbreaks due to non-reportable agents. These systems account for approximately three quarters of all reported foodborne outbreaks. However, many of these investigations do not identify the ultimate source of the outbreak and may result in outbreak misclassification. They are also not sensitive to geographically dispersed, seemingly sporadic cases.

Syndromic surveillance utilizes indicators of health rather than direct reports of illnesses to focus the attention of public health agencies on areas of concern (CIFOR 2014). In theory, data mining from sources with relevant information can act as an early warning system for outbreaks. For example, an increase in persons with diarrheal illness visiting doctors or an increase in sales of antidiarrheal medicines may indicate the emergence of an outbreak. In practice, these systems are costly and have had limited practical value because they typically detect large-scale events, such as influenza outbreaks that are widespread across the community. Most foodborne outbreaks, even relatively large outbreaks, occur at a much smaller scale, though refinements of this newer approach may lead to more effective usage in the future.

13.3 Federal Government Surveillance Systems in the United States

Surveillance is conducted at each of the federal, state, and local levels. Typically, federal systems use data initially collected through state surveillance systems to estimate burden of illness and identify and respond to outbreaks. Federal systems are generally focused on identifying pathogens associated with multistate outbreaks, while state systems are designed to be responsive to both pathogen-based and complaint-based surveillance systems. A summary of the major federal systems for food foodborne illness follows.

13.3.1 Illness Surveillance Systems

Pathogen-specific surveillance systems are designed both to detect clusters of illnesses that may represent an outbreak and as a means of tracking the burden of illness over time. These systems collect data that allow public health officials to detect clusters of cases and to analyze the burden of illness data over time, by location and by the characteristics of those made ill. These surveillance systems collect data either through passive (reliant on health providers, laboratories, and local health departments to make complete and accurate reports) or active (contacting relevant entities) case ascertainment methods.

The National Notifiable Diseases Surveillance System (NNDSS) was the first major surveillance system in the United States (CDC 2015). Today, 57 state, local, and territorial health agencies report to NNDSS. The system tracks nine food-related bacterial illnesses (botulism, brucellosis, campylobacteriosis, listeriosis, salmonellosis, STEC, shigellosis, typhoid fever, and vibriosis), four parasitic illnesses (cryptosporidiosis, cyclosporiasis, giardiasis, and trichinellosis), one viral illness (hepatitis A), and one syndrome (hemolytic-uremic syndrome). Other illnesses not related to food are also tracked by this system. The National Electronic Disease Surveillance System (NEDSS) provides a link between NNDDS and state and local health departments, facilitating data transfer between the two (CDC 2017).

The Foodborne Disease Active Surveillance Network (FoodNet) uses active surveillance techniques to estimate the incidence of illness for nine common foodborne pathogens (*Campylobacter*, *Cryptosporidium*, *Cyclospora*, *Listeria*, *Salmonella*, STEC, *Shigella*, *Vibrio*, and *Yersinia*) and HUS (CDC 2016). FoodNet operates in ten states covering a geographic area that includes 15% of the US population. FoodNet reports genus- and species-specific incidence rates for many pathogens. Hospitalization and death rates are also reported. FoodNet also fields periodic population and physician surveys to assess illness and treatment trends not apparent in reported illnesses, to provide context for translating surveillance results into burden of illness estimates.

13.3.2   Outbreak Surveillance Systems

Outbreak surveillance systems aggregate outbreak data. Knowledge about the agents, foods implicated in outbreaks, and factors contributing to their occurrence help public health officials improve outbreak investigations and are also used to train food service operators to prevent future outbreaks. These systems include outbreaks investigated as a result of identification by both complaint and pathogen-specific surveillance systems.

The Foodborne Disease Outbreak Surveillance System (FDOSS) collects data on outbreaks reported to CDC through the National Outbreak Reporting System (NORS) (CDC 2016). Data on outbreaks are published using the Foodborne Outbreak Online Database (FOOD Tool), which provides information on etiology (confirmed, suspected, or unknown); number of illnesses, hospitalizations, and deaths; outbreak month and year; outbreak location (catered event, restaurant, home, etc.); what food vehicles and ingredients were implicated; and the state the outbreak occurred in.

13.3.3   DNA Profiling Systems

Surveillance systems have increasingly been designed to use DNA profiling to aid in outbreak investigations. These pathogen-specific systems identify new outbreaks by matching seemingly unrelated illnesses in clusters using the pathogens' distinct

DNA fingerprints. This also decreases the likelihood that unrelated cases will be incorrectly associated with outbreaks. Though this type of surveillance can be time-consuming, it is very sensitive, allowing for earlier detection of outbreaks that initially appear as the widespread distribution of sporadic illnesses.

The National Molecular Subtyping Network for Foodborne Disease Surveillance (PulseNet) develops genetic fingerprints from isolates submitted to state and federal public health labs using pulsed-field gel electrophoresis (PFGE) (CDC 2016). PulseNet-affiliated laboratories contribute genetic profiles of tested isolates to a CDC-maintained database of genetic profiles for *Campylobacter*, *Cronobacter*, *Listeria*, *Salmonella*, STEC, *Shigella*, and *Vibrio*. All states have participated to some degree in PulseNet, though not for all pathogens. Recently, PulseNet has adopted whole-genome sequencing (WGS) for *Listeria* leading to faster, more precise outbreak detection. Similarly, PulseNet is preparing for adoption of WGS for routine surveillance of *Salmonella*.

The National Electronic Norovirus Outbreak Network (CaliciNet) is a system that collects genetic and epidemiological data from state public health labs (currently 28 states) on *Norovirus* outbreaks (CDC 2016). It matches *Norovirus* strains and EpiData to link outbreaks to identify a common source that can then be linked to NORS.

FDA's whole-genome sequencing program uses whole-genome sequencing (WGS) to identify genetic profiles more quickly and precisely than PFGE methods (FDA 2017). GenomeTrakr, a network of 15 federal, 28 state and local, and 20 international labs, maintains an open-access database with genetic profiles that can be used to help identify outbreaks or trace contaminated foods to their source (Allard et al. 2016).

13.4   Surveillance and the Economics of Information

The role of information problems in creating food safety deficiencies has been well documented (see, e.g., Elbasha and Riggs 2003; Hobbs 2004). In a perfectly working market without information problems, consumers are able to bargain for safer foods with confidence in their assessments of risk levels associated with foods from different sources. In this perfect world, consumers will make rational choices over trade-offs between risk and price, resulting in optimal levels of safety. The world we live in is not that world. Nevertheless, surveillance can improve outcomes by providing important safety information to both the market and public health agencies (Ford et al. 2015). Specifically, surveillance can be used to create incentives for industry to produce safer foods, identify and remove contaminated food from the market, and provide policymakers with useful burden of illness estimates (Scharff et al. 2016).

A fundamental source of food safety problems is, in many cases, consumers' inability to clearly observe the risk level of a particular food, though industry risk levels may be observable (Unnevehr et al. 2010). For example, in the absence of market-created information, surveillance, or regulations, consumers may know that illnesses from *E. coli* often come from beef but will not know whether a particular brand of beef is safer than another. The safety level of the average beef producer may be attributed to all beef producers, whether they produce beef that is safer or less safe than the average producer. The ability of low-quality (risky) producers to benefit from the reputation of higher-quality (safe) rivals creates a strong incentive to reduce cost by producing riskier beef. Ultimately, in the absence of other sources of information or regulation, the bad will drive out the good, as the model predicts Akerlof (1970). Ironically, one reason that this information problem exists is that food companies do not believe that food safety should be a competitive issue. Thus, companies that invest in food safety systems do not market that information, preventing consumers from distinguishing the relative safety of their products compared to other similar products in the marketplace.

In the absence of government efforts, the market will generate some information on its own. First, individual firms may use third-party audits to confirm brand adherence to quality standards (McCluskey 2000). Industry trade groups may promote this practice through the use of certification programs to promote industry reputation (Henson and Caswell 1999). Even where such programs do not exist, individuals may (imperfectly) correlate consumption experiences with illnesses. Finally, media coverage and investigations may reveal large outbreaks (Chunara et al. 2012). Each of these mechanisms acts to signal brand or industry safety. Nevertheless, in many cases the information content of the signal is weak.

Surveillance programs can act to strengthen signal quality by helping to identify illness sources (Ford et al. 2015). When surveillance systems lead to the rapid and accurate identification of brands (producer, restaurant, and retail) associated with outbreaks, powerful incentives are created for industry actors to reduce risks from the foods they sell. Implication in an outbreak may lead to substantial costs from litigation, recall efforts, plant cleanup efforts, and loss of brand reputation (Buzby and Roberts 2009). The incentive to avoid these costs may lead endpoint retailers, such as Walmart, to implement programs that promote safety across the supply chain (Hobbs 2010).

Surveillance systems can also have ex post effects, providing information about burden of illness to policymakers tasked with mitigating food safety problems. By compiling and aggregating reports of illness, public health officials create a de facto estimate of the burden of illness within a defined geographic area. Simply reporting the apparent burden of disease provides stakeholders (government, industry, and consumers) with a means of prioritizing responses to potential problems when multiple food safety issues are apparent and resources are limited.

The potential effects of foodborne illness surveillance efforts are summarized in Fig. 13.1. Government actions (coded blue) lead to increased information in the system (yellow), which leads to incentive-based reactions by industry (purple) and consumers (green), ultimately creating more information and reducing foodborne illness (red).

Industry incentives to control foodborne illness are enhanced by surveillance, as follows. First, surveillance systems improve outbreak detection through complaint

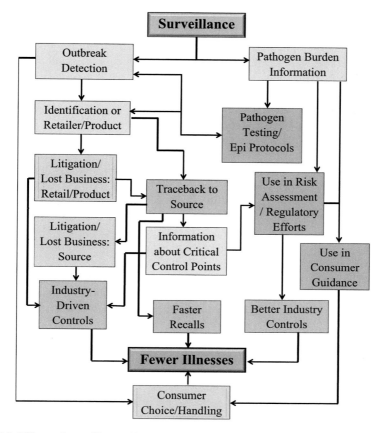

Fig. 13.1 Effects of surveillance efforts on illness outcomes

and pathogen-specific systems. Either system can trigger an outbreak investigation, which may result in the identification of a defined outbreak. Next, a successful investigation will lead to the identification of a branded retail or product source, which will lead to public awareness leading to lost business for the brand, and possibly lead to litigation. If the identified brand was the source of contamination, it (and others like it that observe the costs incurred) will be incentivized to institute new food safety controls in an effort to directly respond to consumer concerns and to avoid these costs in the future. If the brand was not the source of contamination, the implicated business will work with public health investigators to conduct a traceback investigation to identify the actual source (in an effort to transfer costs and either drop the source as a supplier or ensure that the source has taken efforts to reduce risks). This exercise not only imposes costs on the source but also provides information about processing deficiencies that helps to identify critical control points. As a result, enhanced industry controls are identified and incentivized. Costs accruing to end-product retailers will also incentivize the implementation of better traceback systems across the supply chain (e.g., thorough contracts the retailer has with suppliers), reducing recall costs and reducing the time needed to identify products that should be subject to recall. The combination of incentivized enhancement of industry controls and faster recalls leads to a reduction in illnesses.

Figure 13.1 also illustrates the role that surveillance plays in influencing government actions. Both outbreak data and illnesses reported through passive and active surveillance reporting systems add to the body of knowledge about the burden of foodborne illness for identified pathogens. This knowledge affects government efforts to combat foodborne illness in multiple ways. First, food attribution data combined with burden of illness estimates influence protocols for what foods are probed for and what pathogens are tested for in outbreak investigations. Second, burden of illness estimates are used to prioritize food safety efforts and quantify the effects of alternative potential interventions in risk assessments. These risk assessments are also improved as a result of information discovered through successful outbreak investigations that trace contamination to its source. Risk assessments and other regulatory efforts can then be used to design better regulatory controls/guidance that is implemented by industry, leading to fewer illnesses.

As noted above, consumer behavior is also affected by surveillance. Outbreak detection that accurately implicates a retail establishment or specified food may lead consumers (if they are made aware through the media) to avoid those establishments or foods. Similarly, surveillance that improves burden of illness information (coupled with information from other regulatory activities) may lead public health officials to issue consumer guidance related to food consumption or food handling, which some consumers will follow, resulting in illness reductions.

It is important to note that surveillance, by itself, does nothing to improve the safety of the foods we consume. As suggested above, the success of a surveillance system depends critically on the public health infrastructure that surrounds it. Data that is collected, but not shared or used in any other way, will have no effect on illness. Similarly, misclassified illnesses or outbreaks will also undermine the effectiveness of actions taken, in direct proportion to the degree of misclassification. For example, in a 2008 outbreak of salmonellosis caused by contaminated hot peppers, an initial recommendation to avoid eating tomatoes appeared to prevent a number of illnesses because hot peppers and tomatoes are frequently comingled in food items such as salsa (Klontz et al. 2010). Thus, people who tried to avoid tomatoes also frequently avoided the consumption of hot peppers. Finally, good data used in bad analyses will not produce good outcomes. Ultimately, the success of the system is dependent on a combination of the quality of data collected and the way the data is used.

13.5   A Case Study: Assessing the Effectiveness of PulseNet

The PulseNet surveillance system is a network of federal, state, and local public health labs that perform molecular subtyping of foodborne bacteria by pulsed-field gel electrophoresis (PFGE) and upload results into a national database. This allows for earlier and better detection of outbreaks from seemingly sporadic illnesses. In this section, we build on the results of a recent evaluation of PulseNet effectiveness by Scharff et al. (2016).

Following Scharff et al. (2016), PulseNet is evaluated using two basic approaches. A recall model illustrates the effect of earlier outbreak detection on illness due to a recall of a larger portion of contaminated product. A process change model demonstrates how better outbreak detection leads to illness reduction through promotion of industry accountability and the generation of information useful to both industry and public health agencies. The economic benefits from illness reduction are then assessed using a conservative economic model and compared to program costs.

13.5.1 The Recall Model

Results from Scharff et al. (2016) are used to illustrate the effect of PulseNet on illness due to faster recalls. Outbreaks due to *E. coli* O157:H7 and *Salmonella* were examined for the years 2007–2008. Uncertainty in model parameters was assessed through Monte Carlo analysis using @Risk 5.7.1.

The Scharff et al., *E. coli* O157:H7 model used USDA Food Safety Inspection Service data from 15 outbreaks that led to recalls of ground beef. For these outbreaks, recalls led to 0% to 66% of the contaminated meat being recovered leading to 0 to 49 illnesses averted per outbreak. Across all outbreaks, the total number of illnesses averted was 108 (90% CI, 95–266). Adjusted for underdiagnosis, a total of 2819 (90% CI, 2480–6943) illnesses were estimated to be averted as a result of PulseNet's early detection of these outbreaks.

Outbreaks due to *Salmonella* involved recalls led by the FDA, which does not report contaminated product recovery rates. As a result, an alternative model calculated averted *Salmonella* illnesses as the difference between the expected number of illnesses (in the absence of an earlier recall) and the actual number of illnesses that occurred (following a recall conducted due to PulseNet identification of the outbreak). The Scharff study estimated that averted illnesses from five major multistate outbreaks ranged from 0 to 345, totaling 580 (90% CI, 128–1127). Adjusted for underdiagnosis, the expected number of prevented illnesses due to recalls associated with PulseNet identified *Salmonella* outbreaks was 16,994 (90% CI, 3750–33,021).

13.5.2 The Process Change Model

The process change model complements the measurement of direct effects in the recall model by measuring the indirect effects from PulseNet. Specifically, the information created by PulseNet has both short-term and long-term effects on consumers, industry, and government decision-makers. The discovery of more outbreaks that are more accurately tied to specific products and brands leads to more litigation against companies by injured consumers and, when publicized by the media, leads other consumers to respond by reducing purchases of the products and brands implicated. The reputation costs for firms implicated in an outbreak often persist for years. Consequently, by increasing the probability a food firm will be implicated in a costly outbreak, PulseNet incentivizes firms to adopt and improve food safety processes that lead to reduced illnesses. Also, the information created by more accurate outbreak investigations improves the ability of both industry and government decision-makers to craft targeted cost-effective controls to reduce food-borne illness.

The adoption of PulseNet by states to different degrees in different years creates ideal conditions for evaluating the system as a natural experiment. The basic model posited by Scharff et al. (2016), illustrated in Eq. (13.1), estimates reported illnesses from pathogen (p) in a given state (s) at given time (t) as a function of whether PulseNet has been implemented in the state (PulseNet$_{s,t}$), the number of isolate test run (Tests$_{s,t}$), the lagged number of isolate test run (Tests$_{s,t-1}$), and a set of controls ($X_{p,s,t}$):

$$\text{Illnesses}_{p,s,t} = f\left(\text{PulseNet}_{p,s,t},,,,,,,,,,,,,\text{Tests}_{p,s,t},,,,,,,,,,,,,\text{Tests}_{p,s,t-1},,,,,,,,,,,,,X_{p,s,t}\right)$$
$$(13.1)$$

Industry response to existence of PulseNet is expected to negatively affect illnesses due to general expectations that a better food safety system will increase each firm's probability of being implicated in an outbreak. Intensity of PulseNet use has offsetting effects. First, as outbreaks are discovered and publicized, more ill people will come forward to report their illnesses (measured by Tests$_{p,s,t}$). Second, the contribution of the outbreak to expected firm costs and to the knowledge base is expected to lead to longer-term changes in industry processes and fewer reported illnesses (measured by Tests$_{p,s,t-1}$). Multiple empirical specifications were used to test this model.

The influence of spillover effects from testing other pathogens ($-p$) was also examined (Tests$_{-p,s,t-1}$) by Scharf et al.

$$\text{Illnesses}_{p,s,t} = f\left(\text{PulseNet}_{p,s,t},,,,\text{Tests}_{p,s,t},,,,\text{Tests}_{p,s,t-1},,,,\text{Tests}_{-p,s,t-1},,,,X_{p,s,t}\right)$$
$$(13.2)$$

It is expected that controls implemented in response to a perceived increase in outbreak detection likelihood for one pathogen will have some beneficial effect on other pathogens, as well.

Panel data covering the years 1994–2009 was used in the analysis. Illnesses are measured using NNDSS state-level annual estimates for *Listeria*, *Salmonella*, and *E. coli* O157:H7. PulseNet adoption and testing intensity are measured based on annual uploads of isolate test results to the CDC PulseNet database for each pathogen. Controls are included for state population, year, state per capita income, and eight census division dummy variables.

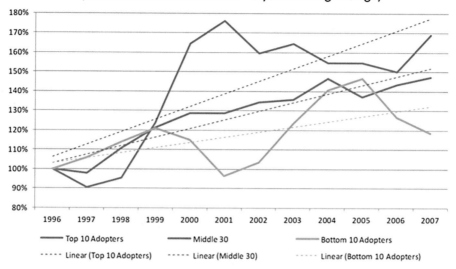

Fig. 13.2 *Salmonella* outbreaks for PulseNet adopters

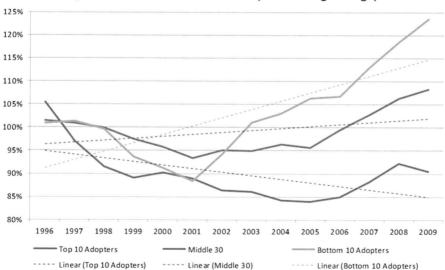

Fig. 13.3 Reported *Salmonella* illnesses for PulseNet adopters. Figure taken from Scharff et al. (2016) (Figure A1)

Table 13.1 Illnesses averted due to PulseNet[a]

	Base models			Spillover effects models		
	(1)	(2)	(3)	(1)	(2)	(3)
Salmonella						
Illnesses	**15,784[b]**	**19,758[b]**	**9096[b]**	**21,249[b]**	**25,181[b]**	**11,291[b]**
90% CI	11,948– 19,623	15,871– 23,662	8504– 9686	16,863– 25,632	20,747– 29,595	5628– 16,948
Adjusted	*462,487*	*578,947*	*266,522*	*622,614*	*737,845*	*330,840*
E. coli O157						
Illnesses	310	**489[c]**	**364[b]**	**2673[b]**	**1597[b]**	**670[b]**
90% CI	−95–717	**48–939**	**274–453**	1718–3627	609–2589	451–889
Adjusted	*8096*	*12,750*	*9489*	*69,755*	*41,684*	*17,475*
Listeria						
Illnesses	**113[b]**	**113[c]**	27	**151[b]**	75	73
90% CI	**39–187**	**31–195**	−38–92	46–256	−46–196	−26–172
Adjusted	*238*	*237*	*52*	*316*	*157*	*153*

[a]Adapted from Table 4 in Scharff et al. (2016)
Note: Boldface indicates statistical significance, [b]significant at 1% and [c]significant at 5%

Figures 13.2 and 13.3 illustrate the relationship between testing intensity and illness. Initially, as Fig. 13.2 demonstrated, the top ten PulseNet adopters (measured by per capita total number of isolates reported) experienced an increase in outbreaks relative to the bottom ten adopters, though the effect eventually dissipated. This is consistent with increased testing revealing more outbreaks, countered by later overall relative declines in illnesses. In fact, as Fig. 13.3 shows, these top adopters eventually experienced significant declines in illness relative to bottom adopters.

A summary of the Scharff et al., empirical results from estimation of the models described in Eqs. (13.1 and 13.2) are presented in Table 13.1. For each set of conceptual models, three empirical models are estimated. The random effects model (1) is a simple OLS model that assumes that each observation is independent to all other observations. The fixed effects model (2) recognizes that there are stable state and year characteristics. The Poisson model (3) is a maximum likelihood model that is preferred for use with count data (such as the number of illnesses). Though estimates from all models are presented as a robustness check, Poisson model estimates are most likely to yield unbiased results. Estimates for the number of illnesses avert due to PulseNet are presented in Table 13.1 based on identified empirical models. More complete empirical results are available in Scharff et al. (2016).

The "base model" (Eq. 13.1) predicts that state involvement in PulseNet leads to large significant reductions of reported illnesses from *Salmonella* and *E. coli* O157:H7. For the Poisson model, 9096 (90% CI, 8504–9686) reported *Salmonella* illnesses, and 364 reported *E. coli* illnesses are averted due to PulseNet. Reported illnesses from *Listeria* are insignificant. Together, these estimates suggest a sizable impact on illness due to the information provided by PulseNet. Adjusted for underreporting and underdiagnosis, foodborne illnesses are reduced by over 276,000.

Table 13.2 Economic benefits[a] from PulseNet (millions of 2017 US $)[b]

	Base models[c]			Spillover effects models[c]		
	CDC	Basic	Enhanced	CDC	Basic	Enhanced
	Process change model benefits					
Salmonella						
Mean	573**	1391**	4061**	711**	1726**	5041**
90% CI	(428–802)	(507–3226)	(937–9565)	(337–1201)	(474–4381)	(984–12,912)
E. coli O157						
Mean	25**	123**	181**	46**	227**	335**
90% CI	(14–44)	(41,316)	(51–440)	(25–84)	(74–579)	(90–813)
	Recall model benefits					
Recall model						
Mean	44**	125**	312**	N/A	N/A	N/A
90% CI	(16–85)	(39–315)	(60–837)	N/A	N/A	N/A

[a]Uses illness reduction estimates from the Poisson model
[b]Based on results from Scharff et al. (2015, 2016), updated as described in Chap. 4
[c]Adjusted for underdiagnosis and underreporting
**significant at 1%; *significant at 5%

The spillover effects model (Eq. 13.2), which allows incentivized control efforts to affect both the pathogen efforts which are targeted toward and other pathogens, increases estimated illness further. Almost 12,000 reported *Salmonella* and *E. coli* illnesses are averted under this model. Adjusting for underreporting and underdiagnosis raises this figure to over 348,000 illnesses averted.

13.5.3   *Economic Benefits and Costs*

To determine economic benefits from PulseNet, Scharff et al. (2016) use a conservative cost of illness model favored by CDC economists. This model includes illness cost values for medical treatment, lost work, and death for both acute illnesses and resulting conditions (sequelae). It is similar to the basic model estimated in Chap. 8, with one exception. Death costs are assessed using age-varying productivity loss estimates. This is one of the more conservative models used by economists, suggesting that true costs are likely higher. Here, for purposes of comparison, we examine benefits from PulseNet using three economic models including (1) the CDC (conservative) model described above; (2) the basic model, which includes a death cost estimate based on preferences toward risk of death; and (3) the enhanced model, which includes both the death estimates from the basic model and monetized lost quality of life losses (described in Chap. 8). Under all methods, PulseNet activities result in sizable economic benefits.

Economic benefits from averted illnesses attributable to PulseNet are summarized in Table 13.2. Process change model benefits are assessed separately for illnesses from *Salmonella* and *E. coli*, while benefits from the recall model are combined. All costs are based on illness estimates that were adjusted for underreporting and underdiagnosis. Benefit estimates for averted *Salmonella* illnesses range from $573 million (90% CI, $428–$802 million) for reported illnesses averted under the base model when the model only examines reported illness to $5.0 billion (90% CI, $1.0–$12.9 billion) for the spillover model with adjusted illnesses. Similarly, benefits attributable to averted *E. coli* illnesses range from $25 million (90% CI, $14–$44 million) to $335 million (90% CI, $90–$813 million). *Listeria* benefits are insignificant and are not reported. Finally, recall model benefits, which include averted illnesses from both *Salmonella* and *E. coli* outbreaks, range from $44 million to $312 million for adjusted illness loss estimates.

For the pathogens examined, the inclusion of more complete death and quality of life estimates has an effect on estimated program benefits larger than would be seen for other pathogens. This occurs for different reasons. The *E. coli* O157:H7 estimates are largely boosted by relatively high death rates for both the acute illness and hemolytic-uremic syndrome. As a result, the inclusion of estimates that include lost welfare (not just lost productivity) has a particularly large effect on these illnesses. For *Salmonella*, the addition of values for lost quality of life has a larger effect. This is driven by the sizable number of persons who are afflicted with reactive arthritis following acute salmonellosis. Though monetary costs are relatively small for arthritis, losses of utility are substantial. In any event, even the most conservative economic estimates demonstrate benefits that exceed measured program costs ($7.2 million).

It is difficult to determine costs associated with PulseNet. While the annualized cost of setting up labs and conducting tests is estimated to be a modest $7.2 million, there are likely to be other unquantified costs. Potentially costly activities that are not quantified include support activities for outbreak investigations that otherwise would not have occurred, industry costs associated with process changes, and costs to regulatory agencies for efforts to interpret and utilize the data to improve food safety efforts. In any event, estimated benefits are large, suggesting that benefits likely justify their costs.

13.6   Conclusion

Surveillance systems are an important part of the food safety establishment. Both state and local complaint systems and state and national pathogen-specific systems play an important role in providing important information to the market and public health agencies. Information provided by surveillance systems improve outbreak detection capabilities and enable industry and government policy decisions that improve social welfare. Our evaluation of PulseNet illustrates the large economic benefits provided by one surveillance program. Furthermore, the large discrepancy in benefits estimated in this model across multiple economic models demonstrates the importance of using estimates including all relevant costs from foodborne illnesses.

References

Akerlof GA. The market for "lemons": quality uncertainty and the market mechanism. Q J Econ. 1970;84:488–500.

Allard MW, Strain E, Melka D, Bunning K, Musser SM, Brown EW, Timme R. Practical value of food pathogen traceability through building a whole-genome sequencing network and database. J Clin Microbiol. 2016;54(8):1975–83.

Buzby JC, Roberts T. The economics of enteric infections: human foodborne disease costs. Gastroenterology. 2009;136(6):1851–62.

Centers for Disease Control and Prevention (CDC). MMWR: summary of notifiable diseases, Atlanta. 2015. https://www.cdc.gov/mmwr/mmwr_nd/index.html. Accessed 7 Jan 2017.

Centers for Disease Control and Prevention (CDC). Surveillance and data systems, Atlanta. 2016. https://www.cdc.gov/ncezid/dfwed/keyprograms/surveillance.html. Accessed 7 Jan 2017.

Centers for Disease Control and Prevention (CDC). Integrated surveillance information systems/NEDSS, Atlanta. 2017. https://wwwn.cdc.gov/nndss/nedss.html. Accessed 8 May 2017.

Chunara R, Andrews JR, Brownstein JS. Social and news media enable estimation of epidemiological patterns early in the 2010 Haitian cholera outbreak. Am J Trop Med Hyg. 2012;86(1):39–45.

Council to Improve Foodborne Outbreak Response (CIFOR). Guidelines for foodborne disease outbreak response. 2nd ed. Atlanta: Council of State and Territorial Epidemiologists; 2014.

Elbasha EH, Riggs TL. The effects of information on producer and consumer incentives to undertake food safety efforts: a theoretical model and policy implications. Agribusiness. 2003;19(1):29–42.

Food and Drug Administration (FDA). Whole genome sequencing (WGS) program. 2017, Silver Spring. https://www.fda.gov/Food/FoodScienceResearch/WholeGenomeSequencingProgramWGS/. Accessed 13 May 2017.

Ford L, Miller M, Cawthorne A, Fearnley E, Kirk M. Approaches to the surveillance of foodborne disease: a review of the evidence. Foodborne Pathog Dis. 2015;12(12):927–36.

Henson S, Caswell J. Food safety regulation: an overview of contemporary issues. Food Policy. 1999;24(6):589–603.

Hobbs JE. Information asymmetry and the role of traceability systems. Agribusiness. 2004;20(4):397–415.

Hobbs JE. Public and private standards for food safety and quality: international trade implications. Estey Centre J Int Law Trade Policy. 2010;11(1):136.

Klontz KC, Klontz JC, Mody RK, Hoekstra RM. Analysis of Tomato and Jalapeño and Serrano pepper imports into the United States from Mexico before and during a national outbreak of Salmonella serotype Saintpaul infections in 2008. J Food Prot. 2010;73(11):1967–74.

McCluskey J. A game theoretic approach to organic foods: an analysis of asymmetric information and policy. J Agric Resour Econ. 2000;29(1):1–9.

Scharff RL. State estimates for the annual cost of foodborne illness. J Food Prot. 2015;78(6):1064–71.

Scharff RL, Besser J, Sharp DJ, Jones TF, Peter GS, Hedberg CW. An economic evaluation of PulseNet: a network for foodborne disease surveillance. Am J Prev Med. 2016;50(5):S66–73.

Unnevehr L, Eales J, Jensen H, Lusk J, McCluskey J, Kinsey J. Food and consumer economics. Am J Agric Econ. 2010;92(2):506–21.

Chapter 14
Economic Rationale for US Involvement in Public-Private Partnerships in International Food Safety Capacity Building

Clare Narrod, Xiaoya Dou, Cara Wychgram, and Mark Miller

Abbreviations

APEC	Asia-Pacific Economic Cooperation
ARS	Agricultural Research Service
BPCS	Better Process Control School
CDC	Centers for Disease Control and Prevention
CSREES	Cooperative State Research, Education, and Extension Service
FAS	Foreign Agricultural Service
FDA	Food and Drug Administration
FSIS	Food Safety and Inspection Service
FSMA	Food Safety Modernization Act
FSPCA	Food Safety Preventive Controls Alliance
FSVP	Foreign Supplier Verification Program
GAP	Good agricultural practices
GFSI	Global Food Safety Initiative
GFSP	Global Food Safety Partnership
GMA	Grocery Manufacturers Association
GMP	Good manufacturing practices
HACCP	Hazard Analysis and Critical Control Points
HVA	High-value agricultural
IEC	International Electrotechnical Commission
IFT	Institute of Food Technologists
IICA	Inter-American Institute for Cooperation on Agriculture
IIT IFSH	Illinois Institute of Technology's Institute for Food Safety and Health
ILSI	International Life Sciences Institute
ISO	International Standard Organization
JIFSAN	Joint Institute for Food Safety and Applied Nutrition
M and E	Monitoring and evaluation

C. Narrod (✉) · X. Dou · C. Wychgram · M. Miller
Joint Institute for Food Safety and Applied Nutrition, University of Maryland, College Park, MD, USA
e-mail: cnarrod@umd.edu

© Springer International Publishing AG, part of Springer Nature 2018
T. Roberts (ed.), *Food Safety Economics*, Food Microbiology and Food Safety,
https://doi.org/10.1007/978-3-319-92138-9_14

MIS	Marketing information services
NASA	National Aeronautics and Space Administration
NASDA	National Association of State Departments of Agriculture
NOAA	National Oceanic and Atmospheric Administration
OASIS	Operational and Administrative System for Import Support
OECD	Organisation for Economic Co-operation and Development
PPP	Public-private partnership
SCM	Supply chain management
SPS	Sanitary and phytosanitary measures
SSA	Sprout Safety Alliance
UNIDO	United Nations Industrial Development Organization
USAID	US Agency for International Development
USDA	United States Department of Agriculture
WHO	World Health Organization

14.1 Introduction

Global agricultural trade has increased substantially during the past three decades, especially trade in high-value agricultural (HVA) products such as horticultural produce, dairy, fish, and meat products. Mike Taylor, the Deputy Commissioner of the Food and Drug Administration (FDA), reports in 2013 that "15 percent of U.S. food supply is imported, including 50% of fresh fruit, 20% of fresh vegetables, and 80% of seafood" (FDA 2013). In 2014 the United States imported a total value of $111 billion in agricultural food, which is nearly three times the 1990 value of $39 billion. Though Canada and Mexico remain the largest exporters to the United States in terms of value, the United States is increasingly sourcing from Asia, especially China, India, Indonesia, Vietnam, and Thailand. There are two reasons for this: firstly, there is year-round demand for seasonal foods (foods consumed as close to harvest as possible), which are usually in the HVA category and not domestically available. Secondly, there is a greater supply capacity, thanks to innovations in transportation and communication technology, enabling retailers to satisfy this growing demand through global sourcing (Fagotto 2010).

The World Health Organization (WHO) initiative to estimate the global burden of foodborne diseases (see Chap. 7) looked at 31 global foodborne hazards and estimated that they were responsible for 600 million (95% uncertainty interval [UI] 420–960) foodborne illnesses and 420,000 (95% UI 310,000–600,000) deaths in 2010 (World Health Organization 2016). With the increasing number of imported HVA foods being consumed in the United States, it is inevitable that some of these illnesses will be caused by imported food. The US Center for Disease Control and Prevention (CDC) estimates that each year foodborne diseases lead to roughly 48 million illnesses, 128,000 hospitalizations, and 3000 deaths in the United States (Scallan et al. 2011). To reduce the burden of foodborne illness, many countries, including the United States, are moving to strengthen their food safety systems by shifting the focus from responding to contamination to preventing it.

Food safety capacity building is a measure of preventive control. FDA has historically worked with the US land-grant system to roll out food safety training material to the states and territories. FDA has also worked with the Joint Institute for Food Safety and Applied Nutrition (JIFSAN) to adapt that material to an international audience and roll it out. The Food Safety Modernization Act (FSMA), passed in 2011, formally shifted the focus from reaction to prevention and placed more responsibility on the private sector for preventing hazards from occurring. This move was in recognition that the private sector is in a better position to ensure that preventive control measures are in place by working with its suppliers.

Historically, regulatory tools, such as regulations and laws, have been used by governments to improve social welfare. These regulatory tools are used to correct market failures through requiring or incentivizing the private sector to change their behaviors. These actions may be costly to the private sector but are considered necessary by the public sector to ensure the safety of food the private sector is supplying to consumers. Capacity building, on the other hand, is a nonregulatory tool that the FDA has made available to help strengthen its efforts in preventing food safety problems in both domestically produced and imported food. Instead of telling the private sector what they should do, capacity building improves the private sector's ability take the required actions or achieve desirable outcomes. Prior to FSMA, FDA has been involved in capacity building abroad surrounding several voluntary measures such as good agricultural practices (GAP) and good aquaculture practices (GAqP). Through FSMA, FDA is required to develop an international capacity-building plan that addresses a wider range of stakeholders.

In 2011, the FSMA required FDA to promote food safety capacity building internationally and implement a complementary monitoring and evaluation plan. This plan allows for cost-benefit analyses and helps to make sure the benefits of the capacity-building efforts outweigh their costs. Data are essential to monitoring and evaluation. The public sector collects some data on imports and rejections, but these data are not sufficient. Both import and rejection data and production and compliance data are needed to measure the impact of food safety capacity building. Since the private sector collects production and compliance data to monitor their suppliers and operations, it is difficult to measure the impact of food safety capacity building without involving the private sector. The private sector may be reluctant to share data with a regulatory agency due to negative repercussions such as loosing proprietary data or facing a possible regulatory sanction. A plausible way forward would be to develop a voluntary data process that focuses on whether food producers are delivering safe food to consumers. The mechanism for measuring impact and data sharing still needs to be worked out. If the public and private sector were to work in a complementary manner, they would be in a better position to inform policy involving such efforts.

This chapter is organized as follows. Firstly, it provides some background on the various actors involved in international food safety capacity building. Secondly, it explains the economic rationale for the public sector to invest in food safety capacity building and to form partnerships with the private sector. Thirdly, it discusses capacity-building efforts that involve the public and private sectors and some public-private partnerships (PPPs) that have already emerged. Fourthly, it discusses the importance of monitoring and evaluating the impact of these efforts so that adjustments can be made if goals are not being achieved. Lastly, it discusses the importance of PPPs not only in food safety capacity-building trainings but also in the monitoring and evaluation efforts associated with these trainings.

14.2 Background on Public and Private Sector Actors Involved in International Food Safety Capacity Building

Various organizations, agencies, and industries form PPPs to support international capacity building in developing countries (Fig. 14.1). These parties are driven by mainly three types of interests: aid interests, trade interests, and food safety interests. International organizations (e.g., WTO, WHO, FAO, World Bank, IICA) and some government agencies (e.g., USAID, USDA/FAS) are driven by aid interests. Aid-driven agencies focus on agricultural capacity building in developing countries to increase agricultural output and food security as well as raise awareness of food safety and nutrition (Testimony on Food Aid and Capacity Building Programs 2015). Some government agencies emphasize the importance of technical assistance to developed countries and endeavor to remove inspection and testing technology barriers to trade. For example, the US Department of Agriculture and Foreign Agricultural Services are interested in building international trade capacity as a means to facilitate US agricultural export and to make sure US producers do not face trade obstacles due to poor testing facilities in the global market (USDA/FAS 2015). The public and private sectors in developed countries also choose to invest in international capacity building to further domestic food safety interests and for the private sector to ensure it is providing consumers globally a safe product. We will focus on the economic rationale supporting the behavior of these stakeholders and use the food market in the United States as an example to illustrate why both the public and private sectors in developed countries are needed in international food safety capacity building.

Fig. 14.1 Public-private partnership and international capacity building

14.2.1   Food Safety Interest and Public Sector Intervention

Figure 14.2 is a highly stylized model of how different players interact with each other in the import and domestic food markets. Domestic importers import food products from international suppliers and sell them to domestic consumers. To ensure the safety of their products, some importers choose to adopt third-party private standards, which help to monitor and ensure food safety practices among suppliers. The safety of imported food is also of interest to the domestic government, which relies on regulatory and nonregulatory tools to improve domestic food safety. Some international suppliers encounter technical difficulty in fulfilling the requirements by governments and importers in developed countries. In addition, some developing countries lack the regulatory capacity to manage their food supply chains. In both cases, international capacity building is an effective nonregulatory tool to ensure the safety of imported food.

Because of the nature of food consumption and structures of the food market, the private sector alone cannot achieve the socially optimal level of food safety and quantity of supply, and public sector involvement is required to correct market failures. In this section, we discuss three such market failures. Firstly, food consumption is food safety consumption in nature, which is considered a public good (Holmes et al. 2006; Roberts 2013; Unnevehr 2007). Food safety has public health benefits that cannot be captured by food prices in the free market. Foodborne illnesses, especially those caused by unsafe practices of suppliers and importers, can

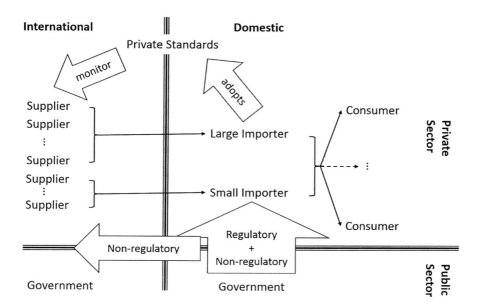

Fig. 14.2 Public and private sector players in food import market

affect large groups of consumers, lead to loss in both output and social welfare, put pressure on the public health system, and sometimes result in loss of lives. In addition, negligence by suppliers or importers can cause significant disruptions in food supply, undermine public confidence, and further reduce social welfare (FDA 2013).

Secondly, the food market suffers from imperfect information and, as a result, underinvestments in food safety. Information problems exist on every link along the food supply chain. For example, consumers are often processors of their own food. Without accurate information on the safety of the food they purchased, they may not take the required actions during preparation to reduce the risk of foodborne illnesses. But it is economically infeasible for the suppliers and importers to sufficiently raise the safety standards of their products because they lack information on consumer behavior (Elbasha and Riggs 2003). In addition, foodborne illnesses are often not recognized or diagnosed, since consumers do not always attribute episodes of illness to their food or seek medical attention. This is especially true when illnesses are caused by chronic exposure. This lack of recognition on the part of consumers leads to their undervaluation of food safety. Even when a food pathogen is identified, it is difficult to traceback to its point of origin due to the technology constraints and limited epidemiological data, especially when products from small-scale suppliers are comingled and sold collectively. The lack of firm-level traceability, then, entails another problem of collective reputation and underinvestment by suppliers in food safety practices (Winfree and McCluskey 2005).

Lastly but not the least, the private sector has limited ability to correct market failures, because each player acts to serve their own interests (Fagotto 2014). In the absence of public sector regulation, private governance did emerge to fill the regulatory gap (Fagotto 2014; Fulponi 2006; Lin 2014). Large importers such as Walmart, Costco, and McDonalds required their suppliers to be certified by private standards. However, the adoption of private standards is insufficient to guarantee a socially optimal level of safety and quantity of supply in the food market. Private standards are voluntary. Smaller importers and their suppliers may not be able to afford to adopt these standards. Moreover, the private standards historically have not been examined or recognized by government agencies to verify that they are sufficient to protect the health of consumers (in the United States, this may change for some of the private certification bodies under the accreditation of third-party certification rule under FSMA). From an efficiency point of view, suppliers certified by private standards are able to differentiate their product from the rest of the industry, implying a less competitive import market (McQuade et al. 2016).

14.2.2 Public Intervention in the Form of International Capacity Building

The market failures discussed above call for actions from the public sector to increase domestic food safety. Traditionally, for imported food, this goal is achieved by inspecting food products at the port of entry, rejecting any unsafe products.

However, this method is insufficient. There is a growing need for new policy tools, given the increasing amount of food being imported. There are three main reasons. Firstly, public resources are limited, while sampling and inspecting are costly. In order to be confident about food import safety, the inspection sample size needs to be large. However, in the United States, the FDA inspects less than 3% of FDA-regulated imports (FDA 2011). Secondly, reaction to foodborne illnesses is insufficient to protect public health and social welfare. Foodborne illness outbreaks are costly to society as they may spread quickly and reduce both public health and confidence in the domestic food system. What is worse is that it is often difficult to detect such outbreaks at their start because many foodborne illnesses are often not recognized or diagnosed. Thirdly, the global food supply chain, made possible by innovations in communication and transportation technologies, is increasingly complex (FDA 2011). The intricate supply chain makes it even harder to trace pathogens to suppliers and hold them accountable. The lack of firm-level traceability implies that it is impossible to deter unsafe suppliers by punishing them. The FDA recognized that it needed to reach beyond US borders and help to ensure the safety of products before they are imported (FDA 2011) and prevent outbreaks of foodborne illnesses arising from imported products.

An important tool of prevention is international food safety capacity building, facilitating the ability of exporting countries to ensure that the food they produce for international markets is safe. Suppliers from developing countries often have difficulty meeting food safety requirements, and developing country governments sometimes lack the capacity to enforce these requirements. For instance, many lack regulatory frameworks to correct market failures, the laboratory infrastructure to identify risks, human capital to conduct risk analysis, and resources to educate and monitor the stakeholders along the food supply chain. Developed countries, with more experience, knowledge, and capacity in food safety, can support international food safety capacity building and secure a sufficient and safe supply of seasonal food domestically, which is mutually beneficial to importers and exporters.

All countries have the right to ensure that the food their consumers eat is safe and to prevent the spread of pests or diseases among animals and plants. Under the WTO Sanitary and Phytosanitary Agreement, countries are allowed to put in restrictions if they are supported by an objective risk assessment that is supported by accurate scientific data. Though many countries do use risk assessment in their regulatory process of reducing the risk of specific diseases, the SPS Agreement also encourages a wider use of risk assessment among all WTO members. Not all countries currently have the human capital to conduct risk assessment, thus the need for capacity building in risk analysis as articulated in the SPS Agreement. Countries under Article 9 of the WTO Sanitary and Phytosanitary Agreement have agreed to facilitate the provision of technical assistance to other members, especially developing country members, either bilaterally or through the appropriate international organization. As countries like the United States and the EU are increasingly reliant on imported HVA from developing countries, they have been providing capacity building to help improve the safety of their imported food.

In 2011, the US Congress passed the Food Safety Modernization Act (FSMA), mandating FDA's participation in international food safety capacity building and enhancing FDA's ability to engage in the global food market. International food safety capacity building is necessary to achieve the goals of FSMA. For example, after FSMA, all suppliers (except for very small farms) need to meet regulatory requirements by the United States, and importers are explicitly responsible for verifying that their suppliers comply (FDA 2013). This policy change requires supplier capacity building on implementing food safety practices and foreign government capacity building on regulating and training their suppliers. In addition, the FDA is tasked to assist the private sector players through the transition brought about by FSMA, which involves helping to develop guidance and training materials on the new requirements.

The scope of FDA's involvement in international food safety capacity building has also broadened over time. In the FDA Global Engagement report (FDA 2011), the FDA summarizes its past efforts in international capacity building as strengthening regulatory capacity building through information provision, training, and exchange programs. The FSMA Section 305 (FDA 2011) charges the FDA to "develop a comprehensive plan to expand the technical, scientific, and regulatory capacity of foreign governments and their respective food industries." In response, FDA's International Food Safety Capacity-Building Plan (FDA 2013) includes enhancing technical assistance as one of the main goals and plans to adapt training materials to "different players along the food supply chain."

A common interest between multiple US government agencies (e.g., the FDA, FSIS, and USDA/FAS) in international capacity building is to promote the use of recognized laboratory methods and testing and detection techniques. Agencies within the US government publish their recommended methods for testing for different food hazards (FDA 2016). There are also ISO documents that can be purchased on different methods. Countries and private suppliers need to test to the requirements of their buyers which may differ depending on which country a supplier is exporting to. If an exporter uses a method to validate the safety of the agricultural product that is different than the one the US government recommends, the exporter needs to show their method is equivalent to the recommended one. Thus, both FDA and USDA/FAS mention in various reports the need for technical assistance and working with local regulatory bodies and industries to develop multilaterally recognized requirements, standards, and methods.

14.2.3 Public-Private Partnership in International Capacity Building

As both the private sector and the public sector have similar interests in food safety capacity building, a PPP surrounding international food safety capacity building makes sense. Rich and Narrod (2010) lay out key processes linking farmers to markets for which PPPs in supply chain management may be optimal given the existence of market failures (see Table 14.1). They also suggest that PPPs have

Table 14.1 Institutional roles in the supply chain management of high-value agriculture: support processes

Supply chain support processes	Traditional institutional role		Needed roles for SCM	Market failures	Possible entry point for PPPs and NGOs
	Public sector	Private sector			
Extension services	Technical assistance to producers in farming practices	Provision of services to farmers and firms linked to private company	Knowledge of specialized techniques for high-value products	Variable smallholder access to public or private extension; limited public knowledge of new techniques; underfunding of services	Creation of partnerships to leverage public and private delivery of specific types of extension services (training, field schools, vaccinations, etc.)
Infrastructure development	Public infrastructure (roads, ports, storage facilities); public distribution of commodities	Private infrastructure (processing, storage); logistics and information services	Manage flows between chain links quickly and efficiently to meet rigid deadlines by buyers; reduce distribution costs to remain competitive with other supply chains	High transportation costs, low access to smallholder areas, poor infrastructure, erratic information flows, crowding out by public sector	Partnerships between public sector and producer groups/NGOs to jointly finance and maintain roads, storage facilities, etc.
Information services	Provision of public statistics on prices, production, etc.; provision of information on varieties through extension	The use of private marketing information services (MIS) and electronic data interchange (EDI)	Integrate information flows across supply chain actors	Imperfect information by smallholders on needs of buyers and customers in HVA	Development of MIS to integrate government statistics agencies with private producer associations, the use of IT to distribute market information

(continued)

Table 14.1 (continued)

Supply chain support processes	Traditional institutional role		Needed roles for SCM	Market failures	Possible entry point for PPPs and NGOs
	Public sector	Private sector			
Certification, grades, and standards	Public certification of seeds and varieties; development and enforcement of public standards and regulations; food safety inspection and monitoring	Private certification of seeds and varieties, development and enforcement of private standards; enforcement of ISO standards	Consistent, credible application of rigid standards on food safety and quality specifications to meet buyer and customer demands	Smallholders' ability to meet public or private standards limited; divergence between public and private standards; low capacity to enforce public standards	Creation of third-party certification agencies that manage quality and food safety in conjunction with government and producer groups
Coordination mechanisms	Creation and enforcement of regulations to ensure competition and market exchanges; mandatory cooperatives (centrally planned economies)	Development of contracts, alliances, and marketing agreements with suppliers	Mechanisms must ensure consistent delivery of high-quality products	Limited enforcement of contracts; divergence in market power between chain actors	Third-party PPP to underwrite and monitor contracts; development and promotion of producer associations to improve enforcement

Source: Rich and Narrod (2010)

advantages over pure public sector intervention in the free market in that, given the public sector's emphasis on social welfare and the private sector's control over its suppliers, PPPs bring forth the best aspects of both sectors (Rich and Narrod 2010).

The public sector's advantage is in its extensive connections with foreign government counterparts and authority to negotiate with other countries, human resources in the agencies and USDA land-grant universities, knowledge on food safety regulatory framework, information on US food safety policy changes and the development of various laboratory methods, and information on agricultural development in foreign countries. It is in the best position to develop guidance and training materials, deploy experts as trainers, and reach out to a wide range of stakeholders. It is also in the best position to identify priorities in international capacity building.

However, public sector involvement alone is not sufficient. Firstly, public sector resources are limited. The budget constraint affects the public sector's capacity in three ways: the number of trainings supported, the number of inspections conducted overseas, and the ability to connect with international suppliers. Secondly, traditionally, extension efforts and research tended to focus less on HVA and more on low value stable products (Rich and Narrod 2010). In addition, though the public sector has historically provided technical assistance in farming practices to producers, public sector extension services have often been criticized as being unresponsive to the diverse needs of farmers. Contracting with private extension service providers could increase responsiveness (Anderson et al. 2007).

The private sector complements the public sector with its existing experience, resources, and infrastructure on monitoring international suppliers. The FSMA, by shifting the responsibility of ensuring international suppliers' compliance with the US food safety regulatory requirements to importers, motivates the importers' participation in international capacity building. Extension service roles that were traditionally played by the public sector can benefit from being transferred to or shared with the private sector providing services to their suppliers. The private sector is able to connect with the private sector players in foreign countries and help to meet FDA's new goals of adapting technical assistance and capacity building to different players along the food supply chain and local needs in different countries. In addition, the private sector has firm-level data that, if made available, could help to evaluate the effectiveness of capacity building, which is essential to prioritizing capacity-building effort and making adjustments as the global food market continues to evolve. The data problem will be discussed in greater detail in Sect. 14.4.

14.3 To Effectively Build International Food Safety Capacity, PPPs Are Needed

The idea of PPP in ensuring food safety is not new. The FDA, under the Federal Food, Drug, and Cosmetic Act of 1938, became responsible for ensuring that foods were unadulterated and truthfully labeled. The FDA built its enforcement activities around pre-market and post-market activities involving the private sector. The FDA also is governed by the Public Health Service Act of 1944 which provides broad authority to protect public health by establishing certain "public-private" cooperative programs, providing authority for emergency authority to prevent the spread of communicable diseases and establishing the role of CDC in public health surveillance.

Hazard Analysis Critical Control Points (HACCP) was the first food safety system to involve training programs. It has its origins with the National Aeronautics and Space Administration (NASA), who had mandated the use of critical control points to ensure the safety of food in flight, and Pillsbury Company, the NASA contractor since the late 1950s(Sperber and Stier 2009). In the early 1970s, the FDA responded to a case of botulism attributed to under-processed, low-acid canned food by reaching out to Pillsbury. Pillsbury organized and conducted a training program for FDA inspectors on how to use critical control points to regulate the production of canned foods (ibid.). With insight from that training program, the FDA published the canned food regulations in 1973, HACCP regulations for seafood in 1995, and subsequently juice HACCP requirements.

In 1998, the FDA published formal guidelines for the microbial safety of fresh produce, suggesting that good agricultural practices (GAP) and good manufacturing practices (GMP) for producers are ways public and private sector entities can ensure the safety of produce (Rushing and Walsh 2006). Later in 1999, the National GAP training program was established at Cornell University through a grant from the USDA Cooperative State Research, Education, and Extension Service (CSREES) and the FDA. The goal of the National GAP program was for Cornell to develop a course material addressing the principles in the 1998 FDA guidelines and to roll out this information through the USDA land-grant extension programs to the fresh produce industry. Although these domestic training programs were effective in the United States, the FDA recognized that they did not address the needs of foreign produce suppliers without a similar extension outreach system abroad. The FDA thus tasked the JIFSAN, one of FDA's Centers of Excellence, to alter the material to the needs of foreign producers and roll out the training internationally.

The JIFSAN, created in 1996, is a PPP between the FDA, the University of Maryland, and the private sector. Its mission is to advance sound strategies that improve public health, food safety, and applied nutrition using risk analysis principles through cooperative research, education, and outreach programs. A major component of its mission is to develop food safety capacity abroad. Initial efforts focused on improving human capital through train-the-trainer programs in good agricultural practices, good aquacultural practices, good fishing vessel practices, food inspection trainings, and commercially sterile packaged food. Much of the JIFSAN's capacity building is funded through an FDA cooperative agreement with a support for specific country programs from the private sector, FDA, USDA-Foreign Agricultural Service (USDA/FAS) and the Food Safety and Inspection Service (USDA/FSIS), and the US Agency for International Development (USAID). JIFSAN trainers are from the industry, are FDA scientist (if an FDA priority country), are retired FDA or USDA scientists, or are the faculty from the University of Maryland or other academic institutions. Starting in 2002, the host countries have cost-sharing agreements supporting JIFSAN International training programs funded through the cooperative agreement. The JIFSAN funds programs up to the port of entry into the host country. The host country and any other partners then provide funding for training activities inside the country. Some governments have also reached out directly to the JIFSAN to request trainings for their food safety specialists; and they either fund themselves or find funding from a donor agency like the World Bank, IDB, and USDA/FAS. Figure 14.3 shows how the effort in capacity building funded through the FDA cooperative agreement with the JIFSAN has increased with increased amounts of imports into the United States. Though the shared funding policy was implemented well before FSMA, it aligns with several of the principles with the FDA International Food Safety Capacity-Building Plan such as ensuring the host country's commitment to the effort while also leveraging JIFSAN resources. Figure 14.4 shows the global reach of all the JIFSAN's training programs.

In 2010 the JIFSAN recognized that one-off training in a country may not be sufficient to reach all the needed stakeholders, which led to the establishment of

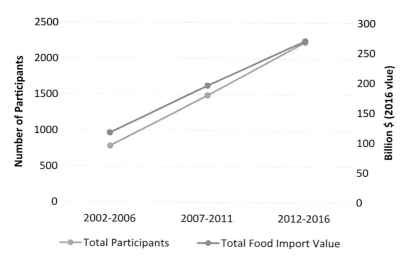

Fig. 14.3 FDA-funded international training and food import values from hosting countries

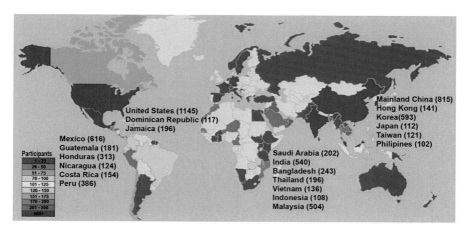

Fig. 14.4 Historical global coverage of JIFSAN's programs

JIFSAN's Global Collaborative Training Center Initiative. The primary goal of the centers is to work with in-country partners to build capacity of both regulators and industry in the use of international best practices in food safety management, enhancing the safety of the food supply in a country or region. The Aquatic and Aquaculture Food Safety Center (AAFSC), in collaboration with the Bangladesh Shrimp and Fish Federation, was established in 2010. The establishment of AAFSC was based on discussions between JIFSAN, the CFSAN's Division of Seafood, and Bangladesh Shrimp and Fish Federation about continuous training needs on good aquacultural practices (GAqP) after an initial training in 2009. Since then, lead trainers have been trained; and the Center has conducted a number of multiplier trainings and has been instrumental in integrating GAqP into university curricula in Bangladesh.

In 2012, the JIFSAN worked with the CFSAN, the FDA's India Office, the Spices Board of India, and the Confederation of India Industry Food and Agriculture Center of Excellence (CII-FACE) to establish the Centre for Supply Chain Management for Spices and Botanical Ingredients in India. Following the same approach as in Bangladesh, lead trainers were identified, trained, and have rolled out multiplier training programs to producers and marketers throughout much of India. In 2013, the JIFSAN, Delta Professional Consultancy, and the Malaysia's Ministry of Health initiated the International Food Safety Training Centre Malaysia, focusing on building laboratory testing capacity, risk analysis capabilities, increasing the skills of the ministry's food inspection staff, and increasing their understanding of global food laws and regulations. Several other initiatives are in the process but have not progressed as much as these three to date.

The JIFSAN, in recognition that the SPS Agreement placed an increased emphasis on risk-based decision-making in facilitating global trade, also established a risk analysis training program in 2002. Though initially the training material was developed with funds from the cooperative agreement, training participation is largely supported through program fees paid for directly by a country's ministry, competitive grants, and funds from the private sector. In 2011, an extended risk analysis fellowship in partnership with International Life Sciences Institute (ILSI) was established. The program is a 3-month program involving 1 month of classroom training followed by a 2-month guided research period. In the guided research period, fellows develop quantitative risk assessments and populated them with data from their countries (or data from nearby regions if country-specific data is currently unavailable). Additionally, the fellows are introduced to various agencies in the US food safety system and participate in several field trips to food companies and retail establishments. Since 2011, 27 fellows from developing countries were trained through this program; and funding came from the ILSI fellowship, USDA-FAS, USDA Borlaug programs, and national governments. In August 2017, a modified extended risk analysis training program began at the MARs Training Center in China.

The laboratory program was established in 2010 as a partnership between the JIFSAN and the Waters Corporation, where it offers hands-on laboratory training in chemical and microbiological food safety analysis. The training courses are "fit for purpose" in that they are designed to teach participants instrument-independent analytical techniques ranging from the most sophisticated to the simplest approach, thereby allowing effective analysis regardless of the facilities available. The focus of the program is on both FDA-recommended methods for sample preparation and analysis required to meet US import standards and the harmonization of methods to ensure food safety worldwide. Participation in this program currently is largely supported through program fees by a country's ministry, competitive grants, and funds from the private sector.

14.3.1 New Era Under FSMA

The abovementioned history illustrates that at each point, the private sector was involved. The FDA is now in a new era under FSMA, the first major change since 1938 in how food is protected in the United States. The goal of the new law is for the FDA to develop a prevention-oriented set of requirements to strengthen accountability of individuals involved in the provision of food and thus ensure high rates of compliance for both imported and domestic foods. The new law created roles for the manufacturers, importers, third-party private standards, foreign regulatory bodies, and the FDA at both the federal and state levels. In 2011, the FSMA formally required the FDA to set requirements; administer training and education programs for the state, local, territorial, and tribal food safety officials; and provide technical assistance so producers and processors know what is expected.

Section 305 of FSMA also charged the FDA to "develop a comprehensive plan to expand the technical, scientific, and regulatory food safety-capacity of foreign governments, and their respective industries, from which foods are exported to the United States." The purpose of the plan is for FDA to be transparent to their stakeholders about FDA's interests and priorities with respect to food safety capacity-building efforts. This was the first time Congress charged the FDA with comprehensively addressing international food safety capacity building. The key principles including ownership, alignment, leverage, managing for results, mutual accountability, and sustainability are articulated in the FDA's International Food Safety Capacity-Building Plan (FDA 2013).

The goals of the international capacity-building plan were to ensure a high level of compliance with the new rules under FSMA; and the FDA recognized that to do this effectively they needed an evidence-based decision-making process. They also needed to coordinate with other US agencies and international organizations and work with partners in the public and private sectors in developing training materials. They needed to prioritize their training and capacity-building efforts based on risk assessments and needs assessment and support the FDA's foreign offices on technical assistance. They also needed to develop a monitoring and impact assessment process.

Under FSMA, the FDA will continue to roll out the Produce Safety Rule internationally through Produce International Partnership (PIP) training program involving the JIFSAN and the Produce Safety Alliance. This is largely in recognition that it will be difficult to get producers trained in these areas without continual support from the public sector. This, however, is not the case with some of the other rules such as the Preventive Controls Rule where there is a large number of lead trainers emerging in both the public and private sectors who can help disseminate the materials internationally and bring all firms up to speed in time to meet the implementation deadlines for the new rule.

The implementation of the FSMA in terms of both domestic and international capacity building is being done in three phases. Phase 1 sets the requirements and develops the regulations and guidance documents, when a series of new rules were

Box 14.1 New Rules Under FSMA

Preventive Controls Rule for food—requires a food facility to have and implement preventive controls to significantly minimize or prevent the occurrence of hazards that could affect food manufactured, processed, packed, or held by the facility.

Preventive Controls Rule for animals—establishes requirements for good manufacturing practices and requires that certain facilities establish and implement hazard analysis and risk-based preventive controls for animal food, including ingredients and mixed animal feed.

Produce safety rule—establishes the science-based minimum standards to reduce the risk of foodborne hazards associated with the production and harvesting of raw fruits and vegetables that are marketed as raw agricultural commodities.

Foreign Supplier Verification Programs rule—describes what a food importer must do to verify that its foreign suppliers produce food that is as safe as food produced in the United States.

Accreditation of third-party certification rule—establishes a voluntary program for the accreditation of third-party certification bodies to conduct food safety audits and issue certifications of foreign facilities and the foods they produce.

Sanitary Transportation of Human and Animal Food rule—establishes requirements for shippers, loaders, carriers by motor vehicle and rail vehicle, and receivers engaged in the transportation of food, including food for animals, to use sanitary transportation practices to ensure the safety of the food they transport.

Intentional adulteration rule—requires facilities to implement a food defense plan to prevent actions intended to cause large-scale public harm.

developed (see Box 14.1). Between 2010 and 2011, the FDA facilitated the creation of the alliances, which are public-private alliances composed of the food industry, academia, and representatives from federal, state, and local food protection agencies. The Produce Safety Alliance was established in 2010 as a collaboration between Cornell University, the FDA, and USDA. In 2011, the FDA provided a grant to Illinois Institute of Technology's Institute for Food Safety and Health (IIT IFSH) for the development of the Sprout Safety Alliance (SSA) and the development of the Food Safety Preventive Controls Alliance (FSPCA). These alliances are responsible for developing a core curriculum for the training and outreach programs. Lead trainers are selected and trained and are then responsible for the training delivery and issuance of certificates of completion to participants around the world.

Phase 2 of FSMA focuses on designing strategies to promote and oversee industry compliance and developing a set of performance metrics. Working groups are developing plans for larger outreach programs to provide the industry with commodity- and sector-specific guidance, education, and technical assistance. These working groups are coordinating with the alliances to get the materials ready as a teaching guidance.

Phase 3 of FSMA focuses on designing an operational plan, implementing the plan, and setting up a monitoring and evaluation approach that focuses on public health impacts. Currently, the FDA is working to develop a set of performance metrics to measure the impact of the training efforts both in the short term and in the long term. A review of public sector data sources, discussed in the next section, indicates that there are limits to publicly available data and partnering with the private sector in the voluntary sharing of potential agreed-upon indicators may provide improved insight to the impact of capacity-building efforts.

14.3.2 Private Sector Involvement in the International Capacity-Building Efforts

In addition to these public sector capacity-building efforts, there are a number of complementary private-sector capacity-building efforts. Several public-private partnerships have emerged to further food safety capacity building globally. Some of the more prominent ones are summarized below.

The Global Food Safety Initiative (GFSI) was formed in 2000 by a group of global food companies that came together at the Consumer Goods Forum and agreed that the way to improve consumer trust in the private sector's efforts to maintain safe supply chains was for the private sector to harmonize their food safety standards and maintain the safety of food along the supply chains they worked in. The GFSI developed a benchmarking model that defined the key elements in food safety schemes for the production of safe food and feed, packaging process, and service provision. With these elements, the GFSI could recognize existing food safety schemes if they are equivalent to the benchmarking model. The recognition of equivalent schemes allowed for flexibility in the private standards marketplace. The GFSI encouraged companies to accept certificates issued during third-party audits against the GFSI-recognized schemes with the goal of enabling their suppliers to work more effectively through fewer audits. The standards currently recognized by the GSFI include requirements about incident management food defense and allergens that go beyond the general principles of food hygiene costs of practice laid out in *Codex Alimentarius* (Fagotto 2014).

The GFSI's Global Markets Program, a food safety capacity-building program created in 2008, established how small- and less-developed food companies can reduce food safety concerns and improve market access in the areas of primary production and manufacturing through certification to one of the GFSI-recognized schemes. This was done because it was recognized that market opportunities may exist for small-scale producers, but these small businesses often lacked access to the expertise, technical, and financial resources that would allow them to meet all necessary food safety requirements (Rey 2016). The GFSI does not carry out training programs, nor does it develop training materials, but relies on a number of organizations which have already developed training manuals and courses for suppliers wishing to implement the Global Markets Program. In 2009 in a PPP with the United Nations Industrial Development Organization (UNIDO), several companies and groups of companies such as Metro, Aeon, Danone, Cargill, and Coca-Cola have rolled out the Global Markets Program. In 2016 UNIDO expanded its partnership with the GFSI to advance food safety using UNIDO Sustainable Supplier Development Program and GFSI's Global Market Program to parts of Africa, China, the Middle East, and Southeast Asia.

The Grocery Manufacturers Association (GMA) through their Science and Education Foundation offers international trainings in HACCP and plant-specific training in better processing controls (BPCS) tailored to the needs of different facilities using the FDA-approved BPCS text. Their mission in regard to training is to deliver training and education programs to the food industry and consumers. They are also working with the Inter-American Institute for Cooperation on Agriculture (IICA) to build the capacity of sector professionals and private sector stakeholders in HACCP to support implementation of SPS measures and increase trade opportunities in the Caribbean countries. This effort is a PPP in essence.

The Food Safety Cooperation Forum was formed in 2007 within the Asia-Pacific Economic Cooperation (APEC) with the goal of building robust food safety systems so as to accelerate progress toward harmonization of food standards internationally to improve public health and facilitate trade. The Partnership Training Institute Network (PTIN) was formed in 2008 to improve food safety practices and technical processing expertise in the Asia-Pacific region utilizing a network of decision-makers and experts from the regulatory, agriculture, and trade agencies. It also includes industry and academia from APEC member economies who help prioritize and coordinate capacity-building activities within in APEC, taking into account the needs of developing member economies and other capacity-building activities in the region. To date, trainings have taken place on developing food laws, standards, enforcement systems, risk analysis, supply chain management, and export certificates and assessing food safety capacity-building needs of food control systems and food safety incident management, including development of food recalls.

The Global Food Safety Partnership (GFSP) housed at the World Bank grew from the APEC forum and emerged in 2012 as a PPP aimed at improving the safety of food in middle-income and developing countries through capacity-building efforts. The program has struggled in its infancy and has undergone a major reorganization to implement sound monitoring and evaluation strategy without duplicating other efforts. To date the partnership has provided trainings in laboratory capacity building, HACCP food safety, and seafood disease management training.

In addition to these collaborative efforts, a number of food companies have increased their involvement in food safety capacity building through social stewardship programs so as to improve environmental, economic, and social impacts of sustainable sourcing. For instance, Cargill, through their Rural Development Initiative partnership with CARE, a humanitarian organization, works with Cargill's local teams to provide training and skills development to improve market access for smallholder farmers, enhance education and nutritional support for children, and provide access to social services for communities they are working with. Similarly, General Mills has developed a Supplier Engagement Program and works with their suppliers to implement these requirements so as to enhance the livelihoods of farming communities, improve yields, and protect natural resources across the supply chain. All these programs have discussed the difficulty in measuring the impact of their efforts.

14.4   The Importance of Monitoring and Evaluation Efforts Associated with International Capacity Building

Integral to FSMA is the need to develop a monitoring and evaluation approach to measure the impact of training efforts. In 2011, prior to the finalization of the new rules under FSMA and the publication of the FDA's International Food Safety Capacity-Building Plan, the FDA's International Program asked the JIFSAN to develop and pilot a set of evaluation/self-assessment tools to measure the effectiveness and impact of JIFSAN's international capacity-building training programs that were already in process. The approach uses a modification of Kirkpatrick's (Kirkpatrick 1959a, b, 1960a, b) "Hierarchical Model of Training Outcomes," one of the most popular methods for assessing behavioral change in training evaluation. The "hierarchy" has four levels. Firstly, the trainer gauges the reaction of trainees to the training program. The idea is that trainees who are satisfied with a training program will get more out of it. Secondly, the trainer determines how much learning actually occurred. Learning can be quantified based on the knowledge or skills acquired or changes in attitudes. Thirdly, the trainer assesses how this learning affects actual job performance. This step is a measure of how behavior on the job changes as a consequence of the training. Finally, the trainer measures the impact of the training on the ultimate outcomes of interest (e.g., increased sales or productivity, improved market access, etc.).

The program was piloted in 2012, and primary data were subsequently collected at each international training session for program evaluation (see Fig. 14.5). First, questionnaires were used to collect participant feedbacks. Secondly, pre- and post-training factual tests (pretests and posttests from here on) were administered to provide a quantitative measure on knowledge gain in the training. These data enabled the JIFSAN to evaluate the immediate training effects and improve future trainings (Kirkpatrick levels 1 and 2). Approximately a year after training, another survey instrument was disseminated to collect information on medium-term effect of the training (Kirkpatrick level 3). Several years after training has taken place, secondary data sources, including FDA refusal data, the FDA inspection data, trade data, and CDC traceback data, are used to determine if there has been any long-term changes

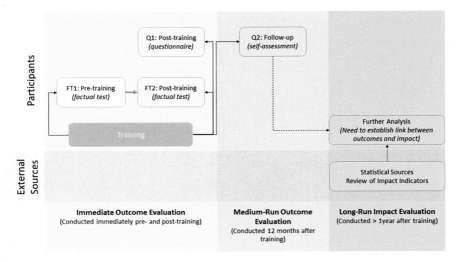

Fig. 14.5 JIFSAN's metrics approach

associated with rejections of a product or in trade patterns from a country in which training has occurred (Kirkpatrick level 4).

The approach was adopted by the GMA and IICA who used it in their recent trainings on HACCP in the CARIFORM countries. Maryland Extension programs have also used it in GAP trainings with local farmers. The GFSP laboratory capacity-building training in China also made use of the approach. Recently, it was also adopted as a way to measure the impact of international produce safety trainings that use alliance material to teach the Produce Safety Rule.

The JIFSAN still is in the early process of measuring impact. The goal is that several years after training has taken place, JIFSAN's monitoring and evaluation team will use secondary data sources, including the FDA refusal data, FDA inspection data, trade data, and CDC traceback data to determine if there has been any change associated with rejections of a product from a country in which training has occurred and to identify changes in trade patterns. All this secondary data was collected for specific purposes, which were not measuring the impact of food safety capacity building. So often it is in a form that does not really facilitate attributing changes to a specific training. For instance, the FDA refusal database does not provide data on the volume of product refused; thus it is difficult to know to the full cost of a rejection to the supplier. Further, the FDA's commodity codes do not match the trade data collected by the Department of Commerce, which makes it difficult to understand the value of trade affected. Similarly, CDC's outbreak data have limited entries on actual tracebacks, as many countries are still developing their monitoring programs to conduct actual tracebacks. Table 14.2 describes the different sets of secondary data available that might point to impact. As noted in the table under each potential secondary dataset, there are limitations to much of the publicly available data. This points to the need to engage the private sector in helping the FDA measure the impact of capacity-building efforts.

Table 14.2 Potential useful secondary datasets

Description of database	Possible limitations
FDA's Operational and Administrative System for Import Support (OASIS) database Information on product that FDA detained on regulated products that are out of compliance with the Food, Drug, and Cosmetic Act. Information of the products, country of origin, and reason for refusal are entered into and is publically available. Predict (described below) will replace it.	The difficulty in using this is it provides data indicating that a product from a specific country and from a specific firm was refused. It does not provide data on amount of product refused. The FDA commodity codes used in the refusal database and the codes of the trade data collected by Department of Commerce do not match, making it difficult to estimate the financial impact of that turned away or destroyed due to a food safety hazard
FDA's Inspection Classification Database— Results of the FDA's inspections of regulated facilities to determine if a firm's compliance with regulations and the Food, Drug, and Cosmetic Act. For this dataset, FDA is disclosing the final inspection classification for inspections conducted of clinical trial investigators, Institutional Review Boards (IRB), and facilities that manufacture, process, pack, or hold an FDA-regulated product that is currently marketed	Inspection classifications listed in this report reflect the compliance status when the report was generated and may not represent the final agency determination. The disclosure of the information is not intended to interfere with planned enforcement actions; therefore some information may be withheld from posting until such action is taken. The database does not represent a comprehensive listing of all conducted inspections, and the FDA states that the database should not be used as a source to compile official counts
CDC National Outbreak Reporting System (NORS) which contains traceback data on foreign sources of foodborne illness outbreaks in the United States exists	Currently there are limited entries of actual tracebacks, as many countries are in the process of still developing monitoring programs to conduct trackbacks
European Union's Rapid Alert System for Food and Feed contains monitoring reports on problems associated with imported foods	These are reported problems once the product has entered the EU and are not associated with the amount, preventing the researcher from calculating real trade impact
National Oceanic and Atmospheric Administration (NOAA) data. NOAA provides training programs on seafood HACCP; they certify establishment as being capable of producing safe, wholesome products in accordance with specific quality regulations promulgated by the US Department of Commerce. There may be some country data information that they collect associated with training	Currently we are unable to find publicly available data but expect that NOAA has such databases where they keep track of such information

(continued)

Table 14.2 (continued)

Description of database	Possible limitations
PREDICT (Predictive Risk-Based Evaluation for Dynamic Import Compliance Targeting) is an electronic screening tool that the FDA uses to flag high-risk imports of food products for additional monitoring and inspection. PREDICT uses a variety of assessments including information on the product, information on weather conditions during shipment, country of origin, and manufacturer's safety record to rank and score shipments according to risk. Based on the risk score, inspectors will target higher-risk shipments for examination	Whether the public version of the tool will facilitate better understanding of changes in trade based on capacity-building efforts is yet to be seen, given the current limitation of the public version of the OASIS system, but we plan to also see what other information can be gleaned
FDA-TRACK program may prove useful in the future. For instance, the FDA collects on the number of inspections completed by investigators based on in country, the data on total number of inspections completed in the month, and number of verifications of foreign firm registrations with their China, India, and Latin America offices	Currently, the public version is in the aggregate, thus somewhat limiting. If more detailed data was available, it could help facilitate impact evaluations associated with food safety capacity-building efforts
Possible new data associated with new rules under FSMA—as FSMA is rolled out, one might be able to also look at increases in the numbers of participants in Foreign Supplier Verification Program (FVSP), the voluntary qualified importer program (VQIP), and third-party auditors. Increases in the number of foreign laboratories accredited, increases in country system recognition or equivalence assessments of foreign food safety systems, and increases in the number of foreign inspections and facilities registered	The ability to measure impact based on this data will depend on what the FDA makes publicly available

14.4.1 Engaging Private Sector in Monitoring and Evaluation Efforts

Stakeholders involved in food safety capacity building have different interests in improving public health, livelihoods, and financial measures. Their interests in measuring the effectiveness of capacity-building efforts also differ. Understanding the interests of these stakeholders is crucial when designing monitoring and evaluation programs. This is because different stakeholders may be better motivated to fund different capacity-building efforts. Here it would be helpful to refer back to Fig. 14.1 where we identified the stakeholders involved in international food safety capacity building. The FDA is a public health agency whose main goal in capacity building is to improve health outcomes (a health measure). There are limits to the capacity building and impact evaluation that the FDA can do based on their mission and the fact that they are a regulatory agency. USAID's Feed the Future initiative on the other hand is focused on reducing global hunger and improving food security; thus the outcomes they would be interested in examining would include the health and livelihood improvements among the poor in Feed the Future countries. The private sector and industry organizations are interested in capacity-building efforts primarily for financial reasons such as preventing the production of defective products and the consumption of unsafe food products and improve economic and social outcomes through sustainable sourcing.

The private sectors may be interested in several potential measurements of production outcomes, for example, the changes in (1) the number of products going through the "first-pass" quality check without having to be reworked or diverted to a lesser value stream, (2) the number of products on hold, (3) the number of marketplace actions taken based on customer complaints or recalls, and (4) the ability to attract new customers and enter new markets. Potential internal control measures for a company include (a) the development of facility internal control measures, (b) increased number of analytical test results within acceptable values, (c) improved audit scores through internal or third-party audits, (d) improved "risk" score among those companies who create risk scores for their plant and/or suppliers, (e) external certification of the facility/operation, (f) decreases in frequency of required audits, and (g) reductions in regulatory violations (Geisert 2014). Whether the private sector would share such data with the public sector is unclear without some sort of novel PPP aimed at measuring the combined effect of capacity-building efforts.

Sharing such data can be difficult, due to confidentiality concerns and worries over possible regulatory sanctions. Feedback from a product-tracing study in 2011 suggested there was a concern from the industry that data collection efforts would be costly and it was unclear if all industry would share data unless it was through a voluntary approach (Institute of Food Technologists 2012). This does not have to be the case; a public-private partnership can be made that facilitates the sharing of data in a way that blinds or aggregates the data that some companies may voluntarily share through a third party, so that more fruitful impact analysis can be done. A mechanism might be for an industry group to work with a university who can blind the data received.

Partnering with the private sector and forming such partnerships for data sharing is not new and is increasingly looked upon as a positive way to achieve improved public health outcomes. The recently released USDA Branded Food Products Database is the result of a successful PPP with USDA/ARS, International Life Sciences Institute (ILSI), GS1 US., 1WorldSync, Label Insight, and University of Maryland's JIFSAN. Through this initiative, a number of private companies who work with ILSI voluntarily chose to submit data to the JIFSAN. The goal of the PPP is to enhance public health and the sharing of open data by complementing the USDA National Nutrient Database for Standard Reference with nutrient composition of branded foods and private label data provided by the food industry. This partnership and the development of the mechanisms for sharing data came out in the 2011 Presidential Memorandum from President Obama that directed agencies to develop public-private partnerships in areas of importance to the agency's mission (Kretsera et al. 2015).

14.5   Conclusion

This chapter examines the evolving rationale promoting PPPs in food safety capacity-building efforts. It discusses how the private sector developed both voluntary mechanisms to improve food safety and ways to audit such approaches among their suppliers. It discusses how the public sector has altered their regulatory mechanism to embrace private-sector efforts and how the public sector may want to focus their training efforts on those who were not aligned to these private mechanisms. It suggests that, in addition to PPPs in capacity-building efforts, PPPs in monitoring and evaluation efforts are needed to guide public and private actions and deliver capacity-building outcomes more effectively in the future. It is argued that in order for the FDA to achieve sustained public health outcomes, it will be necessary to work beyond traditional methods to deliver food safety capacity building. It will also be necessary to evaluate outcomes of interest to other stakeholders investing in international capacity-building efforts. This will include measuring outcomes that go beyond the FDA's mission and looking at some of the spillover effects such as improved livelihoods, which are of interest to the aid community, and production measures that are of interest to the private sector and industry organizations. If suppliers knew that behavioral changes had positive livelihood and health impacts, they would be more likely to sustain these changes. Currently, to our knowledge there are no studies measuring spillover effects, and this is worthy of research.

References

Anderson JR, Feder G, Agricultural Extension. In: Evenson RE, Pingali P, editors. Handbook of agricultural economics, Agricultural development: farmers, farm production and farm markets, vol. 3. Amsterdam: Elsevier; 2007. p. 2343–78.

Elbasha E, Riggs T. The effects of information on producer and consumer incentives to undertake food safety efforts: a theoretical model and policy implications. Agribusiness. 2003;19:29–42.

Fagotto E. Governing a global food supply: how the 2010 FDA food safety modernization act promises to strengthen import safety in the US. Erasmus Law Rev. 2010;3:257–73.

Fagotto E. Private roles in food safety provision: the law and economics of private food safety. Eur J Law Econ. 2014;37:83–109.

FDA. Global engagement. 2011. https://www.fda.gov/downloads/aboutfda/reportsmanualsforms/reports/ucm298578.pdf. Accessed 15 Aug 2017.

FDA. FDA's International Food Safety Capacity -Building Plan, Food Safety Modernization Act Section 205, 2013. U.S. Department of Health and Human Services. 2013. https://www.fda.gov/downloads/Food/GuidanceRegulation/UCM341440.pdf. Accessed 15 Aug 2017.

FDA. Laboratory methods. 2016. https://www.fda.gov/Food/FoodScienceResearch/LaboratoryMethods/default.htm. Accessed 15 Aug 2017.

Fulponi L. Private voluntary standards in the food system: the perspective of major food retailers in OECD countries. Food Policy. 2006;31:1–13.

Geisert S. former Sr. Director, global product safety and regulatory, General Mills. Personnel communication. 2014.

Holmes P, Lacovone L, Kamondetdacha R, Newson L. Capacity-building to meet international standards as public goods. UNIDO; 2006. https://www.unido.org/fileadmin/import/60028_03_international_standards_public_goods.pdf. Accessed 15 Aug 2017.

Institute of Food Technologists. Pilot projects for improving product tracing along the food supply system—final report. 2012. https://www.fda.gov/downloads/food/guidanceregulation/ucm341810.pdf. Accessed 15 Aug 2017.

Kirkpatrick DL. Techniques for evaluating training programs: Part 1—reactions. J ASTD. 1959a;13:3–9.

Kirkpatrick DL. Techniques for evaluating training programs: Part 2—learning. J ASTD. 1959b;13:21–6.

Kirkpatrick DL. Techniques for evaluating training programs: Part 3—behavior. J ASTD. 1960a;14:13–8.

Kirkpatrick DL. Techniques for evaluating training programs: Part 1—results. J ASTD. 1960b;14:28–32.

Kretsera A, Murphya D, Finleyb J, Brennerc R. A partnership for public health: branded food products database. Procedia Food Sci. 2015;4:18–26.

Lin CF. Public-private interactions in global food safety governance. Food Drug Law J. 2014;69:143–60.

McQuade T, Salant SW, Winfree J. Markets with untraceable goods of unknown quality: beyond the small-country case. J Int Econ. 2016;100:112–9.

Rey Y. Leading the charge for food safety capacity building: four global initiatives: part one: the global food safety context and the GFSI mission. 2016. http://globalfoodsafetyresource.com/food-safety/food-safety/capacity-building. Accessed 15 Aug 2017.

Rich K, Narrod C. The role of public–private partnerships in promoting smallholder access to livestock markets in developing countries: methodology and case studies. IFPRI; 2010.

Roberts T. Lack of information is the root of US foodborne illness risk. Choices. 2013;28:1–6.

Rushing J, Walsh C. The United States Food and Drug Administration's approach to training international suppliers in food safety and security. Hortic Technol. 2006;16:4566–9.

Scallan E, Hoekstra R, Angulo RV, Tauxe R, Widdowson M, Roy S, Jones J, Griffin P. Foodborne illness acquired in the United States—major pathogens. Emerg Infect Dis. 2011;17:7–15.

Sperber W, Stier R. Happy 50th birthday to HACCP: retrospective and prospective. Food Safety Magazine. 2009;15(42):44–6.

Testimony on Food Aid and Capacity Building Programs: Testimony to the U.S. House of Representatives Committee on Agriculture Cong (testimony of Philip Karsting). 2015. https://www.fas.usda.gov/sites/default/files/2015-06/karsting_testimony_-_final.pdf.

Unnevehr L. Food safety as a global public good. Agric Econ. 2007;37:149–58.

USDA/FAS. U.S. Department of Agriculture Foreign Agricultural Service Strategic Plan FY 2015–2018. 2015. https://www.fas.usda.gov/sites/default/files/2015-07/administaror_approved_fas_2015-2018_strategic_plan_150128.pdf. Accessed 15 Aug 2017.

Winfree J, McCluskey J. Collective reputation and quality. Am J Agric Econ. 2005;87:206–13.

World Health Organization. WHO estimates of the global burden of foodborne diseases: foodborne disease burden epidemiology reference group 2007–2015. 2016. http://apps.who.int/iris/bitstream/10665/199350/1/9789241565165_eng.pdf. Accessed 15 Aug 2017.

Chapter 15
The Political Economy of US Antibiotic Use in Animal Feed

Walter J. Armbruster and Tanya Roberts

Abbreviations

ADAA	Animal Drug Availability Act
AHI	Animal Health Institute
AMS	Agricultural Marketing Service
APHIS	Animal and Plant Health Inspection Service
AR	Antibiotic resistance
ARDs	Antibiotic-resistant determinants
ARMS	Agricultural and Resource Management Survey
CDC	Centers for Disease Control and Prevention
CR	*Consumer Reports*
CSPI	Center for Science in the Public Interest
ERS	Economic Research Service, USDA
FACT	Food Animal Concerns Trust
FAO	United Nations' Food and Agriculture Organization
FDA	US Food and Drug Administration
FSIS	Food Safety and Inspection Service, USDA
GFI	Guidance for Industry
IOM	Institute of Medicine
MRC	UK Medical Research Council
NAE	No Antibiotics Ever
NAHMS	National Animal Health Monitoring System
NARMS	National Antimicrobial Resistance Monitoring System—Enteric Bacteria
NASS	National Agricultural Statistics Service
NRDC	Natural Resources Defense Council

W. J. Armbruster (✉)
Farm Foundation (Retired), Darien, IL, USA
e-mail: waltja@live.com

T. Roberts
Economic Research Service, USDA (retired), Center for Foodborne Illness
Research and Prevention, Vashon, WA, USA
e-mail: tanyaroberts@centurytel.net

© Springer International Publishing AG, part of Springer Nature 2018
T. Roberts (ed.), *Food Safety Economics*, Food Microbiology and Food Safety,
https://doi.org/10.1007/978-3-319-92138-9_15

OIE	World Organization for Animal Health
OTC	Over-the-counter
RWA	Raised without antibiotics
UK	United Kingdom
US	United States
UCS	Union of Concerned Scientists
USDA	United States Department of Agriculture
VCPR	Veterinarian-client-patient relationship
VFD	Veterinary Feed Directive
WAHIS	World Animal Health Information System
WHO	World Health Organization

15.1 Introduction

Antibiotic resistance has been widely recognized as a serious public health problem. Hence, there is a major public good to be realized in safeguarding the effectiveness of existing antibiotics and creating new ones. Antibiotics are used to treat human infections and used in animal agriculture. While many drugs are dual-use, others are animal- or human-use specific. The production benefits of sub-therapeutic levels of antibiotics in animal agriculture have been recognized since the late 1940s (CAST 1981). In animal agricultural production, antibiotics are used at therapeutic levels to treat infections and at sub-therapeutic levels to prevent infections and promote animal growth (Sneeringer et al. 2015; Van Boeckel et al. 2017; WHO 2017).

As the organizational complexity of the animal agricultural supply chain increased, the number of economic stakeholders in on-farm antibiotic use has also increased. The major stakeholders include pharmaceutical companies, production integrators, feed suppliers, farm groups, producers, restaurants, food retailers, the public, the medical community, the scientific community, government regulators and policy makers. Each of these stakeholders faces a different set of incentives and disincentives related to on-farm use of antibiotics in animal agriculture. Knowledge of these incentives and disincentives has evolved with the accumulation of scientific and economic research. To understand the regulatory outcomes governing antibiotic use in agriculture, it is important to recognize the political economy context in which they are developed. The various stakeholders are driven by the relative benefits they receive under policies as they affect their industry segment (Zilberman et al. 2014).

15.2 Context of Antibiotic Resistance

Alexander Fleming, who discovered penicillin, warned that "...misuse of the drug could result in selection for resistant bacteria" (Rosenblatt-Farrell 2009). Antibiotic resistance (AR), a term sometimes used interchangeably with antimicrobial resistance, occurs when bacteria change in ways that make antibiotics less effective in treating infections, thereby allowing the bacteria to survive, multiply, and cause additional harm. AR has been recognized as a serious public health problem among the medical and scientific communities. Antibiotics are used to treat human infections and used in animal agriculture. Particularly concerning is resistance for those antibiotics that are of value in treating human health issues, the so-called medically important antibiotics.

The use of antibiotics along with other advances in agricultural technology has facilitated the concentration of animal production on farms in the United States (US) and elsewhere. For example, in 2012, 88% of all US sales of hogs and pigs were by the 13% of farms with 5,000 or more head, and 66% of all layers were produced on the less than 1% of farms that sold 100,000 or more to egg producers (NASS 2014). The majority of the production of hogs, broilers, and eggs occurred under contractual arrangements between growers and integrators, with the integrators prescribing certain production practices, including the use of antibiotics for treating infections, for disease prevention and for promoting growth.

Many of the antimicrobial drugs administered to food-producing animals are also important in treating humans, worldwide. Domestic sales of medically important antimicrobial drugs for use in food-producing animals in the United States accounted for 62% of the domestic sales of all antimicrobials approved for use in food-producing animals. And, 28% of domestic sales of all medically important antimicrobials approved for use in food-producing animals are labeled for therapeutic use only (FDA 2015). Importantly, animal drug sales data represent products sold or distributed by manufacturers through various outlets for intended sale to the user. Since veterinarians and others in the supply chain may have substantial inventory on hand for possible use, these numbers do not accurately reflect the amount of product ultimately administered to animals. Given the number of humans versus a much larger number of animals in each of the species, as well as other confounding factors, no definitive conclusions from any direct comparisons between the quantities of antimicrobial drugs sold for use in humans versus animals can be drawn (FDA 2016a).

There are obvious situations where antibiotics are required to treat sick animals in agriculture, but the proper therapeutic use versus prophylactic use remains in question among stakeholders. Farm groups and others in the food animal supply chain recognize that antibiotics in animal feed keep animals healthy and meat costs down. But over 1000 medical doctors and other healthcare providers signed petitions to Congress asking for new legislation to reduce non-therapeutic antibiotic use in food animals (Miller 2011). The animal health industry is very concerned that needed preventative use will be threatened by the recent FDA ban on use of medically important antibiotics for growth. FDA classifies as therapeutic those antimicrobials targeted for treatment, control, and prevention of bacteria or disease identified on the product label. FDA explicitly states that the use of antibiotics in animal feed for growth promotion is not allowed.

Those who characterize preventative use as routine overlook the difference between treating animals versus humans. If preventative measures are not taken and a disease outbreak occurs and spreads rapidly within a flock or herd, it risks large numbers of animals developing a deadly, high mortality disease. Waiting until a disease is clearly evident makes successfully treating the active infections very difficult due to the large number of animals involved. By contrast, a human patient can generally be quickly diagnosed and treated. While some are concerned that producers will continue to use antibiotics for growth under the guise of prevention, the FDA-approved label is specific about dose and duration for a specified bacterium or disease. Off-label use of antibiotics in animal production is illegal, and FDA only allows a veterinarian to decide whether to use or not to use a preventative treatment based on their judgment of a disease threat (Carnevale 2016).

In an economic framework, antimicrobial resistance can be considered as an unwanted side effect, or externality, associated with the use of antibiotics. The efficacy of antibiotics can be considered as a public good that must be managed with government involvement. This is because the costs of overuse by any single individual are borne by society and, in the case of antibiotics, globally. Hence, not only is there a role for government involvement with the animal agriculture industry in managing the stock and use of antibiotics as an important public good, but it must be done cooperatively across countries.

15.3   Challenges in Recognizing the Problem

In 1969, the United Kingdom's (UK) Parliament received the Swann Report, which concluded that using antimicrobials at sub-therapeutic levels in food-producing animals created risks to human and animal health (Joint Committee on the use of Antibiotics in Animal Husbandry and Veterinary Medicine 1969). It noted a dramatic increase in numbers of animal-origin bacteria strains which showed resistance to one or more antibiotics and that these strains could transmit resistance to other bacteria. It recommended that only antimicrobials that are not medically important for humans should be used without prescription in animal feed and that antimicrobials should only be used for therapeutic purposes under veterinary supervision. The primary reason that producers were using these sub-therapeutic doses of antibiotics was to promote faster weight gain in the animals.

In 1970, a US Food and Drug Administration (FDA) task force was charged to do a comprehensive review of antibiotic use in animal feed (FDA 2012). Its report found that sub-therapeutic use of antimicrobials in food-producing animals was associated with development of resistant bacteria and that treated animals might provide a reservoir of antimicrobial-resistant pathogens capable of causing human disease. The task force recommended that medically important antimicrobial drugs meet certain guidelines they identified or be prohibited from growth promotion or other sub-therapeutic use by certain dates. Further, antimicrobials not meeting the guidelines should be limited to short-term therapeutic use only under veterinarian control.

In the 1970s, the Animal Health Institute (AHI), a US trade association for the animal drug industry, funded an on-farm study to determine the impact of adding low-dose antibiotics to chicken feed. Within 1 week of adding tetracycline, the intestinal flora in the chickens "…contained almost entirely tetracycline-resistant organisms" (Levy et al. 1976). The antibiotic resistance was not located in the DNA of the bacteria which is hard to transfer among bacteria but in plasmids located on the outside surface of the bacteria. Plasmids are easily exchanged among bacteria living in the intestine. Importantly, the tetracycline-resistant bacteria in the chicken's intestines were resistant to multiple antibiotics. Furthermore, some members of the farm families began to harbor these same antibiotic-resistant bacteria in their intestines within 6 months.

In 1977, the FDA proposed withdrawing the new animal drug approvals for the sub-therapeutic uses of human medically important penicillin and tetracycline in animal feed based on lack of evidence to show they were safe. However, the US Congress intervened and asked for more research first. The AHI was one of the groups advocating in Congress to delay regulation pending additional research, then and now. In 2010 Congressional Testimony, Richard Carnevale, vice-president at AHI, testified that while it is possible for human antibiotic resistance to be caused by antibiotic use in farm animals, "…it does not happen enough that we can find it and measure it" (Carnevale 2010). This statement contradicted the data produced by the AHI-funded study by Levy (Levy et al. 1976) that was published in the prestigious *New England Journal of Medicine* in 1976.

Richard Carnevale also mentioned in his 2010 testimony that prior to joining AHI he was Deputy Director of New Animal Drug Evaluation in FDA and had worked at USDA in the Food Safety and Inspection Service (FSIS). His testimony illustrates two points in the political economy of food production: (1) how industry has an opportunity to influence regulators' decision-making via the revolving door of employment and (2) how industry carefully selects its facts to present a point of view that bolsters their profits, namely, for drug companies in this case (Oreskes and Conway 2010).

Another example of the political economy in action involved USDA prohibiting an agency research microbiologist from talking about the significant levels of antibiotic-resistant bacteria detected in the air near Midwest hog confinement operations (Union of Concerned Scientists 2004). A third element of the political economy is shown by industry efforts to influence policy makers through campaign contributions and strong lobbying of proposed legislation which may affect their bottom line. Pharmaceutical companies spent at least $135 million and agribusiness companies another $70 million during 2009, in large part to fight possible limits on antibiotic use in animal feed (Mason and Mendoza 2009).

In response to Congressional pressure in the late 1970s, FDA withdrew its proposal and instead funded three studies to determine the impact of using low levels of antibiotics in animal feed (industry won this round, obtained a delay in regulations, and funded more reports):

1. In 1980, the National Academy of Sciences reported that there was limited epidemiological research on the topic. Available evidence at that time did not prove nor disprove dangers of seven therapeutic antimicrobials in animal feed, but that did not preclude the existence of hazards (National Academy of Sciences 2009).

2. In 1984, the FDA funded the Seattle-King County Department of Public Health to analyze *Salmonella* and *Campylobacter*, which were chosen as models to estimate the flow of potentially pathogenic bacteria from animals to humans through the food chain. Their report was based on sampling retail meat and poultry and investigating *Salmonella* and *Campylobacter* enteritis cases in humans. Isolates from human illness cases and retail foods were analyzed for antibiotic resistance of these pathogens, using plasmid analysis and serotyping. The report found that *Campylobacter* was a more common cause of enteritis than *Salmonella* and appeared to flow from chickens to humans through consumption of poultry products, with tetracycline resistance being plasmid-mediated (Seattle-King County Department of Public Health 1984).
3. In 1988, the Institute of Medicine (IOM) undertook a FDA-requested independent quantitative risk assessment of human health impacts from sub-therapeutic use of penicillin and tetracycline in animal feed. Based on a risk-analysis model of *Salmonella* infections that resulted in human death, the IOM did not find substantial direct evidence that sub-therapeutic use in animal feed posed a human health hazard. However, they found a considerable body of indirect evidence implicating both sub-therapeutic and therapeutic use of antimicrobials as a potential health hazard and strongly recommended additional study of the issue (Institute of Medicine 1988).

15.4   Global Concern About Antibiotic Resistance

Numerous research results quantifying the extent of the antimicrobial resistance problem have been published in the scientific literature and indicate a growing and serious threat to human health. The many channels for AR to affect humans are shown in Fig. 15.1. The two main channels for food animals are (1) AR bacteria in the food animal's gut can contaminate the meat or poultry eaten and (2) environmental contamination, such as manure used to fertilize fields that contain AR bacteria, may contaminate the environment and some of the food crops grown on these fields. *Consumer Reports* (CR) tested products sold in US supermarkets and found resistance to multiple antibiotics in the following percent of samples: beef 14%, shrimp 14%, turkey 83%, and chicken 57% (Consumer Reports 2016). CR also found that ground beef from conventionally raised cows was twice as likely to contain antibiotic-resistant pathogens as ground beef from cows raised without antibiotics.

Like other threats to human health, AR is best managed across national boundaries. Increasing international trade may spread antibiotic resistance through imported food products as more trade agreements are approved. This scenario could be exacerbated to the extent FSIS approves additional international facilities, local regulations, and inspections as "equivalent to the United States." Future trade agreements will need to include provisions which address reduced use of medically important antibiotics in producing food animals.

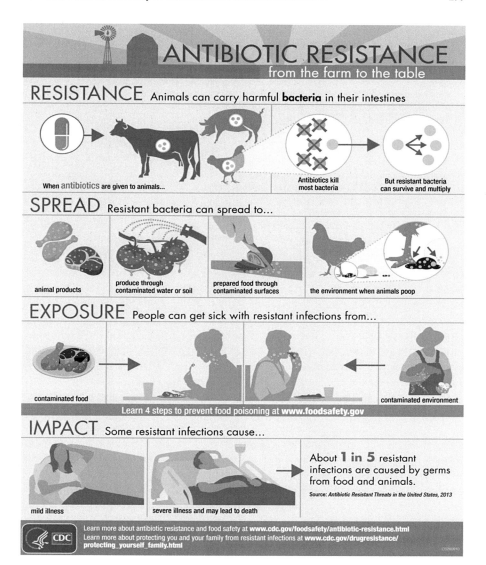

Fig. 15.1 How antibiotic resistance happens and spreads. Source: The US Centers for Disease Control and Prevention, AR-infographic.508c.pdf

Numerous trusted institutions from the United States (US) and the United Kingdom (UK) as well as international organizations such as the World Health Organization (WHO), the United Nations' Food and Agriculture Organization (FAO), and the World Organization for Animal Health (OIE) have acknowledged the threat of antibiotic resistance related to use in producing food animals. The following excerpts from a few recent reports highlight the role that low-dose antibiotic use in animal feed plays in spreading AR.

The Centers for Disease Control and Prevention (2013b) reported that:

Each year in the United States, at least 2 million people acquire serious infections with bacteria that are resistant to one or more of the antibiotics designed to treat those infections. At least 23,000 people die each year as a direct result of these antibiotic-resistant infections. Many more die from other conditions that are complicated by an antibiotic-resistant infection.

Antibiotic-resistant infections add considerable and avoidable costs to the already overburdened U.S. healthcare system. In most cases, antibiotic resistant infections require prolonged and/or costlier treatments, extend hospital stays, necessitate additional doctor visits and healthcare use, and result in greater disability and death compared with infections that are easily treatable with antibiotics. The total economic costs of antibiotic resistance to the U.S. economy has been difficult to calculate. Estimates vary but have ranged as high as $20 billion in excess direct healthcare costs. Adding on the costs for lost productivity brings the total societal costs (sic) for AR to $35 billion a year (2008 dollars). (CDC 2013a, p. 11)

This CDC report also indicates that foodborne cases are responsible for 20% of human AR infections (Fig. 15.1). Thus, societal costs of these AR foodborne illnesses could total $7 billion annually of the $35 billion/year total costs to the US economy. These societal costs could be prevented if the foods were free of contamination with AR pathogens. There may be additional costs associated with environmental pathways of human contamination from use of antibiotics in meat production, such as exposure to contaminated water.

In 2014, WHO stated: "Antimicrobial resistance (AR) is an increasingly serious threat to global public health. AR develops when a microorganism (bacteria, fungus, virus or parasite) no longer responds to a drug to which it was originally sensitive. This means that standard treatments no longer work; infections are harder or impossible to control; the risk of the spread of infection to others is increased; illness and hospital stays are prolonged, with added economic and social costs; and the risk of death is greater—in some cases, twice that of patients who have infections caused by non-resistant bacteria. The problem is so serious that it threatens the achievements of modern medicine. A post-antibiotic era—in which common infections and minor injuries can kill—is a very real possibility for the 21st century" (WHO 2014, p. 3).

In 2015, OIE noted: "Today, in many countries, including developed countries, antimicrobial agents are widely available, directly or indirectly, practically without restriction. Of 130 countries recently evaluated by the OIE, more than 110 do not yet have relevant legislation on the appropriate conditions for the import, manufacture, distribution and use of veterinary products, including antimicrobial agents. Consequently, these products circulate uncontrolled like ordinary goods and are often falsified."

To date, there is no harmonized system of surveillance on the worldwide use and circulation of antimicrobial agents. That information is necessary, however, to monitor and control the origin of medicines, obtain reliable data on imports, trace their circulation, and evaluate the quality of the products in circulation. It is in this context that the OIE was mandated by its member countries to gather that missing information and create a global database for monitoring the use of antimicrobial agents, linked to the OIE's World Animal Health Information System (WAHIS). That mandate is also supported by FAO and the WHO within the framework of the WHO's global action plan on antimicrobial resistance. The database will form a solid basis for the three organizations' work to combat antimicrobial resistance (OIE 2015, p. 2).

In 2016, a UK evaluation of 139 academic, peer-reviewed research articles addressing antibiotic use in agriculture determined that only 5% found no link and 75% found a positive link between antibiotic use in animals and antibiotic resistance (AR) in humans (O'Neill 2016). Taken together, these and numerous other scientific findings show an indisputable relationship between antibiotic use on farms and drug-resistant infections in people (Van Boeckel et al. 2017; Silley and Stephan 2017; Tang et al. 2017; Webb et al. 2017).

15.5 Farm-Level and Public Health Economic Impacts

To evaluate proposals to ban the use of growth-promoting or sub-therapeutic levels of antibiotics in food animals, USDA's Economic Research Service (ERS) economists added questions on antibiotic use to the Agricultural and Resource Management Survey (ARMS). ARMS is a nationally representative survey of farms administered jointly by ERS and USDA's National Agricultural Statistics Service (NASS). Hog producers were surveyed in 2006 and 2009, and broiler producers were surveyed in 2006 and 2011. ERS also drew upon their research using data in the National Animal Health Monitoring System (NAHMS) to develop a model to estimate the impacts of withdrawing antibiotics for other than therapeutic use in food animals. Using Monte Carlo simulations, ERS estimated the impacts of eliminating antibiotic use for growth promotion of poultry and pork, not just the FDA-specified "medically important" antibiotics (Table 15.1). Simulation results showed less than 0.5% reduction in output and an approximate 0.75% increase in wholesale prices, netting pork producers greater total revenue of 0.29% and poultry producers 0.42%. ERS concluded that these small effects were not statistically significant (Sneeringer et al. 2015).

Table 15.1 Ban on antibiotics used for growth promotion has statistically insignificant impact

Impact of Ban on Growth Promotion Antibiotics (%)	Hogs	Broilers
Percent change in output	−0.47	−0.31
Percent change in wholesale price	0.77	0.73
Percent change in value of production	0.29	0.42

Source: Data from Sneeringer et al. (2015)

These ERS results are consistent with research studies post-2000 indicating that productivity gains from using antibiotics for growth promotion were lower than earlier research had found (Teillant and Laxminarayan 2015). Another report suggested that phase out of growth promotion use in food animals over a 5-year period would avoid most of the 67% projected global growth in such use and cost agricultural sectors a small portion of the costs of AR in each country. Further, reduced infection risk and costs of medications would cover most farm-level costs of improving animal husbandry practices to offset loss of antimicrobials for production purposes (Laxminarayan and Chaudhury 2016).

Presuming that any new antibiotic classes probably will not be made available for veterinary medicine, it is important to preserve the effectiveness of existing antibiotics which are necessary for treatment of infectious diseases to maintain animal health (Teillant and Laxminarayan 2015). An alternative to encourage development of new antibiotics would be to delay or not approve drugs which mimic others, but for which the applicant company has not performed antibiotic research (Amábile-Cuevas 2016). Even better, several production practices may be used to enhance animal health in the absence of using antimicrobials for growth or for prophylactic disease prevention (Sneeringer et al. 2015; WHO 2017; MacDonald and Wang 2011). These include:

- Improved management practices, such as more space per animal and better control of the housing environment
- Tightened biosecurity to prevent diseases and improve productivity by avoiding introduction of infectious agents by wild animals, domestic pets, and nonessential workers or other humans; through increased cleanliness of production facilities; and from timely removal of dead animals
- Optimized nutrition to increase growth and mitigate stress-related factors and provide vitamin and mineral supplements to reduce disease susceptibility
- Improved gut microflora to improve feed efficiency by providing enzymes, organic acids, prebiotics, probiotics, and immune modulators
- Vaccinations to prevent some diseases
- Hazard Analysis Critical Control Point plans to improve productivity in the absence of using sub-therapeutic antibiotics in animal production

Generally, these practices may raise production costs modestly at the farm level because of the need for more resources required to successfully manage them. Since ERS found no statistically significant evidence that antibiotics reduce the costs of producing pork or broilers, we conclude that there are small or no costs to producers from withdrawing growth-promoting or prophylactic uses of antibiotics in production of food animals.

In contrast, the public health benefits of withdrawing these antibiotics from agriculture are significant. As reported above, CDC estimates that the medical costs and productivity losses of AR illnesses attributed to agriculture are $7 billion US dollars annually. The benefit/cost analysis becomes $7 billion in public health protection benefits vs. the very small costs to animal production from withdrawing antibiotics from non-therapeutic use. In other words, the protection of the public health will come at little or no cost to agriculture. Furthermore, this benefit/cost analysis provides a conservative estimate of public health protection benefits. The CDC public health protection benefits do not include estimates for protection from an increasing number of "superbugs" that would be created if low-level antibiotics would continue to be used. And CDC does not include the costs of long-term health outcomes caused by foodborne pathogens (see Chap. 8).

15.6   Other Societal Costs

Aside from costs to agricultural producers, there are also other societal costs related to AR and connected to antimicrobial use in animal production, both in their production and use/misuse, affecting human and environmental health. In economic terminology, these costs are considered negative externalities to society from the individual use of antibiotics. Moreover, since the science of AR is unfolding, there may be additional unknown human health and environmental risks associated with the use of antibiotics in food animal production.

Pharmaceutical Production. A major issue involved with manufacturing of active ingredients for antibiotics and the effluent from factories producing them is the potential to contaminate nearby water systems. Pharmaceutical factories often contaminate the environment, since guidelines for pharmaceutical waste discharge focus on chemicals used in manufacturing, rather than active pharmaceutical ingredients. This is a primary concern in countries outside the United States, but international trade makes it a worldwide problem.

Use and Misuse. Worldwide, antibiotics are used heavily in animal agriculture. This practice has created resistance problems transmissible from animals to humans. For example, China has mrc-1 colicin resistance in pork and *Salmonella* resistance to cephalosporins at higher levels than in the United States (Zhang et al. 2016). Their practice of applying human waste on fields and the closeness of population centers to agriculture contribute to cross-mixing of pathogens in China. Parasites are common in Chinese soil and can contaminate pork. And low levels of chlorine in Chinese water supplies allow accumulation of biofilms containing antibiotic-resistant pathogens in water pipes. In India, manufacturing of pharmaceuticals and waste disposal practices lead to contamination of water and soil. Further, over-the-counter antibiotics are available and heavily used there.

Farm antibiotic use is of concern in India and China in poultry and pigs (APUA Newsletter 2016). The threat of superbugs via food is worldwide, due to the distribution of animal food products from China (Zhang et al. 2016; Zhu et al. 2013). Rosenblatt-Farrell (2009) drew upon existing literature to identify additional environmental paths to exposure to antibiotic resistance. Veterinary antibiotics are frequently excreted intact from food animals (Table 15.2). For the widely used tetracycline, 60–80% of the antibiotic is excreted in the feces or urine and not metabolized by the food animal. The transfer of this animal waste to croplands may transfer antibiotics and possibly AR pathogens. In one study, AR genes in soil

Table 15.2 Antibiotics used in US animal agriculture and percent not metabolized and discharged into feces and urine

Drug class and example	Quantity sold for veterinary use (kg)	Fraction not metabolized (%)
Aminoglycosides/neomycin	270,342	80–90
Cephalosporins/ceftiofur	28,337	<10
Ionophores/monensin	4,434,657	50–80
Lincosamides/lincomycin	236,450	10–50
Macrolides/tylosin	563,251	10–80
Penicillins/penicillin G	828,721	80
Sulfonamides/ sulfadimethoxine	384,371	20–50
Tetracyclines/chlortetracycline	6,514,779	60–80

Source: Data from Aga et al. (2016)

increased fourfold after manure from hog and dairy farms was applied to the soil (Moyer 2016). Runoff from farms, feedlots, or cropland can lead to antimicrobial resistance problems in soil, surface water runoff, and groundwater. Animal waste held in lagoons allows birds and insects to become contaminated with antibiotic-resistant bacteria, and flies around food animal facilities can carry antibiotic-resistant enteric bacteria which increases potential human exposure. Migratory birds and seagulls which become infected with antibiotic-resistant bacteria or viruses can widely transmit resistance to other birds as well as marine life.

Others note that antibiotics should never be used to compensate for poor hygiene and husbandry practices or conditions in livestock production (Van Boeckel et al. 2017). Veterinary medicine should only use antibiotics to treat diagnostically determined bacterial infections not otherwise treatable and only those antibiotics authorized for the diagnosed pathogenic indication and the specific bacteria involved. Further, given the potential for acute diseases that require immediate treatment, it is important that routine testing (surveillance) be carried out for farm-specific pathogens for all relevant antibiotic classes (Silley and Stephan 2017). WHO also emphasizes the need for surveillance and monitoring antimicrobial use in food-producing animals to evaluate the extent to which their guidelines are implemented.

15.7 US Policy Response

FDA has increased regulation of antibiotic use in food animals. As noted in Sect. 15.3 above, FDA attempted to withdraw new animal drug approvals for sub-therapeutic uses of human medically important penicillins and tetracyclines in animal feed based on lack of evidence to show they were safe. After industry opposition and Congressional intervention to require further study, this early policy response was withdrawn. Subsequently, the US Congress gave something to each group when it enacted the Animal Drug Availability Act (ADAA) in 1996. This Act both facilitated the approval and marketing of new animal drugs and medicated feeds and gave FDA new regulatory controls. The Act created a new category, Veterinary Feed Directive (VFD), for drugs used in animal feed that could only be used under the professional guidance of a licensed veterinarian. FDA implemented the ADAA VFD provisions through final regulations published in 2000. However, subsequent feedback from various stakeholders led FDA to seek public input on improvements needed in the regulations. Following lengthy delays as shown in Table 15.3, FDA in recent years issued three core documents to implement a policy framework for judicious use of medically important antimicrobial drugs in food animals:

Table 15.3 US policy actions to reduce antibiotics in animal feed

Year	Policy action	Comments
1951	FDA approved antibiotics in livestock feed	Production purposes and therapeutic uses
1970s	FDA proposed ending production-purpose use	FDA 1970 report: antibiotic use might pose human health threat
1975	Antibiotic drug sponsors required to show that products not a human health threat	Result of the 1970 FDA report
1977	FDA proposed withdrawal penicillin, tetracycline sub-therapeutic feed use	Congress directed FDA wait for further study to be conducted
1980	FDA-contracted National Academy of Science report issued	Available data could neither prove nor disprove human health hazards
1980–2003	FDA continued research support into safety of sub-therapeutic antibiotic use	Reviews of research that associated antimicrobial livestock use and human disease resistance
2003	FDA guidance, requiring risk assessment for any new antibiotics for livestock agriculture	Not applicable to majority of antibiotics used in meat production, approved before 2003
2005	FDA withdrew enrofloxacin (type of fluoroquinolone) for poultry production	Lengthy and challenging process to withdraw approval of specific drug use
2010	FDA voluntary guidelines: medically important antibiotics in livestock limited to nonproduction use, with veterinary oversight	Allowed disease treatment, control, or prevention purposes; FDA to rely on voluntary industry response
2011	FDA draft guidance: "judicious use" of antimicrobial drugs in livestock production	Concern that production-purpose use of medically antimicrobial drugs adversely affects human public health
2012	FDA Guidance for Industry # 209 finalized	Limit medically important antibiotic use to assuring animal health; veterinary oversight
2013	FDA guidance document # 213	Detailed information to pharmaceutical manufacturers on withdrawing production uses; veterinary oversight of remaining OTC uses
2014	As of June 2014, all 26 producers of antibiotics used in livestock feed agreed to FDA's request in guidance # 213	It only took 44 years, starting in 1970, to stop production use of antibiotics in food-producing animals

Source: Sneeringer et al., pp. 12–14

- Guidance for Industry (GFI) #209 "The Judicious Use of Medically Important Antimicrobial Drugs in Food-Producing Animals" was published in April 2012, citing primary scientific literature they considered (FDA 2012). It identified several steps for ensuring appropriate judicious use by eliminating feed and water use of medically important antimicrobial drugs for production purposes and bringing remaining therapeutic uses under the oversight of licensed veterinarians.
- GFI # 213 "New Animal Drugs and New Animal Drug Combination Products Administered in or on Medicated Feed or Drinking Water of Food-Producing Animals: Recommendations for Drug Sponsors for Voluntarily Aligning Product Use Conditions with GFI # 209" was published in December 2013. It detailed a process and timeline for implementing GFI #209 measures. When fully implemented, feed use of antimicrobial drugs would transition from over-the-counter (OTC) to VFD marketing status (Veterinary Feed Directive 2015).
- The third element in this process is the VFD regulation issued in June 2015. The FDA amended regulations for Veterinary Feed Directive (VFD) drugs used in animal feeds to improve the efficiency of the program but still protect human and animal health. VFD drugs are new animal drugs for use in feed, allowed only under professional supervision of a licensed veterinarian who issues an order (VFD) under a valid veterinarian-client-patient relationship (VCPR). The VFD rule is the final document to implement FDA's policy framework for judicious use of medically important antimicrobial drugs in food-producing animals. It eliminates feed and water use of medically important antimicrobial drugs for production purposes. Remaining therapeutic uses of these drugs are brought under oversight of licensed veterinarians. The final rule provides accountability through important controls regarding distribution and use of VFD drugs. It facilitates transition of the currently large number of over-the-counter (OTC) feed-use antimicrobial drugs to a new VFD status (Veterinary Feed Directive 2015).

On January 3, 2017, FDA announced that it had completed implementation of the Guidance for Industry #213. This means that medically important antimicrobials provided to food-producing animals may no longer be used for growth promotion purposes and may be used to treat, prevent, or control animal illnesses only under direction of a veterinarian. FDA worked with industry participants to implement this voluntary compliance to slow development of antimicrobial resistance and preserve effectiveness of medically important antibiotics. More than 70 percent of 292 new drug applications subject to GFI #213 were converted from over-the-counter to prescription status, 84 applications were withdrawn, and all 31 applications indicating production use withdrew that specified use. FDA also indicated plans to work with industry stakeholders to support antimicrobial stewardship in food animal production and to evaluate the effectiveness of strategies to reduce antimicrobial resistance development under the allowed uses (FDA 2017).

15.8   Industry Stakeholders and Responses

Some industry stakeholders in the supply chain are actively engaged in responding to consumer and general public health concerns about AR in the food supply chain amidst mounting scientific evidence, but responses vary considerably by country, place in the supply chain, and individual company. Aside from farm groups, stakeholders include feed companies, pharmaceutical companies, integrators or meat processors, restaurant chains and other retail outlets, and consumer and other interest groups.

Pharmaceutical Companies. In the case of pharmaceutical companies, little evidence exists that they are responding to the AR problem yet. As described earlier, most antibiotics are produced in India and China, and their production has resulted in significant risk, especially environmental risk. Regulators need to set minimum standards for the treatment of manufacturing waste before it is released into the environment. Other industries which purchase these pharmaceuticals need to establish higher standards through their supply chains to help correct this environmental pollution (O'Neill 2016).

Furthermore, the drug companies are not required to compensate victims who become ill or die from either the environmental or food exposure. The drug companies and their trade associations have resisted more regulation to prevent misuse of antibiotics. The companies therefore have been getting a "free ride" at the expense of the ill consumers and the general public.

Integrators and Meat Processors. Some chains and food retailers have recently responded to customer concerns by restricting the use of antibiotics in their food supply chains. Large meat processors committing to judicious use of antibiotics have already led many producers to eliminate the use of antibiotics for production enhancement purposes.

In a case study of voluntary labeling in the broiler industry, "Raised without Antibiotics" (RWA) label claims by Tyson Foods and by Perdue Farms in 2007, respectively, numbers one and three in total broiler production, resulted in mixed outcomes. At that time, USDA FSIS had not published a standard for such claims, nor was a clear definition established. Perdue and Tyson developed their own standards and submitted the label claims to FSIS for approval along with supporting documentation. After initially approving both firms' label claims, FSIS determined in September 2007 that Tyson's claim was false and misleading and gave them the opportunity to submit a revised label claim. However, Tyson continued their advertising of the RWA claims. The diverse label claims in which Tyson and their competitors were using different standards for their claim resulted in consumer confusion, and eventually court challenges were filed jointly by Sanderson Farms, the fourth largest producer, and Perdue against Tyson. The suit was upheld in court in April 2008. Tyson was found not to have delivered the RWA attribute promised to the marketplace and to thereby have harmed competitors, while Tyson profited from introducing a false and misleading claim. In June 2008, FSIS rescinded Tyson's qualified RWA label claim and required its removal within 2 weeks, after the claims and advertising had continued for more than a year. The authors found no evidence that the events had any impact on Tyson's brand, suggesting that companies may have incentives to introduce misleading label claims since the size of penalties is uncertain (Bowman et al. 2016).

Perdue Farms Inc. was the only major chicken producer to eliminate all medically important and animal-specific antibiotics from use in its chicken production as of 2016. By replacing antibiotics with vaccines and improving its production facilities and practices, it has been able to produce chicken at virtually the same cost as when using antibiotics. Perdue estimates that its conventional chicken sales are increasing by not more than 3% annually, while sales for product raised without antibiotics are growing 15–20% annually (Bunge 2016).

GNP Company, a leading provider of premium natural chicken products, is adopting antibiotics-free production of chicken products. Its Gold'n Plump brand will feature a "No Antibiotics-Ever" claim. This will go well beyond what many companies are currently focusing on—eliminating the use of medically important antibiotics, rather than all antibiotics. USDA regulations allow this label claim only for chicken never having received antibiotics, even inside the egg. The company will continue to treat flocks for illness as necessary, but not market them under their premier Gold'n Plump brand. The company plans extensive media and in-store support to educate consumers about the transition to its chicken products raised totally without antibiotics (GNP Company 2015).

Tyson Foods, a leading producer of chicken, pork, and beef and products thereof, adopted a position to eliminate the use of human-use antibiotics in broiler production by September 2017. They stopped the human antibiotic use in their hatcheries and reduced usage in producing broilers by 80% since 2011. They also have worked with farmers and others in the beef, pork, and turkey supply chains to explore ways to reduce human antibiotic use at the farm level. Tyson is employing alternative husbandry strategies such as use of probiotics and essential oils, improved housing, and selective breeding to offset the potential impact of eliminating the use of the antibiotics. They are also interacting with the food industry and other involved supply chain participants, as well as academics, to increase research on disease prevention and alternatives to replace antibiotics (Tyson Foods 2017).

Feed Companies. The feed companies are also getting into the discussion to address public health concerns about antibiotic resistance and the relationship to livestock production uses of antibiotics. Phibro Animal Health Corporation recently launched a website AnimalAntibiotics.org to "…provide accurate and credible information while still creating open dialogue about animal agriculture in the use of antibiotics." It will address all issues involving animal antibiotics and changes underway within the industry to promote responsible use of antibiotics in livestock (Johansen 2016). This is very consistent with the historical pattern of the animal agriculture industry making its case in the political economy in reaction to the strong push to limit use of antibiotics to help quell rising antibiotic resistance of medically important drugs.

Restaurant Chains. An interesting example of restaurant chains and poultry producers working together is provided by Panera Bread Co. and Perdue Farms Inc.

Panera is one of the restaurant companies for which Perdue supplies chickens raised without antibiotics. When Panera pioneered antibiotic-free chicken in its restaurant products over 10 years ago, they paid a 50% premium versus chicken produced using antibiotics. With improved production practices, the cost differential has virtually disappeared (Bunge 2016) and is thus consistent with the ERS estimates cited earlier.

Consumer and Other Interest Groups. In the process of developing these new FDA regulations, activist groups petitioned the Federal Courts. For example, in May 2011, the Natural Resources Defense Council (NRDC), Center for Science in the Public Interest (CSPI), Food Animal Concerns Trust (FACT), Public Citizen, and Union of Concerned Scientists (UCS) filed a case against the FDA. They charged that FDA failed to ban penicillins and tetracyclines used at low doses in animal feed for growth promotion, despite evidence FDA put forth in 1977 that penicillin and most tetracyclines were not shown to be safe and may pose a risk to human health (APUA 2016). In 2012, the Federal Court ruled in favor of these petitioners. In a later ruling in 2012, the Federal Court directed FDA to reexamine its decision on five other classes of "medically important drugs" used as growth promoters addressed in two citizen petitions (filed in 1999 and 2005) to ensure the safety and effectiveness of all drugs sold in interstate commerce (Ibid).

Given that most governments have neglected to acknowledge and address the problem of increasing antibiotic resistance, international organizations with a role in health issues have been stymied from doing so. It will take more concerted action by societies around the globe to successfully address this cross-border issue (Amábile-Cuevas 2016).

15.9   Consumer Demand for Labeling of Antibiotic Use

US consumers, in general, have much less information about the product than does the seller (Chaps. 2 and 3). This asymmetric information can offer opportunity for the selling firm behavior that is detrimental to the interests of the consumer, as when a product is labeled as containing or not containing certain desirable or undesirable attributes. In the case of many products known as credence goods, it is impossible to determine whether the attributes are as stated, even when the product is used or consumed. This market failure can be addressed either through government regulations or through voluntary steps by the sellers to assure that the stated attributes are factual. The latter could be accomplished through advertising to build and maintain the firm's brand and reputation, and competition with other sellers could result in consumers having increased variety of product choices. However, some consumers may not trust private companies' word about product attributes and prefer certification programs which monitor products against some standard established either by the private sector or by government agencies.

Lusk (Lusk 2013) argues that voluntary labels are dynamically efficient in responding to changes in market conditions and encouraging innovation more than mandatory labels implemented through regulations, since the latter are more subject to manipulation by those with vested interests. Further, USDA's Agricultural Marketing Service (AMS) process-verified and certification programs are very effective in helping to assure the credibility of voluntary labeling, while accommodating innovation from the private sector. GNP's adoption of antibiotics-free production discussed in 15.8 is an example of dynamic market efficiency through use of voluntary labeling to innovate in response to changing consumer demand

USDA's Food Safety Inspection Service (FSIS) currently employs an animal production claims protocol for evaluating and allowing or denying labeling claims. Labeling applications must provide supporting documentation such as operational protocols detailing production practices and affidavits or testimonials about production practices. FSIS then evaluates whether protocols support the accuracy of the proposed label. Also, feed formulations must be provided and reviewed to ensure they do not include substances not permitted by the claim. Commonly approved claims relevant to the use of antibiotics include "raised without added hormones" (only allowed for use in beef cattle and lamb production) and "raised without antibiotics." Claims not allowed include that animal products are antibiotic-, hormone-, or residue-"free" (FSIS 2016). Given the current trend among meat producers, restaurants, and retail livestock product marketers, it can be anticipated that there will be increasing attention to labeling the lack of antibiotic use for other than therapeutic purposes. This will likely result in animals that have been raised with antibiotics to promote growth and uniformity of size consistent with processor contract agreements being diverted to marketing outlets where such promises do not exist. The impact of labeling in this manner will vary according to how much consumers know about the use of antibiotics in livestock production and their ability to currently purchase antibiotic-free livestock products (Lusk et al. 2006).

O'Neill and his British colleagues emphasize improving transparency as a major step in addressing antimicrobial resistance related to the livestock production. Recent attention by companies such as food retailers, wholesale producers, and fast-food chains, as well as investors, for reducing antibiotic use in their supply chains, has been in response to consumer pressure. Providing greater transparency through voluntary approaches is helpful in the short term, but it may be necessary to mandate transparency requirements about how antibiotics are used in the supply chains to have longer-term impacts. Labeling that refers to antibiotic use could improve consumer knowledge to allow them to make better informed decisions. They also argued that third-party validation of support from independent institutions to monitor progress may be beneficial (O'Neill 2016).

15.10 Surveillance, Data Gaps, and Transparency

Improved transparency by food producers about antibiotics used in producing meat could help consumers make better informed purchasing decisions. But there are large gaps in data needed to allow monitoring of types and quantities of antibiotics used in animal agriculture and their impacts (CFI 2016), as well as on emergence and spread of resistance in animals. The WHO also identified major gaps in surveillance and data sharing on emergence of antibiotic resistance in bacteria and its impact on animal and human health. WHO called for integrated surveillance systems harmonized across countries to enable better comparison of data from food-producing animals, food products, and humans (WHO 2014).

In the United States, FDA requires drug companies to voluntarily submit data on drugs sold for use in food animals. The publicly available data are not detailed, and 97% of the sales of medically important antimicrobials are over-the-counter (OTC). Tetracyclines are primarily added to feed and accounted for 71% of domestic sales of animal drugs that are "medically important" to human medicine in 2015 (FDA 2016b). From 2009 to 2015, domestic sales and distribution of tetracycline products approved for use in food-producing animals increased by 31%. While Levy et al. (1976) discovered how rapidly tetracycline created antibiotic resistance in the gut of chickens, 40 years later, the public does not have access to information on what antibiotics are used in which food animals at what stage of life.

This will change somewhat in FDA implementation of GFI # 213 (FDA 2017) that will identify whether the sales are intended for use in cattle, sheep, hog, or poultry. FDA (2016b) issued a final rule amending an existing requirement that sponsors of drug products containing antimicrobial active ingredients report annually the amount of each such ingredient in the drug products sold or distributed for use in food-producing animals. Effective July 11, 2016, drug sponsors were required to submit species-specific estimates of product sales as a percent of their total sales. Additional reporting requirements are expected to facilitate better understanding of antimicrobial drug sales for specific food-producing animal species and the relationship between such sales and antimicrobial resistance. As reported above, drug sponsors have all adopted voluntary revision of FDA-approved labels for use of new medically important antimicrobial animal drugs administered through feed or water. Under this rule, sponsors all voluntarily removed the growth promotion and feed efficiency uses and brought the remaining therapeutic uses under veterinarian oversight by the end of December 2016. The rule makes it illegal to use medically important antibiotics for production purposes. Despite the scientific and economic evidence, many comments to the proposed final regulation reflected ongoing resistance to the elimination of food animal production use of medically important antibiotics.

Data available on antibiotics used in the US livestock industry is derived primarily from two nationally representative surveys of farms conducted by the USDA's Economic Research Service (ERS) and National Agricultural Statistics Service (NASS). The Agricultural and Resource Management Survey (ARMS) is designed to collect information on farm finances, production practices, and resource use focuses on three commodities annually, livestock included. Different types of livestock are resurveyed every 5–6 years and represent commercial producers in states producing 90% of production for that livestock type. Some questions have been included in these surveys on antibiotic use for hogs and broilers. The hog surveys ask about use of antibiotics in feed or water for growth promotion, disease prevention, and/or disease treatment in breeding, nursery, and finishing hogs. Given the widespread use of hog production contracts under which farm operators may receive feed from integrators, the surveyed operators may not know if antibiotics are included in it. For broilers, there is only a single question about whether they were raised without antibiotics in feed or water other than for therapeutic treatment of illness. Production contracts dominate the broiler industry, so surveyed farm operators are in a similar situation as hog producers in not necessarily knowing whether antibiotics are included in the feed provided. A further complication is that ARMS does not separate traditional antibiotics and ionophores, which are not used in human medicine (Sneeringer et al. 2015).

The National Animal Health Monitoring System (NAHMS) consists of national studies to provide essential information on livestock and poultry health and management. Major food livestock species are surveyed about every 5 years to provide current and trend information important to industry participants, researchers, and policy makers. Each study includes states that represent at least 70% of the targeted animal population and at least 70% of the farm operations involved and provides statistically sound information for decision-making. A NAHMS study is a collaborative, voluntary, confidential, scientifically sound product. Descriptive reports are prepared along with information sheets which briefly address very specific topics, such as biosecurity practices (APHIS 2016). The NAHMS focuses on animal health and management, providing information on disease occurrence and disease prevention practices, as well as more detailed information on antibiotics used in production, including by specific purpose. However, the information collected on antibiotics varies greatly across commodities, as well as over time with the same commodities. Further, ARMS focuses on hog production operations with 25 or more head versus NAHMS focus on 100 or more head. This complicates comparison of statistics across surveys, assuming smaller operations may have different characteristics than larger ones (Sneeringer et al. 2015).

To track antimicrobial resistance changes over time, the National Antimicrobial Resistance Monitoring System—Enteric Bacteria (NARMS) was established by CDC in 1996. The program is a collaboration between state and local public health departments and three federal agencies to monitor changes in antimicrobial susceptibility for certain enteric bacteria from ill people (CDC), retail meats (FDA), and food animals (USDA) in the United States. It provides information about emergent bacterial resistance, the ways resistance is spread, and how resistant infections differ from susceptible ones (NARMS 2016).

15.11 Global Responses to Antimicrobial Use in Livestock

The World Organization for Animal Health (OIE) plans to address antimicrobial resistance as a major risk to the international community, in the face of concern about agriculture's role in increased antimicrobial resistance. The goal is to preserve effectiveness of antimicrobials used in animal medicine, protect animal welfare, and help maintain important antimicrobials for use in human medicine. OIE has already developed international standards, most recently revised in 2015. The new strategy introduced at the 84th OIE General Session in May 2016 outlines plans to help nations improve legal frameworks to preserve antibiotics, communicate about the AR problem, train animal health workers, and monitor antibiotic use in animals. They are currently working to create a database of information on the use of antimicrobial agents in animals and develop performance indicators to assist countries by increasing information flow and transparency in their use of antimicrobials. Further, the OIE expert network is working to reinforce scientific knowledge about new technologies and replacement solutions for current antimicrobials (Mitchell 2016).

The EU has banned the use of antimicrobials in food animal production, other than by veterinarian prescription for specific therapeutic use. Some other countries have adopted similar bans, and, as discussed above, the United States fully implemented voluntary guidelines in 2016 requiring current drug sponsors to withdraw antibiotics for growth promotion. However, the Animal Health Institute's Carnevale has said that the new FDA guidance on antibiotics may not decrease the total quantities of antibiotics used in animal food production (Moyer 2016). Generally, variations among countries in implementing regulations have resulted in the spread of resistance. There is ample evidence to support the need for global coordination to prevent continued spread of antimicrobial resistance, and elements of a framework to make such global coordination effective have been posited (So et al. 2015).

The UK Review on Antimicrobial Resistance Final Report proposes three broad steps to deal with reducing unnecessary use of antibiotics in animals. First, establish 10-year targets for reduction in use, with milestones to support progress consistent with countries' economic development. This could encourage farmers to reduce non-therapeutic use to be able to allocate the resulting reduced amounts of antibiotics to treating sick animals. Second, implement restrictions or bans on certain types of highly critical last-line antibiotics for humans from being used in agriculture. This would require a harmonized approach to identify the most important human health antimicrobials across countries and good systems of veterinarian oversight to assure compliance. Third, improve transparency from food producers on antibiotic use in meat production to allow consumers to make better informed buying decisions. Voluntary industry efforts may be one of the most practical approaches to reduce antibiotic use in the near term, but third-party validation to monitor progress would be beneficial (O'Neill 2016). Generally, voluntary industry approaches require monitoring by an outside party to assure both industry participants and consumers that standards are being met as required.

The UK Medical Research Council (MRC) recently made three large grants focused on antimicrobial resistance through an initiative established in 2014 to address the growing AR issue. The projects will use new technology to exploit natural compounds, develop a better and faster diagnostics tool, and study how the body's immune system can be harnessed to better fight infections. The goal is not only to develop antibiotics but also explore alternatives to antibiotic use, working with other UK research councils to bring to bear a wide range of disciplines to tackle AR (MRC 2016).

The need to focus increased attention to developing new antibiotics is supported by CDC data which shows that many of the most widely used drugs have developed resistance. The number of years to develop resistance varies greatly but never extends more than a couple of decades, and more recent antibiotic introductions have been resistant for only a year or two. For example, the widely used tetracycline was introduced in 1950 and developed resistance to *Shigella* by 1959. This is near the midrange of years to resistance reported (CDC 2013a). Given this scientific fact, the slow pace of adopting policies to proscribe use of human-use antibiotics in animals and to encourage greater investment in developing newer antibiotics or alternatives is unacceptable. Increasing detection of bacteria resistant to last-resort drugs has driven stakeholders to countenance accelerating government efforts to increase surveillance of drug use and to develop new antibiotics (FDA Week 2016). Promising approaches which provide more rapid assessment utilizing newer technologies such as genomics are now being utilized by scientists.

Microbiologists are embracing high-throughput genomics to quickly examine individual organisms or entire microbial communities. A project underway at the University of California, Davis, the *100 K Foodborne Pathogen Genome Sequencing Project*, will sequence 100,000 foodborne isolates for the most important worldwide foodborne illness outbreak organisms. It involves a consortium of academic, government, and industry to create a massive database of genome signatures for the most significant foodborne disease-causing microbes. The goal is to allow public health agencies and the food industry supply chain to trace any foodborne illness outbreaks to their source. By comparing the pathogen genome to the database which includes millions of pieces of information on previously detected strains, including their exact location, the contamination source will be positively identified. Bioinformatics and the analytics involved can be used to turn the vast amount of genomic information into actionable knowledge. These event sequencing approaches will enable new diagnostic and public health approaches to manage foodborne disease to facilitate improved public health. The database will increase ability to detect and mitigate pathogenic organisms in food, the environment, and livestock. That capacity is now constrained by continual genetic evolution of pathogens which hinders the ability to defend the food supply. This project will facilitate speedy testing of raw ingredients and finished products from outbreak investigations with precision and accuracy unparalleled using existing methods of analysis. Genomics enabled diagnostics with molecular tools will allow surveillance, risk assessment, and diagnosis of foodborne pathogens directly throughout the global food chain. The result will be a genetic catalog for some of the most important outbreak organisms impacting human health. The database will provide insights into molecular methods of infection and drug resistance for use in creating new vaccines and therapies. And importantly, it will assist in systematic definition of biomarker gene sets associated with antibiotic resistance (Weimer 2016).

A recent innovative metagenomics study also provides new insights on possible impacts of antibiotic use in food animal production and AR in humans. The research investigated antimicrobial resistance potential—the resistome—by tracking specific pens of intensively managed cattle from feedlot through slaughter to market-ready beef products. Study results found no antibiotic-resistant determinants (ARDs) in the beef products beyond the slaughter facility. This suggests that intervention during slaughter minimizes potential for antibiotic-resistant determinants passing through the food chain. The results also highlight potential risks through indirect environmental exposures to the feedlot resistome through wastewater runoff, manure application on cropland, and wind-borne particulate matter. The insights provided can be used to better inform future agricultural and public health policy. However, this first of its kind study suggests the scientific community must develop a better understanding of the risk of different resistomes and resistance genes. It also identifies a pressing need to standardize ARD nomenclature so that databases and analyses are comparable across studies (Noyes et al. 2016).

The World Health Organization has recently developed guidelines to mitigate human health consequences from use of medically important antimicrobials in food-producing animals (WHO 2017). The guidelines are evidence-based recommendations and include best practices for use of medically important antimicrobials in food-producing animals, especially antimicrobials deemed critically important to human medicine. They also can help preserve effectiveness of antimicrobials for veterinary medicine. The recommendations include:

- An overall reduction in use of all classes of medically important antimicrobials in food-producing animals.
- Complete restriction for use in growth promotion.
- Complete restriction of use to prevent infectious diseases that have not yet been clinically diagnosed.
- Antimicrobials designated as critically important for human medicine should not be used to control spread of clinically diagnosed infectious disease identified within a group of food-producing animals, nor for treatment of food-producing animals with a clinically diagnosed infectious disease.
- For best practices, any new class of antimicrobials for use in humans will be considered critically important for human medicine unless otherwise categorized by WHO. Further, medically important antimicrobials not currently used in food production should not be so used in the future.

These guidelines apply universally, and improved animal health management can be used to reduce the need for antimicrobials including improvements in disease prevention strategies, housing, and husbandry practices as noted in Sect 15.5 above.

Economic incentives in regulations were addressed in a recent article. In some European countries, capping total antimicrobial use per animal through regulations has been successful in reducing use by more than half while maintaining competitive livestock sectors. The second option was to impose user fees on veterinary antimicrobials, applied at the point of manufacture or wholesale purchases for imported products, which could also reduce use significantly. As a policy option, some combination of these two strategies would significantly reduce antimicrobial use in food animal production (Van Boeckel et al. 2017). Finally, as discussed in Chap. 12, Sweden does not allow use of antibiotics in broiler production. If there is the political will, strong regulations can provide strong economic incentives to control antibiotic use.

15.12   Need for Producer Education

To promote the understanding and implementation of the FDA's new Veterinary Feed Directive, the Farm Foundation and the Pew Charitable Trusts sponsored a series of meetings with livestock and farming communities throughout the United States. Twelve educational workshops provided livestock producers, feed suppliers, veterinarians, and support service organizations information and insights on the new policies. The workshops also provided opportunity for participants to interact with FDA and USDA's Animal and Plant Health Inspection Service (APHIS) personnel about implementation challenges.

Among livestock producers attending the workshops, small- and medium-sized operators, as well as many veterinarians, were unaware of the pending requirements. Lack of understanding about responsibilities under the revised VFD rule means that producers and veterinarians need education. Some land grant university extension services are now offering balanced education programs to inform these audiences about their obligations going forward, rather than having interested parties in the food animal industry be the primary source of information to producers and veterinarians about the requirements.

While seeing positives of improved public perception and livestock management as result of the new rules, workshop participants were concerned about increased costs in animal health due to restrictions on access to antibiotics and lack of veterinary services. Perhaps the biggest challenge is that many small producers do not have established relationships with veterinarians needed to establish a veterinarian-client-patient relationship (VCPR). This may be particularly challenging in remote rural and urban fringe areas where fewer veterinarians are available to treat food-producing animals.

In sum, workshop participants saw a need for education and outreach; continuing dialogue between industry representatives, consumers, and state or federal regulators; and the need to provide better access to veterinary services for food animals.

15.13   Conclusions

There is widespread agreement that the scientific evidence indicates a global human health and environmental crisis due to antibiotic resistance, in part resulting from production practices in animal agriculture. Government action in regulating the animal agriculture industry, to date, has done little to slow the advance of AR. Most countries still need to pass legislation to establish appropriate conditions for the import, manufacture, distribution, and use of veterinary products, including antimicrobial agents. Continued easy access to antimicrobial drugs for use on the farm is not acceptable. Important stakeholders in the animal production industry include pharmaceutical companies, feed companies, livestock production integrators, and some farm groups, each with their own set of incentives and supporters. They must be engaged in the effort to reduce agriculture's role in contributing to development of AR, which CDC estimates at 20% (Fig. 15.1). Even so, other major industry groups must be engaged to significantly reduce their 80% contribution to resistance development.

In the United States, some progress was made with the passage of the Animal Drug Availability Act in 1996 and its very gradual implementation through various regulations over the past two decades. However, there are serious gaps in these regulations. Given the gridlock that has prevailed in the US Congress and the power of the pharmaceutical lobby at the national level, state actions are leading the way to responsive regulation in the public interest. For example, California is the first US state to prohibit all human antibiotic use in food animal production.

In contrast to the halting actions of governments and industry, consumer and interest group actions are being at least partially successful in getting fast-food and retail establishments to not market animals fed human-use antibiotics for growth-promoting purposes. This suggests that a productive approach may be finding ways to provide information to and educate consumers about the risks of antibiotic resistance to enable them to make better informed decisions. There is an important role for educators to extend scientific information in a nontechnical way to the lay public.

The drive to use antibiotics more responsibly and in the public interest may be facilitated by recent economic results that show that reducing antibiotic use in animal production need not come at a significant economic cost to producers or consumers. Since the benefits of using antibiotics for livestock growth promotion appear to have resulted in increasingly smaller productivity gains, independent producers where input mix is a farmer-driven choice based on farm-level economics may be better off to substitute good management practices rather than using antibiotics for prophylactic disease prevention. However, much of meat animal production on US farms is produced under contract, where the integrator provides inputs, often including antibiotics, that the grower is required by contract to use. Recent actions by integrators and meat processors to reduce antibiotic use and substitute alternative strategies to protect animals from diseases and maintain productivity are an important development, especially since production-purpose use of antibiotics is now prohibited.

Presuming that any new human-use antibiotic classes will probably not be made available for veterinary medicine, it is important to preserve the effectiveness of existing antibiotics. Some policy makers and industry now recognize the urgency to identify new antibiotics. This will require increased antibiotic research funding and judicious use of existing antibiotics. The ban of human-use antibiotics in animals for production purposes is expected to help slow the growth of antimicrobial resistance, giving more time to discover new antibiotics for animal uses and for human health uses. To the extent they can be developed and used separately, the potential for animal antibiotic use leading to antimicrobial resistance for important human antibiotics will be mitigated.

The ban on antibiotics used for humans also being used for animal production purposes will necessitate adopting improved cultural practices to reduce the potential for disease and to increase feed efficiency. This calls for research on best management practices to accommodate today's supply chain requirements for food safety, production efficiency, and attribute verification. Moreover, there is a need to educate producers—for example, through the USDA-State Cooperative Extension Service—about safe production practices for managing AR. This will allow producers to maintain efficiency in their operations and assure that they comply with current regulations to address the growing concern about antimicrobial resistance in the food supply chain. Improved data collection and analysis to allow tracking of potential antimicrobial resistance development are essential to facilitate the food animal industry implementation of cultural practices to reduce the potential for contributing to antimicrobial resistance. It would also allow policy makers to better understand the need for any necessary interventions. These investments in the public good can be very cost-effective, though not without additional public investment or internal agency budget reallocation.

Increasing international trade may spread antibiotic resistance through imported food products as more trade agreements are approved. This scenario could be exacerbated to the extent FSIS approves additional international facilities, local regulations, and inspections as "equivalent to the United States." In many developed and developing countries, antimicrobial agents are readily available. Policies need to be implemented establishing appropriate conditions for use of veterinary products, including antimicrobial agents. Future trade agreements will need to include provisions which address reduced use of medically important antibiotics in producing food animals.

To date, there is no harmonized system of surveillance on the worldwide use and circulation of antimicrobial agents. That information is necessary to monitor and control the origin of medicines, obtain reliable data on imports, trace their circulation, and evaluate the quality of the products in circulation. The OIE initiative to create a global database for monitoring the use of antimicrobial agents is an important step that can provide valuable information for private sector and public policy leaders worldwide.

The serious implications of growing antibiotic resistance require a concerted effort across all stakeholders and society generally. Increased attention to this issue is emerging in the medical community where overuse of existing drugs and inadequate sanitary precautions account for 80% of the resistance. Lack of development of new antibiotics exacerbates the problem, and industry focus and perhaps government policy are needed to improve this situation. Animal agriculture stakeholders need to improve production practices to reverse the other 20% of resistance attributable to foodborne sources. Government policy and agencies have been slow to acknowledge the seriousness of antibiotic resistance and appropriately address it. Public and private sector collaboration internationally is necessary to successfully deal with this critical societal issue.

Acknowledgments We deeply appreciated the conversations with and review comments from Mary Ahearn in developing and writing this chapter. Any remaining errors, mistakes, or omissions are of course ours.

References

Aga DS, Lenczewski M, Snow D, Muurinen J, Sallach JB, Wallace JS. Changes in the measurement of antibiotics and in evaluating their impacts in agroecosystems: a critical review. J Environ Qual. 2016;45:407–19.

Amábile-Cuevas C. Society must seize control of the antibiotics crisis. Nature. 2016;533:439. https://doi.org/10.1038/533439a. Accessed 26 May 2016.

APHIS. National Animal Health Monitoring System (NAHMS). Animal and Plant Health Inspection Service, United States Department of Agriculture. 2016. https://www.aphis.usda.gov/aphis/ourfocus/animalhealth/monitoring-and-surveillance/nahms/ct_national_animal_health_monitoring_system_nahms_home. Accessed 10 May 2016.

APUA. Major developments in U.S. policy on antibiotic use in food animals. Alliance for the prudent use of antibiotics. http://emerald.tufts.edu/med/apua/policy/policy_antibiotic_food_animals.shtml, Accessed 26 Dec 2016.

APUA Newsletter. Farm antibiotic use remains worrisome in India. Alliance for the prudent use of antibiotics. APUA Newsletter. 2016;34(1(Spring)):15.

Bowman M, Marshall KK, Kuchler F, Lynch L. Raised without antibiotics: lessons from voluntary labeling of antibiotic use practices in the broiler industry. Am J Agric Econ. 2016;98(2):622–42. https://doi.org/10.1093/ajae/aaw008.

Bunge J. Perdue chickens now free of antibiotics. Wall St J. 2016:B3.

Carnevale R. Antibiotic resistance and the use of antibiotics in animal agriculture: hearing before the subcommittee on health of the house committee on Energy & Commerce, 1__th Cong., 1st Sess. (?) (2010). http://www.ahi.org/wp-content/uploads/2011/06/July-14-2010-Testimony-by-Richard-Carnevale.pdf. Accessed 17 July 2016.

Carnevale RA. Across the divide: the importance of antibiotics for animal health. 2016. http://asmcultures.org/3-1/7/. Accessed 3 Apr 2017.

CDC. Antibiotic resistance threats in the United States, 2013. Centers for disease control and prevention. 2013a. http://www.cdc.gov/drugresistance/threat-report-2013/pdf/. Accessed 29 Sep 2016.

CDC. CDC 2011 estimates of foodborne illness in the United States. Centers for disease control and prevention. 2013b. http://www.cdc.gov/foodborneburden/pdfs/factsheet_a_findings_updated4-13.pdf. Accessed 22 May 2016.

CFI. Tracking animal antibiotic use in food animals. The center for foodborne illness research and prevention. 2016. https://www.foodborneillness.org/cfi-library/2016_CFI_Fact.Sheet_Track.animal.antibiotic.use_FINAL.pdf. Accessed 18 May 2016.

Consumer Reports, America's Antibiotic Crisis. Part 3. 2016. http://www.consumerreports.org/media-room/press-releases/2015/11/consumer-reports-meats-produced-without-antibiotics-harbor-fewer-superbugs/. Accessed 6 Oct 2016.

Council for Agricultural Science and Technology (CAST). 1981. Antibiotics in animal feeds. Report No. 88. Ames, Iowa.

FDA. The judicious use of medically important antimicrobial drugs in food-producing animals. Guidance for Industry #209. 2012. https://www.fda.gov/downloads/AnimalVeterinary/GuidanceComplianceEnforcement/GuidanceforIndustry/UCM216936.pdf. Accessed 3 June 2016.

FDA. 2014 Summary report on antimicrobials sold or distributed for use in food-producing animals. 2015. https://www.fda.gov/downloads/ForIndustry/UserFees/AnimalDrugUserFeeActADUFA/UCM476258.pdf. Accessed 3 June 2016.

FDA. 2015 Summary report on antimicrobials sold or distributed for use in food-producing animals. FDA, Department of Health and Human Services. 2016a. https://www.fda.gov/downloads/ForIndustry/UserFees/AnimalDrugUserFeeActADUFA/UCM534243.pdf

FDA. Anti-microbial animal drug sales and distribution reporting. FDA, Department of Health and Human Services. 2016b. https://www.gpo.gov/fdsys/pkg/FR-2016-05-11/pdf/2016-11082.pdf. Accessed 22 May 2016.

FDA. FDA announces implementation of GFI #213, outlines continuing efforts to address antimicrobial resistance. 2017. https://www.fda.gov/AnimalVeterinary/NewsEvents/CVMUpdates/ucm535154.htm. Accessed 27 Feb 2017.

FDA Week. Stakeholders: colistin-resistance shows need for congressional action. 2016. http://insidehealthpolicy.com/fda-week/stakeholders-colistin-resistance-shows-need-congressional-action. Accessed 5 June 2016.

FSIS. Animal production claims: outline of current process. FSIS, U.S. Department of Agriculture. 2016. http://www.fsis.usda.gov/OPPDE/larc/Claims/RaisingClaims.pdf. Accessed 06 Oct 2016.

GNP Company. Gold'n Plump To Add 'No Antibiotics-Ever' and Humane Certified Attributes. 2015. Perishable News.com. http://www.perishablenews.com/print.php?id=0050530. Accessed 6 Oct 2016.

Institute of Medicine. Report of a study: human health risks with the subtherapeutic use of penicillin or tetracyclines in animal feed. Committee on human health. Washington, DC: National Academies Press; 1988.

Johansen J. Accurate, credible info on animal antibiotics. Phibro Animal Health. 2016. AnimalAntibiotics.org. Accessed 5 Aug 2016.

Joint Committee on the use of Antibiotics in Animal Husbandry and Veterinary Medicine. 1969. Report presented to parliament by the secretary of state for social services, the secretary of state for Scotland, the Minister of Agriculture, Fisheries and Food and the Secretary of State for Wales by Command of Her Majesty.

Laxminarayan R, Chaudhury RR. Antibiotic resistance in India: drivers and opportunities for action. PLoS Med. 2016;13(3):e1001974. https://doi.org/10.1371/journal.pmed.1001974. Accessed 5 Aug 2016.

Levy SB, FitzGerald GB, Macone AB. Changes in intestinal Flora of farm personnel after introduction of a tetracycline-supplemented feed on a farm. N Engl J Med. 1976;295:583–8. https://doi.org/10.1056/NEJM197609092951103. Accessed 5 Aug 2016.

Lusk JL. In: Armbruster WJ, Knutson RD, editors. Consumer information and labeling. New York: Springer; 2013.

Lusk JL, Norwood FB, Pruitt JR. Consumer demand for a ban on antibiotic drug use in pork production. Am J Agric Econ. 2006;88(4):1015–33. https://doi.org/10.1111/j.1467-8276.2006.00913.x. Accessed 29 Sep 2016.

MacDonald JM, Wang SL. Foregoing sub-therapeutic antibiotics: the impact on broiler grow-out operations. Appl Econ Perspect Policy. 2011;33(1):79–98. Accessed 1 June 2016.

Mason M, Mendoza M. Pressure rises to stop antibiotics agriculture. PhysOrg. 2009. http://phys.org/news/2009-12-pressure-antibiotics-agriculture.html. Accessed 24 May 2016.

Miller M. Antibiotic discussions intensify in WDC. Pork Network. 2011. http://www.porknetwork.com/pork-news/antibiotic-discussions-intensify-in-wdc-114024789.html. Accessed 24 May 2016.

Mitchell A. World animal health organization sets out action on antibiotic resistance. Poultry News. 2016. http://www.thepoultrysite.com/poultrynews/37091/world-animal-health-organisation-seeks-action-on-antibiotic-resistance/. Accessed 25 May 2016.

Moyer MW. The looming threat of factory-farm superbugs. Sci Am. 2016;315(6):70–9.

MRC. MRC announces cross-council awards worth nearly £10m to tackle antibiotic resistance. Medical Research Council. 2016. http://www.mrc.ac.uk/news/browse/mrc-announces-cross-council-awards-worth-nearly-10m-to-tackle-antibiotic-resistance/. Accessed 25 May 2016.

NARMS. National antimicrobial resistance monitoring system for enteric bacteria human isolates surveillance report for 2014 (Final Report). Atlanta, GA: U.S. Department of Health and Human Services, CDC; 2016. https://www.cdc.gov/narms/pdf/2014-annual-report-narms-508c.pdf. Accessed 29 Sep 2016.

NASS. 2012 Census of Agriculture. United States summary and state data. Vol. 1, Part 51. AC-12-A-51. U.S. Department of Agriculture, National Agricultural Statistics Service. Washington, DC; 2014.

National Academy of Sciences. The effects on human health of subtherapeutic use of antimicrobial drugs in animal feeds. Committee to study the human health effects of subtherapeutic antibiotic use in animal feeds. Washington, DC; 2009

Noyes NR, Yang X, Linke LM, Magnuson RJ, Dettenwanger A, Cook S. Resistome diversity in cattle and the environment decreases during beef production. eLife. 2016;5:e13195. https://elifesciences.org/content/5/e13195. Accessed 3 Apr 2016.

O'Neill J. Tackling drug-resistant infections globally: final report and recommendations. The review on antimicrobial resistance. 2016. http://amr-review.org/. Accessed 7 July 2016.

OIE. Antimicrobial Resistance. World Organization for Animal Health. 2015. www.oie.int. Accessed 22 May 2016.

Oreskes N, Conway EM. Merchants of doubt: how a handful of scientists obscured the truth on issues from tobacco smoke to global warming. London: Bloomsbury Publishing Plc; 2010.

Rosenblatt-Farrell N. The landscape of antibiotic resistance. Environ Health Perspect. 2009;117(6):A244–50. http://www.ncbi.nlm.nih.gov/pmc/articles/PMC2702430/. Accessed 7 July 2016.

Seattle-King County Department of Public Health. Surveillance of the flow of *Salmonella* and *Campylobacter* in a community. Prepared for United States Department of Health and Human Services, Public Health Service, Food and Drug Administration, Bureau of Veterinary Medicine; 1984. Contract Number.

Silley P, Stephan B. Prudent use and regulatory guidelines for veterinary antibiotics—politics or science? J Appl Microbiol. 2017;123:1373–80. Accessed 30 Nov 2017.

Sneeringer S, MacDonald J, Key N, McBride W, Mathews K. Economics of antibiotic use in U.S. livestock production. U.S. Department of Agriculture, Economic Research Service. ERR-200, November. 2015. http://www.ers.usda.gov/publications/err-economic-research-report/err200.aspx. Accessed 1 June 2016.

So AD, Shah TA, Roach S, Chee YL, Nachman KE. An integrated systems approach is needed to ensure the sustainability of antibiotic effectiveness for both humans and animals. J Law Med Ethics. 2015;43(Suppl 3):38–45. Accessed 30 May 2016.

Tang KL, Caffrey NP, Nóbrega DB, Cork SC, Ronksley PE, Barkema HW, et al. Restricting the use of antibiotics in food-producing animals and its associations with antibiotic resistance in food-producing animals and human beings: a systematic review and meta-analysis. Lancet Planet Health. 2017;1(8):e316–27. https://doi.org/10.1016/S2542-5196(17)30141-9. Accessed 18 Nov 2017.

Teillant A, Laxminarayan R. Economics of antibiotic use in U.S. swine and poultry production. Choices 1st Quarter. 2015;30(1). http://choicesmagazine.org/choices-magazine/theme-articles/theme-overview/economics-of-antibiotic-use-in-us-swine-and-poultry-production. Accessed 5 June 2016.

Tyson Foods. Position statements: antibiotic use. 2017. http://www.tysonfoods.com/media/position-statements/antibiotic-use. Accessed 27 Feb 2017.

Union of Concerned Scientists. Scientific integrity in policymaking: an investigation into the Bush Administration's misuse of science. Cambridge, MA; 2004. http://www.ucsusa.org/sites/default/files/legacy/assets/documents/scientific_integrity/rsi_final_fullreport_1.pdf. Accessed 1 Mar 2017.

Van Boeckel TP, Glennon EE, Chen D, Gilbert M, Robinson TP, Grenfell BT, Levin SA, Bonhoeffer S, Laxminarayan R. Insights: reducing antimicrobial use in food animals. Science. 2017;357(6358):1350–2.

Veterinary Feed Directive A rule by the food and drug administration on 06/03/2015. 2015. https://www.federalregister.gov/documents/2015/06/03/2015-13393/veterinary-feed-directive. Accessed 3 June 2016.

Webb HE, Angulo FJ, Granier SA, Scott HM, Loneragan GH. Illustrative examples of probable transfer of resistance determinants from food animals to humans: Streptothricins, glycopeptides, and colistin [version 1; referees: 2 approved]. F1000Research. 2017;6:1805. https://doi.org/10.12688/f1000research.12777.1.

Weimer BC 100 K genome project 2016. Veterinary Medicine, UC Davis. 2016. Accessed 5 July 2016; OIE. Fact sheet on antimicrobial resistance. World Organization for Animal Health. 2015. http://www.oie.int/fileadmin/Home/eng/Media_Center/docs/pdf/Fact_sheets/ANTIBIO_EN.pdf. Accessed 24 Oct 16.

WHO. Antimicrobial resistance: Global Report on Surveillance. 2014 summary. World Health Organization. 2014. http://apps.who.int/iris/bitstream/10665/112647/1/WHO_HSE_PED_AIP_2014.2_eng.pdf?ua=1. Accessed 30 Sep 2016.

WHO. WHO guidelines on use of medically important antimicrobials in food-producing animals. Geneva: World Health Organization; 2017. License: CC BY-NC-SA 3.0 IGO

Zhang WH, Lin XY, Xu L, Gu XX, Yang L, Li W, et al. CTX-M-27 Producing *Salmonella enterica* serotypes Typhimurium and Indiana are prevalent among food-producing animals in China. Front Microbiol. 2016;7:436. https://doi.org/10.3389/fmicb.2016.00436. Accessed 7 July 2016.

Zhu YG, Johnson TA, Su JQ, Qiao M, Guo GX, Stedtfeld RD, et al. Diverse and abundant antibiotic resistance genes in Chinese swine farms. PNAS. 2013;110(9):3425–40. http://www.pnas.org/cgi/doi/10.1073/pnas.1222743110. Accessed 7 July 2016.

Zilberman D, Hochman G, Kaplan S, Kim E. Political economy of biofuel. Choices. 2014. 1st Quarter 2014;29(1). http://choicesmagazine.org/choices-magazine/theme-articles/theme-overview/economics-of-antibiotic-use-in-us-swine-and-poultry-production. Accessed 7 July 2016.

Chapter 16
The Role of Consumer Advocacy in Strengthening Food Safety Policy

Patricia Buck

Abbreviations

AR	Antibiotic resistant
ARS	Agricultural Research Service
CDC	Centers for Disease Control and Prevention
COB	Congressional Budget Office
ERS	Economic Research Service
FDA	Food and Drug Administration
FOOD	Foodborne Outbreak Online Database
FSIS	Food Safety Inspection Service
FSMA	Food Safety Modernization Act
GAO	Government Accountability Office
HHS	Health and Human Services
IFSAC	Interagency Food Safety Analytics Collaboration
LTHO	Long-term health outcomes
MOFS	Make Our Food Safe coalition
MT	Mechanically tenderized
NGO	Nongovernmental organization
NNDSS	National Notifiable Diseases Surveillance System
OIG	Office of Inspector General
OMB	Office of Management and Budget
PR/HACCP	Pathogen Reduction/Hazard Analysis and Critical Control Points
SFC	Safe Food Coalition
USDA	US Department of Agriculture

P. Buck (✉)
Center for Foodborne Illness Research & Prevention, Grove City, PA, USA
e-mail: buck@foodborneillness.org

© Springer International Publishing AG, part of Springer Nature 2018
T. Roberts (ed.), *Food Safety Economics*, Food Microbiology and Food Safety,
https://doi.org/10.1007/978-3-319-92138-9_16

16.1 Introduction

Safeguarding America's food supply is a public health ("general welfare") responsibility for the US federal government, the individual 50 states, and their local jurisdictions. Previous to the twentieth century—when our country was largely rural and information on food/foodborne pathogens was limited—the USA had a local-regional food supply. Later, as transportation opportunities expanded and the science about food and its role in the transmission of disease grew, the federal government enacted laws[1] to protect consumers from contaminated food. By the end of the twentieth century, two major federal agencies—namely, the US Department of Agriculture's Food Safety Inspection Service (USDA/FSIS) and the Food and Drug Administration (FDA)—were the major food oversight agencies, and they are responsible for most of America's food regulations pertaining to food inspection and enforcement procedures, while other food-related areas, such as international law, food trade agreements, state inspections, and Internet sales, are activities handled by a variety of governmental entities (Fortin 2016). Today, many federal agencies[2] have some food safety responsibilities, while the states' departments and localities enact their own rules and regulations to meet various agricultural and food industry activities within their borders. As a result, the US food safety system has become highly fragmented, both within the national framework and also between the federal and state systems. In 2017, the Government Accountability Office issued a report on the challenges associated with this fragmented food safety system (GAO 2017) and concluded that "there is a compelling need to develop a national strategy to address ongoing fragmentation and improve the federal food safety oversight system." Unfortunately, changing America's disjointed approach will be difficult and will require years of advocacy, as well as a large commitment to increased funding.

A major food safety challenge has been identifying which foodborne pathogens (bacteria, viruses, and parasites) are associated with sporadic and outbreak cases of foodborne illnesses. Currently, the USA has several national surveillance programs (CDC 2011), most of which are led by the Centers for Disease Control and Prevention (CDC). As with any surveillance, these programs are only as good as the data entered into the system, yet obtaining timely foodborne illness data is difficult.

---

[1] Major federal food laws: Meat Inspection Act (1906), Pure Food and Drug Act (1906), Food, Drug and Cosmetic Act (1938), Poultry Products Inspection Act (1957), Egg Products Inspection Act (1970), and most recently, the FDA Food Safety Modernization Act (2011). In addition to these laws, there is the Pasteurized Milk Ordinance (1927) and the Pathogen Reduction; Hazard Analysis and Critical Control Points Final Rule (1996).

[2] Federal agencies—in addition to FDA and USDA/FSIS—with food safety responsibilities: Animal and Plant Inspection Services; Bureau of Alcohol, Tobacco, Firearms and Explosives; Centers for Disease Control and Prevention; Cooperative Research and Extension Services; Environmental Protection Agency; Federal Trade Commission; Grain Inspection, Packers & Stockyards Administration; National Agricultural Statistics Services; USDA/Agricultural Research Services; USDA/Economic Research Services; U.S. Customs & Border Protection; US Department of Homeland Security.

According to the 2010 FDA Food Survey (Lando 2010), the public has the following misconceptions about foodborne disease:

- Most of the 4237 survey participants (83%) said that they did not believe that anyone in their household had experienced a foodborne illness in the past year, even though CDC reports that 1 in 6 Americans are sickened each year, which is a very high incidence rate for any disease.
- Most participants said that foodborne illness is a temporary, nuisance illness that does not need treatment or investigation. When asked if the sickened individual received medical attention, 86% of the participants said that the individual did not seek medical attention, which means that medical providers—who are required to report specific infectious diseases—did not see a majority of the foodborne illness patients, and could not send information about the confirmed foodborne illness cases to the state and national foodborne illness surveillance systems.

Obviously, the public's lack of knowledge about foodborne disease and their unwillingness to seek medical attention are major gaps in our efforts to control the spread of these infectious diseases (see Chap. 2 for more information).

Another food safety challenge is that reporting a foodborne illness is a slow and labor-intensive task. First of all, when foodborne illness is suspected, both the victims and their medical providers (as well as some insurance providers) regard investigative procedures, like stool testing, to be unnecessary since most cases of foodborne illness resolve themselves quickly. It is only when there are a lot of cases (an outbreak) or when the case is severe (generally when a victim is hospitalized) that stool testing is ordered. If the result is a confirmed positive, then (finally) those results are reported, but only to the state/local public health departments. After the local/state departments are notified, most states have laws/regulations mandating them to (1) report the disease—if it is "nationally reportable"—to the National Notifiable Diseases Surveillance System and (2) conduct an investigation into possible sources for the foodborne illness.

The first step in a foodborne illness investigation is the public health food history interview. Conducting these food history interviews are time-consuming and frequently yield little information because victims do not always remember the food that they consumed in 3–4 days prior to the onset of symptoms (this is especially true if the disease has a lengthy incubation period, such as with hepatitis infections and listeriosis). Not surprisingly, the majority of the state/local public health food history interviews are inconclusive, meaning that no food source is identified and the investigation ends. However, sometimes the state asks the CDC to help with a particular investigation, or more commonly, the investigation shifts to the federal agencies if evidence—from testing or surveillance efforts—indicates that the event is a national public health concern. Depending on the circumstances, the federal actions are orchestrated by the CDC, FDA, or USDA and include (1) publicly announcing a foodborne illness outbreak, (2) issuing a public health alert, or (3) recalling contaminated food. Generally, for public health alerts and food recalls, the federal agencies' response is ordered after positive results are found through the government's microbiological testing programs at food facilities, which means that most alerts and recalls have source information. However, in the case of outbreaks (or recalls associated with outbreaks), the federal agencies involved in the investigation must have two confirmed positive tests results—from a victim and from a food—that match the outbreak strain. Further, in order to ensure accuracy, testing on a food associated with an outbreak must be performed on an unopened food package found in the possession of a foodborne illness victim or a food retailer. Given these requirements, it is not surprising that many foodborne illness outbreak investigations stall because the federal food safety agencies cannot confirm a food source. For example, in late 2016 through August 2017, CDC identified a large cluster of illnesses involving 100 cases in multiple states that had the same strain of antibiotic-resistant *Salmonella*. This event was discussed multiple times at meetings between consumer groups and FSIS because the victim food histories showed that 87% of the victims had consumed ground beef prior to the onset of symptoms, meaning that there was enough evidence to connect these illnesses to a food product. However, despite this finding, neither CDC nor FSIS would take action to declare an outbreak or issue a recall because they had not found an unopened package of ground beef at a victim's home or at a food retailer establishment to test for the antibiotic-resistant strain. As a result, this investigation was closed without any notification to the public.

Currently, the USA is working on better ways to connect foodborne illnesses to specific foods. PulseNET, CDC's large national foodborne illness database, which matches PFGE patterns of victims to contaminated food products, has been our major tool for gathering information on "clusters" and "outbreaks" of foodborne illnesses. Recently, additional initiatives—CDC's Foodborne Outbreak Online Database (FOOD) tool, the Interagency Food Safety Analytics Collaboration (IFSAC), and CDC's use of whole genome sequencing technology—have also spurred progress in more rapid detection while expanding our understanding about which foods harbor dangerous pathogens. Everything from dairy to apple cider, from raw protein products to produce, and from processed flour to peanut butter has been associated with major foodborne illness outbreaks (Donald 2010). The impacts of these outbreaks, in terms of the human toll, are documented in CDC's estimates of the US burden of foodborne disease reports (Scallan et al. 2011), while the economic costs are reported to be in billions of dollars (Scharff 2012). Unfortunately, the recent proposed budget for FY 2018 is suggesting a $1.2 billion cut to CDC's budget (Achenbach and Sun 2017), and this action—if implemented—will impact on CDC's ability to continue with its new initiatives, thereby making it more difficult for our nation to respond to foodborne disease outbreaks and food recalls in a timely fashion.

One good example of how contaminated food impacts our society is the 2009 Peanut Corporation of America—*Salmonella* outbreak. During this large national outbreak, over 700 were sickened, 116 were hospitalized, and 9 people died. FDA food recalls escalated to include over 3900 products made by 360 companies. At the 2009 US House of Representatives' hearing on the *Impact of Food Recall on Small Businesses*, peanut producers testified that their losses for sales and production *during the outbreak* were at least $1 billion (Koehler 2009). As a result, the PCA outbreak and recalls became a turning point in US food safety reform and led to changes in FDA's approach to their food inspection strategies. Later, in 2014, another impact was felt when three PCA employees were found guilty of criminal charges because they knowingly introduced both adulterated and misbranded food into interstate commerce (Flynn 2014).

In addition to large outbreaks, other factors have also contributed to our nation's heightened interest in food safety. Over the past 20 years, US consumers have exhibited new food preferences, including increased consumption of fresh produce (Produce for Better Health Foundation 2015) and a trend to eat more meals away from the home (ERS 2016a). To meet the "fresh produce" preference, the USA has been importing more fruits and vegetables (ERS 2016b), but until the recent passage of FSMA, the FDA has not had any legal authorities to ensure that foreign suppliers are meeting US standards in produce production. This absence of imported food protection for produce has heightened consumer concerns, especially since raw produce has been implicated in multiple foodborne illness outbreaks (CDC 2013).

With regard to the consumer preference to eat more meals away from home, a solution will not be easy since restaurants are inspected by local public health departments, which typically have a long list of unfunded public health mandates (NACCHO 2011). Further, the impact of insufficient resources for the public health infrastructure—at the national, state, and local levels—has a rippling effect on our food safety system. For example, in 2015, CDC reported that sit-down restaurants are the most common *setting* for foodborne illness outbreaks (Heiman et al. 2015), yet given the workload and limited resources of the local public health departments, the frequency of restaurant inspections varies considerably across the country. Further, given the restaurant industry's low profit margin, most food service workers have poor or nonexistent sick leave, which means that workers come to work sick, thereby spreading preventable diseases (CDC 2016). Finally, and again largely due to the low profit margin, restaurant employers have insufficient resources to fully train all their employees, so they rely on the supervisors, who attend the formal training sessions, to instruct the other employees.

Given this backdrop, it is not surprising that American consumers continue to be concerned about food safety and are eager to have more information and transparency on food safety regulations and policies. In the 2016 International Food Information Council Foundation's (IFICF) Food and Health Survey, the data shows that 29% of the participants regarded pathogenic food contamination as the number one food safety issue (IFICF 2016). In another study, conducted by the Center for Food Integrity, consumers responded to questions about food safety. This study (Beck 2016) found the following:

- 65% of US consumers are either not confident or unsure that their food is safe to eat.
- 76% of US consumers lack confidence in our government's ability to protect us from unsafe food.
- Over 50% of US consumers do not trust the safety of imported food.
- Over 50% of US consumers want more accurate labeling information, including details about food sources and food preparation.

Today, government policies about food-related topics range from large farm subsidies to the minutia on labeling food products. In addition, while most food-related policies begin at the domestic level, they almost always have a global impact, with trade agreements and developmental aid being huge factors. To handle this ever-expanding arena, the USA will need to focus more effort on the drivers of food safety, i.e., the public health and economics that are embedded into our food safety policies and practices.

16.2 Consumer Food Safety Advocacy: The Challenges, Strategies, and Tools

Food safety policies and practices are currently hot topics for discussion, not only here in the USA but throughout the world. In the USA, the large, multistate foodborne illness outbreaks and high-profile media exposures have fueled national demands for change. In particular, the foodborne illness outbreaks associated with ground beef (1993), spinach (2006), pet food (2007), and peanut butter (2007 and 2009) helped to spur reform initiatives, while news media coverage, such as *Philadelphia Inquirer's* 2003 series on "tainted meat" (Surendran 2003) and the documentary *Food Inc.*,[3] personalized the victims and discussed the impacts that food production and food contamination have had on environments, animals, and people, e.g., One Health (King L. Institute of Medicine 2012). New challenges, such as antibiotic-resistant (AR) strains of foodborne pathogens, are being addressed by both international initiatives (World Health Organization 2015) and national programs (Obama, President 2014). As a result, food safety has become an important issue for consumers, public health officials, governmental regulators, and agricultural enterprises.

As a general rule, food policies are difficult to change for four reasons: (1) the farm-to-fork continuum has many participants and perspectives making agreement difficult; (2) food safety issues are complicated because there are so many product and/or regional variables; (3) the proposed solutions can be costly; and (4) the general public is far removed from food production realities, so they underestimate the importance of science- and risk-based food policies. Two of the drivers of food policy reform—innovations in science/technology or foodborne illness crisis events—disrupt the status quo and lead to a revision of food safety goals, as well as a shift in food product availability and/or the need to change food handling behaviors. The third and fourth food policy drivers—economics and litigation—are hindered by a lack of information, in particular for attribution data that links foodborne illness to a specific food or pathogen. In other words, as stated by the economist Dr.

---

[3] *Food Inc. is* a documentary that raises awareness about the changes in agriculture since the 1960s. It focuses on large corporate agricultural enterprises and how the new farming techniques impact on people, animals and the food we eat. 2010. http://www.pbs.org/pov/foodinc/film-description/. Accessed 12 Dec 2016.

Tanya Roberts, "the imperfect information about the risk associated with food means that neither the legal system nor the marketplace may be able to provide adequate economic incentives for the production of safe food" (Roberts 2013).

Food safety reforms are generally initiated by food and food safety professionals, including researchers, investigative reporting groups, business leaders, and crisis management experts. They urge change based on the results of their work but are frequently frustrated when governmental policy makers or local/state/federal regulators—the groups most responsible for devising and implementing a policy change—fail to appreciate the importance of their findings. As a result, food and food safety stakeholders have vigorous debates as they work independently or in collaborations to develop effective and science-based food safety, public health, and agricultural policy changes. Meanwhile, the public—which has a weak understanding of the underlying issues or the political process to implement change—struggles to make sense of the conflicts.

Another important factor about food safety change proposals is that they always have an element of uncertainty, which means that all proposals are subject to a high level of scrutiny. To prepare for that accountability, food safety policy change proponents conduct literature reviews on the available research and prepare detailed metrics outlining the goals, resources, outcomes, and evaluation measurements for a proposed change. Food safety proposals may also include scientific studies, details about innovative technologies, economic cost-benefit analysis, government reports on specific topics, investigative summaries on crisis events, media clips/publications, as well as polls/surveys to determine the public's awareness/concerns about the topic or the need for change. As a result, change proposals for food safety are lengthy and resource intensive to prepare.

In addition to preparing the food safety proposal, change proponents must also develop a base of supporters, which can be difficult given the diversity of the food stakeholders. Sometimes large coalitions form around a single issue or plan, but generally, the collaborations are limited to a few like-minded academia, health, industry, media, or consumer advocates. Once the proposal has been prepared and a support base has been identified, the proponents must then present the idea in a clear and concise manner to the appropriate policy change personnel. To achieve that end, food and food safety advocates develop briefings, handouts, and letters that highlight key aspects of the proposed change and provide their audience with references for more detailed information. For optimum results, change proponents must follow up their meetings and presentations with timely summaries and, if necessary, arrange additional opportunities to discuss the benefits of their plans.

Two final notes on food safety advocacy:

1. Change proponents should use a variety of actions to increase their likelihood of success. Besides coalitions, these efforts have also been successful: demonstrations, economic analysis, litigation, media exposure, partnerships, petitions, public comments, public meetings (including state or national Congressional hearings), requests for governmental investigations, and scientific reports/white papers.

2. Change proponents should consider all mechanisms to achieve their goals. Developing a new law is appealing because this type of reform generates new programs and has a wider impact. However, law-making is an onerous task, so reformers should consider other options. Memorandums of understanding, cooperative agreements, judicial actions, state or federal executive orders, and regulatory initiatives/reforms can also be effective in securing change.

16.3   New Regulations for US Department of Agriculture (USDA): The Pathogen Reduction/Hazard Analysis and Critical Control Points Final Rule (PR/HACCP) and Other Notable USDA/FSIS Regulations (1996–2016)

After the 1993 Jack-in-the-Box outbreak associated with *E. coli* O157:H7, proponents of food safety policy change wanted federal food oversight of meat and poultry inspection to shift from a reactive approach to one that was more preventive. At the time of the outbreak, federal meat and poultry oversight was largely rooted in the "poke and sniff" inspection approach, even though these organoleptic inspections were ineffective in detecting most of the microbiological agents that cause foodborne illness. The development and eventual adoption of UDSA's 1996 PR/HACCP Final Rule was not smooth and easy, but the consumer advocates were certain that a new approach was necessary. The leaders of the consumer movement felt that there were several "bad actors," within the meat/poultry industry, and that these entities needed mandatory regulations to improve the safety of their products (Tucker-Foreman 2002).

Prior to the Jack-in-the-Box outbreak, USDA had been interested in changing its meat and poultry inspection, but those proposals met with failure. However, after the outbreak, it became obvious that something had to be done, so in September, 1994, Mike Taylor, then employed as FSIS's Administrator, announced that *E. coli* O157:H7 would be an adulterant in ground beef (Bottemiller 2011). Opponents responded by suing USDA/FSIS—they maintained that a product contaminated with *E. coli* O157:H7 was only injurious to health if the product was not cooked properly. The court disagreed saying that many Americans have cooking practices for ground beef (rare, medium-rare, and medium) that are insufficient to kill the pathogen. Further, given *E. coli* O157:H7's very low infectious dose (10–100 microbes), and its potential for person-to-person transmission, it was clear that food contaminated with *E. coli* O157:H7 should not be allowed in household settings. Still, the opponents—mainly the meat industry—worked to end the development of the PR/HACCP system; their primary position was that the Federal Meat Inspection Act gave USDA the authority to protect the public from sick animals, not deadly bacteria (Nestle 2010). These industry groups lobbied Congressional Members to introduce anti-HACCP legislation, and there were several incidents of hostility between industry and USDA/FSIS field inspectors. This discord between FSIS and its field inspectors was very intense, and according to some accounts, the pressures imposed on the field inspectors fueled the animosity between industry and USDA inspectors (Organic Consumers Association 2000), thereby contributing to the fatal shoot-out in California, which killed three inspectors (Nestle 2010).

Despite all of this turmoil, in 1996, Secretary of Agriculture Dan Glickman used his authority, under the Federal Meat Inspection Act and the Poultry Products Inspection Act, to issue the PR/HACCP Final Rule. The new rule demanded meat and poultry companies to develop a comprehensive food safety plan that included identifying hazards, mitigating those hazards by designing and using scientifically proven interventions, and then verifying the food safety plan's effectiveness by using microbial testing. The purpose of this new approach was to improve the industry's sanitation practices to the point that a meat or poultry facility would successfully meet PR/HACCP's generic *E. coli* sanitation standard, as well as the new end-product *Salmonella* performance standard. It was an entirely different way of inspection and its focus was to reduce foodborne illnesses by carefully verifying an industry's practices based on a science-detailed plan, developed by the industry, but approved by the government.

In the final PR/HACCP rule, the costs and the benefits of implementing this new inspection system were examined. For many owners of small meat/poultry businesses or companies that produced multiple species and/or products, the cost of daily microbiological testing (one of the verifying mechanisms) was seen as an onerous burden. However, in the discussion about the cost versus benefits for implementing HACCP over 20 years, the Final Rule's analysis showed a huge public health benefit that could not be denied (Federal Register 1996).

PR/HACCP's implementation began in 1998 with large ground meat processors, but within a year, the *Salmonella* performance standards were challenged. In 1999, Supreme Beef Inc., a Texas-based meat processor and grinder, failed the PR/HACCP's *Salmonella* performance standards three times in 8 months, meaning that the end-product *Salmonella* standard had not been met. As a result, USDA informed Supreme Beef of its intent to close the plant. Supreme Beef responded by securing a court injunction to remain open and then filed a suit against USDA. Subsequently, the court decided in favor of Supreme Beef, so USDA appealed the ruling in the US Fifth Circuit Court of Appeals. In its 2001 ruling, the Fifth Circuit Court of Appeals said that USDA had no statutory authority in the Federal Meat Inspection Act to suspend operations at a grinding plant solely on noncompliance with the PR/HACCP *Salmonella* standard. Further, the court accepted Supreme Beef's argument that the company's failed tests were due to the condition of the incoming raw materials, not their lack of sanitation control. USDA did not file for an appeal of this Fifth Circuit Court ruling (Supreme Beef Processors Inc. v. United States Department of Agriculture 2001).

As a result of the Supreme Beef case, USDA changed its strategy and started to develop compliance and enforcement regulations—other than suspension—which would improve the agency's ability to control for dangerous pathogens in its meat and poultry products. For example, in early 2008, USDA began posting completed verification sample set results from establishments in Categories 2 and 3.[4] This action was initiated because FSIS testing results from young chicken slaughter establishments, during 2000–2005, showed a 71% increase of *Salmonella enteritidis* isolates in its samples. Further, as a result of the web posting, FSIS later reported that the number of establishments not meeting the standard fell by 50% in the 2-year period following the time when FSIS began posting category information (USDA/FSIS 2014a).

Over the past 15 years, USDA/FSIS has also used partnerships and meetings (both public and technical) with academia, food industry leaders, and consumer groups to explore food safety issues related to USDA-inspected products. For example, during the Bush and Obama Administrations, both industry leaders and the Safe Food Coalition[5] met monthly with USDA/FSIS to discuss important inspection and oversight issues. In addition to these meetings, the government has provided opportunities for public comments on proposed regulations posted in the Federal Register. Based on these efforts, there has been some important progress in meat and poultry safety (Table 16.1, Fig. 16.1).

While this is an impressive list of USDA accomplishments, USDA's authority to enforce PR/HACCP's *Salmonella* performance standards has not been fully resolved, so in the future, there may be efforts to further reform America's meat and poultry inspection systems. Meanwhile, the high incidence of *Salmonella*, which is the nation's leading foodborne illness killer, continues to worry food safety experts. According to CDC's FoodNet surveillance, the incidence of *Salmonella* has basically stalled over the past 15 years, while *E. coli* O157:H7 and *Listeria monocytogenes*, which were declared adulterants in the 1990s, have shown a decline (Henao et al. 2015) (Fig. 16.2).

When food safety experts discuss the US *Salmonella* problem, they generally agree that pathogen standards for food products, along with animal management programs, are of upmost importance. In the 1980s, Denmark had a high rate of human salmonellosis, which they were able to reduce by developing integrated, preharvest, animal management programs (Wegener et al. 2003). Unfortunately, in the USA, our meat/poultry laws and regulations do not provide USDA with oversight of farms, so developing these types of programs is currently not an option. Further, many other experts maintain that the problem with *Salmonella*'s persistence is not due to a lack of performance standards or animal management programs. Instead, they feel that the pervasive nature of the pathogen—it is in everything!—is the major issue. At the National Food Policy Conference in 2015, Dr. David Goldman (FSIS Office of Public Health Science Administrator) reported that while 60% of *Salmonella*'s 200+ strains

[4] Category 2 establishments are characterized as having variable process control and Category 3 as having very variable process control, while Category 1 establishments are characterized as having consistent process control.

[5] Current members of the Safe Food Coalition, coordinated by Consumer Federation of America, include Center for Food Safety; Center for Foodborne Illness Research and Prevention; Center for Progressive Reform; Center for Science in the Public Interest; Consumers Union, Food, and Water Watch; Government Accountability Project, National Consumers League; STOP Foodborne Illness; US PIRG; and United Food Commercial Workers.

Table 16.1 Selected USDA food safety policy changes, 2001–2016

Timeline	Food safety situation needed to be changed	Actions taken
2001– present	*Food safety education and outreach programs*: Educational materials were limited to the general public. In addition, there was no central website for food safety.	New campaigns within USDA and its partners were developed to target appropriate messages for the vulnerable populations. The "foodsafety.gov" website was launched in 2012
2003 and 2015	*Listeria control* needed to be improved, especially for ready-to-eat (RTE) products. *Listeria monocytogenes* is a very dangerous pathogen. It has a high death rate, and survivors frequently experience long-term health problems.	In 2003, USDA developed an interim rule to control for *Listeria monocytogenes* in ready-to-eat meat and poultry products (USDA/FSIS 2003). In 2015, USDA updated the rule's guidance to include environmental testing of surfaces (USDA/FSIS 2015a).
2007– present	*Data sharing*: USDA's *Salmonella* verification testing data from meat and poultry products was not being shared with other federal agencies. CDC's PulseNet wanted the *Salmonella* isolate information so it could match human cases of illness to the positive samples that FSIS was finding in meat and poultry products.	In 2007, a cooperative agreement to share information about FSIS' *Salmonella* isolates was implemented between USDA/FSIS and USDA/ARS and CDC/PulseNet (see Fig. 16.3). Since then, both USDA and CDC have developed more "data-sharing" agreements, thereby allowing more researchers access to food safety information.
2008	*Control of Salmonella in poultry slaughter and processing facilities*: After the Supreme Beef ruling in 2001 stating that enforcing the PR/HACCP *Salmonella* performance standards were not under USDA's authority, *Salmonella* control was difficult to achieve. Alternative methods for control needed to be developed, especially since *Salmonella* has never been declared an adulterant in meat and poultry products.	USDA's web posting of verified *Salmonella* results (see above, p. 15) became an effective tool for controlling this pathogen. Note: In 2016, USDA stopped its web posting as the poultry industry transitioned into using a new "sampling broth." Once the transition was over, the Safe Food Coalition members repeatedly asked USDA to resume its web posting. Finally, in late 2017, USDA told SFC that it would resume web postings in January 2018.
2011- 2015	*Gaps in ground beef safety*: • Under PR/HACCP, all hazards have to be identified, yet many beef producers were not listing *E. coli* O157:H7 as a hazard likely to occur in ground beef. • USDA could only request beef grinders to not ship product until after the results from FSIS' *E. coli* testing were completed. • *E. coli* O157:H7 is not the only *Shiga* toxin-producing *E. coli* to cause serious human disease. The World Health Organization and CDC have identified at least six other strains. • FSIS had no program for testing beef "trim" that was intended for use in raw ground beef production. • USDA did not have a rule requiring beef grinders to maintain accurate records.	New regulations for beef: • Directive 6410.1 Rev. 1 provided improved guidance by describing points in beef slaughter where contamination is most likely to occur (USDA/FSIS 2011a). • USDA mandated "Test and Hold" for beef grinders, requiring them to hold shipment until FSIS' testing was completed (USDA/FSIS 2011b). • USDA announced that non-intact raw beef products containing *E. coli* O26, O103, O45, O111, O121, and O145 were also considered adulterated (USDA/FSIS 2012). • USDA Directive 10,010.1 Rev. 4 changed the sampling protocols to include testing of beef trim intended for ground beef production (USDA/FSIS 2015b). • USDA required grinders to keep accurate supplier records, thereby improving product tracing (USDA/FSIS 2015c).

(continued)

Table 16.1 (continued)

Timeline	Food safety situation needed to be changed	Actions taken
2012–2014	*Poultry pathogen standards: Salmonella and Campylobacter*, two leading foodborne illness pathogens, are frequently associated with raw poultry. According to CDC, the incidence of *Salmonella* illnesses has showed little change between the 1999 baseline and 2015 (see Fig. 16.2).	*USDA improves poultry standards* by updating its *Salmonella* reduction performance standards and adding *Campylobacter* to its microbial testing programs (USDA/FSIS 2016a).
	Labeling gaps: Meat and poultry products injected with solutions or treated with mechanical tenderization are not intact products and could cause illness if the product is not cooked thoroughly (USDA/FSIS 2016b). Labels to identify and provide instructions are needed.	*USDA finalizes two labeling rules* mandating that all products injected with added solutions (USDA/FSIS 2014b) or treated with mechanical tenderization (USDA/FSIS 2013) be identified. In addition, these labels must provide appropriate cooking instructions.

cause illness, the virulence of the strain may be the biggest factor. To control for the most deadly strains, Goldman says, we must become better at identifying, in a timely manner, the most problematic *Salmonella* serotypes. He is hopeful that whole genome sequencing, a new surveillance technology, will help (Clapp 2017).

Moving forward, USDA should declare deadly antibiotic-resistant *Salmonella* strains as adulterants in meat and poultry products and expend more resources on important foodborne illness surveillance efforts. Similarly, worker safety and fatigue issues—both of which can impact food safety—will not be improved until more research is conducted on controversial practices, such as the use of high-speed slaughtering and processing lines. Finally, there are concerns about the shortage of FSIS inspectors. In 1981, FSIS employed about 190 workers per billion pounds of meat and poultry inspected and passed, but in FY 2011, FSIS employed 88 inspection workers per billion pounds, a 54% decrease (Center for Effective Government 2012). Unfortunately, the resulting shortages have been routinely handled by assigning inspectors to work overtime, thereby creating a situation where inspector fatigue can pose a real problem, causing inspector error or an inability to complete all of the required food safety tasks.

16.4   A New Law for the Food and Drug Administration: FDA Food Safety Modernization Act (FSMA)

In 2006, a large recall of fresh spinach contaminated with *E. coli* O157:H7 rocked consumer confidence in America's food supply. Spinach, a plant food that was considered to be "very healthy," was found to contain a deadly pathogen that was

United States Department of Agriculture
Office of the Secretary
Washington, D.C. 20250

October 24, 2007

Ms. Patricia Buck
Center for Foodborne Illness
Research & Prevention
Post Office Box 206
Grove City, Pennsylvania 16127

Dear Ms. Buck:

Thank you for your June 1, 2007, letter to former Agriculture Secretary Mike Johanns regarding microbiological subtype data sharing within and among food safety surveillance agencies. The Acting Secretary asked me to respond to your letter.

Regarding your concerns about the sharing of data between the Agricultural Research Service (ARS) and the Food Safety and Inspection Service (FSIS), let me assure you that these agencies have a long history of working collaboratively and sharing data on issues of public health. In the specific area of sharing the subtype data from *Salmonella* isolates, the exchange of data has been limited as in a foodborne outbreak investigation, or as part of a collaborative study examining trends or attribution issues related to a specific serotype.

Thanks to your input, FSIS has determined that future risk management strategies must incorporate the full range of data available from subtype analysis. Moreover, the Agency recognizes that these data may play a critical role in further reducing the uncertainty associated with attribution of foodborne illness to the products that we regulate. To this end, FSIS established a workgroup that is researching methods for using subtype data. Another group is working with both ARS and the Centers for Disease Control and Prevention (CDC) to improve or modify the existing information technology infrastructure so that FSIS may compare data available from meat, poultry, and processed egg product isolates with data on human clinical isolates maintained by CDC. To ensure that the risk management needs of the Agency are met, FSIS and ARS recently implemented a cooperative agreement, which is enclosed.

The attached addendum represents our best efforts to provide insight into some of the specific concerns raised in your letter. We appreciate your thoughtful input on this issue and we hope that this information is helpful to you.

Sincerely,

Richard A. Raymond, M.D.
Under Secretary
Office of Food Safety

Fig. 16.1 Letter about data use protocols for USDA *Salmonella* verification program

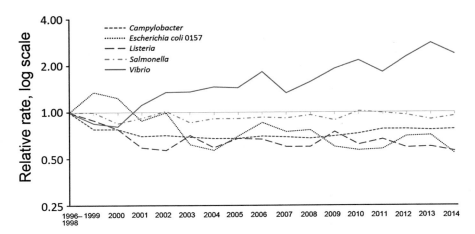

Fig. 16.2 CDC comparison of major foodborne pathogens, 1996–2015. Relative rates of culture-confirmed infections with *Campylobacter, Escherichia coli O157, Listeria, Salmonella,* and *Vibrio,* compared with 1996–1998 rates, Foodborne Diseases Active Surveillance Network, USA, 1996–2014. The position of each line indicates the relative change in the incidence of that pathogen compared with 1996–1998. The actual incidences of these infections cannot be determined from this graph. Data for 2014 are preliminary (CDC 2015)

generally associated with raw meat. To add to the apprehension, FDA told consumers to simply "Throw it out!" since washing would not remove the dangerous bacteria and handling the spinach could lead to more illnesses. As a result, consumer confidence in the food supply sank and multiple concerns about FDA's oversight of food were raised. The following year, there were two other large outbreaks: one involving melamine in pet food imported from China (FDA 2007) and the other involving a common food item, namely, peanut butter, which previously had been regarded as a low-risk food (CDC 2009).

Americans were shocked by this steady stream of outbreaks and recalls, but then in the early part of 2009, a second multistate outbreak of contaminated peanut butter occurred, killing nine people. This outbreak, so close on the heels of the other food safety events, generated public outrage when it was discovered that Peanut Corporation of America (PCA) allowed peanuts, contaminated with *Salmonella*, to be sent to customers on multiple different occasions between 2007 and 2009 (Goetz 2012). Consumers were also aghast when they learned that FDA had not inspected the PCA plant since 2001 and that the last state inspection was conducted 3 years before the outbreak (Leighton 2015). Obviously, consumers had mistaken ideas about the inspection frequency at FDA-inspected facilities.

When FDA finally reinspected the Blakely GA plant in 2009, inspectors found "a range of unsanitary conditions, including dead roaches, mold on the ceiling and walls in the cooler where finished products were stored, and rainwater leaking from skylights into the production room" (Layton 2009). The PCA recalls involved almost 4000 products produced by over 300 companies—many of which used PCA's peanuts or peanut products as an ingredient in other food items (Leighton

2015). Without a doubt, the PCA outbreak and recalls highlighted multiple weaknesses in America's fragmented (and under-resourced!) food safety system. The time was ripe for reform (Produce News 2009).

Further, but not surprising, the impact of this outbreak/recall extended into the financial and criminal areas. PCA declared bankruptcy in 2009, and in 2010, the court allowed bankrupt PCA "to settle tort claims with more than two dozen victims" (Hawkins 2010). Six years later, the US Department of Justice sentenced three PCA executives to prison terms and lamented that "the case was never just about shipping tainted peanut product; it was about making sure individual wrong doers were held accountable and the losses suffered by the victims and their families are never forgotten" (U.S. Department of Justice 2015).

Those involved in the efforts to pass the FDA Food Safety Modernization Act (FSMA) credit a "perfect storm" to the law's enactment. Starting in the 1990s, various bills to modernize FDA's federal food safety authorities had been introduced, but none had advanced. However, after the PCA outbreak, the mood of the country changed. President Obama—in the middle of this food crisis—faulted the FDA for its poor food safety record (Blum 2009), and shortly after his statement, leading Democrats in the Senate and House called for committee hearings. In addition, the groups most commonly associated with food safety issues became mobilized: the food producers wanted improved oversight to restore the consumers' shaken confidence in the safety of the food supply, while the food safety and public health advocates wanted FDA to transform its food oversight to be less reactive and more proactive. Taken together, this diverse coalition of food stakeholders led to a hospitable environment for food safety modernization.

After the introduction of food safety legislation to reform FDA in 2009, proponents of the legislation successfully worked to gain bipartisan support. On the side of the reformers, there was a large group of food industry leaders, health professionals, and consumer advocates. Opposition consisted of a mix of local, sustainable, and organic food/agriculture sectors, along with libertarian advocates endorsing a free market view. To help organize the legislative effort, The Pew Charitable Trusts initiated the development of the Make Our Food Safe (MOFS) coalition (Pew Charitable Trusts 2009a, b), since at that time, there was no coalition focused on FDA reform.[6] In this new venue, the various food safety stakeholders found areas of common ground, and they led the efforts to inform the public about FDA's deficiencies (Table 16.2).

As the process to pass new legislation began, it was very clear that a review of the existing laws, as well as analytic economic studies/reports, would be crucial in moving the bill forward. Since 1975, the Congressional Budget Office (COB) has provided economic analyses to Congressional committees showing how the legislation might financially impact state, local, and tribal governments, as well as the private sector (Congressional Budget Office 2016). The initial COB cost estimates for FSMA were extremely high, so change proponents had to consider ways to

---

[6]The MOFS coalition, formed in 2009, focuses on FDA-regulated food, while the Safe Food Coalition, formed in 1997, works to improve USDA inspection programs for meat and poultry.

Table 16.2 Consumer, health, and industry nonprofit organizations promoting the passage of the FDA Food Safety Modernization Act (FSMA)

American Public Health Association www.apha.org	International Association for Food Protection www.foodprotection.org
Association of Food and Drug Officials www.afdo.org	National Association of Country and City Health Officials www.naccho.org
Center for Foodborne Illness Research & Prevention www.foodborneillness.org	National Association of Manufacturers www.nam.org/
Center for Science in the Public Interest www.cspinet.org/	National Confectioners Association http://www.candyusa.com/
Consumer Federation of America www.consumerfed.org	National Consumers League http://www.nclnet.org/
Consumers Union www.consumersunion.org	National Restaurant Association http://www.restaurant.org/Home
Federal Emergency Management Agency Biological Threats www.ready.gov/biological-threats	Produce Marketing Association http://www.pma.com/
Food & Water Watch www.foodandwaterwatch.org	United Fresh Produce Association www.unitedfresh.org
Food Marketing Institute www.fmi.org	SNAC International, the Snack Food Association http://snacintl.org/
Grocery Manufacturers Association www.gmaonline.org	STOP Foodborne Illness www.stopfoodborneillness.org
Institute of Food Technologists www.ift.org	The Pew Charitable Trusts www.pewtrusts.org/en/projects/food-safety
International Association for Food Protection www.foodprotection.org	Trust for America's Health www.healthyamericans.org
International Bottled Water Association http://www.bottledwater.org/	US Chamber of Commerce https://www.uschamber.com/
International Dairy Foods Association http://www.idfa.org/	US Public Interest Research Group www.uspirg.org

reduce those impacts. In the end, a compromise about FDA's inspection frequency rates, as well as the addition of the Tester-Hagan Amendment (See *Footnote 8*), was added to the Senate version of the bill.

Another factor in FSMA's development was the willingness of foodborne illness victims to meet with policy makers and their staff about the life-altering changes they experienced as a result of foodborne disease. These victim volunteers and their family members[7] traveled to Washington, D.C.—some repeatedly—to engage legislators and their staff with their stories and concerns. It was inspiring to listen to

---

[7] Most of these volunteers were associated with three consumer group organizations, namely, the Center for Foodborne Illness Research and Prevention, Consumers Union, and STOP Foodborne Illness.

the accounts told by the survivors of foodborne illness, and it was heartbreaking to listen to those who had lost loved ones. These victim stories were powerful, and they motivated Congressional Members to regard food safety reform as a top priority. The stories also played another role: they helped the proponents for change to highlight the loopholes and gaps that existed in America's food inspection and foodborne illness surveillance systems.

Despite all of these pluses—powerful motivating events, a strong collaborative approach, meaningful victim involvement, and strong Congressional supporters—the journey to passage was not easy. Each week new challenges, requiring a unified position, arose. To develop appropriate responses, MOFS leaders scheduled regular weekly meetings to discuss developments and strategies. Collaborations that previously were considered impossible were created as industry and consumer groups fought to gain support for the pending legislation. In January 2010, seventeen organizations representing the food industry, consumers, and public health organizations sent a letter to the Senate urging the passage of S. 510. In this letter, they advocated for "food safety legislation that better protects consumers, restores their confidence in the safety of the food they eat, and addresses the challenges posed by our global food supply" (Fig. 16.3).

Two of FSMA's nagging problems were related to the small business concerns and the high inspection costs. Anti-FSMA activists staged large and persistent efforts to derail the legislation. These challenges primarily came from two agriculture stakeholder groups—the small local, sustainable, organic agriculture movement (organized by the National Sustainable Agriculture Coalition) and the anti-big-government group, the Farm-to-Consumer Legal Defense Fund. Before each vote, Congressional offices received thousands of emails and phone calls voicing concerns that the new law would cause small food producers to fail or initiate huge increases in federal spending. In an effort to resolve some of these concerns, a variety of food stakeholders endorsed the Tester-Hagan Amendment,[8] in order to address the concerns of the small and sustainable farmers. Meanwhile, other groups, fearful of the huge inspection costs that FSMA could impose on the federal budget, asked for and received a compromise on FSMA's "frequency of inspection" plans. Basically, FSMA's final inspection provision stipulates that FDA will apply its inspection resources in a risk-based manner, meaning that the agency will adopt a variety of inspection approaches. As a result, the frequency of inspection will vary according to the function and/or the scope of the food establishment. For example, facilities that serve high-risk foods or prepare food for vulnerable populations are

---

- 8 Tester-Hagan Amendment stipulates that a facility is allowed exemptions to FSMA when:The facility has, on a 3 year average, annual gross revenues of less than $500,000, including all subsidiaries and affiliates of a business.
- The facility sells more than half of its products directly to consumers or other qualified end users that are in the same state or within 275 miles of the facility.
- The facility grows, harvests, packs, or holds produce for personal or on-farm consumption.

Under the amendment, FDA retains its authority to withdraw an exemption from a farm or facility that has been associated with a foodborne illness outbreak.

January 21, 2010

Office of the Senate Majority Leader   Office of the Senate Minority Leader
S-221 Capitol Building   S-230 Capitol Building
Washington, D.C. 20510-7020   Washington DC, 20510-7010

Dear Majority Leader Reid & Minority Leader McConnell:

Our organizations are writing to urge you to schedule a vote on S. 510, the FDA Food Safety
Modernization Act of 2009, at the soonest possible date. The HELP Committee approved a
strong, bipartisan bill in November, and we believe that a vote in the first weeks of 2010
would keep the momentum going for enactment of landmark food-safety legislation.

Strong food-safety legislation will reduce the risk of contamination and thereby better protect
public health and safety, raise the bar for the food industry, and deter bad actors. S. 510 will
provide the U.S. Food and Drug Administration (FDA) with the resources and authorities the
agency needs to help make prevention the focus of our food safety strategies. Among other
things, this legislation requires food companies to develop a food safety plan; it improves the
safety of imported food and food ingredients; and it adopts a risk-based approach to
inspection.

Our organizations – representing the food industry, consumers, and the public-health
community – urge you to bring S. 510 to the floor early this year, and we will continue to
work with Congress for the enactment of food safety legislation that better protects
consumers, restores their confidence in the safety of the food they eat, and addresses the
challenges posed by our global food supply.

Sincerely,

American Public Health Association
Center for Foodborne Illness Research & Prevention
Center for Science in the Public Interest
Consumer Federation of America
Consumers Union
Food Marketing Institute
Grocery Manufacturers Association
International Bottled Water Association
National Association of Manufacturers
National Confectioners Association
National Consumers League
National Restaurant Association
Produce Marketing Association
The PEW Charitable Trusts
Trust for America's Health
Snack Food Association

Fig. 16.3 Letter to US Senate in support of Food Safety Modernization Act

ranked high risk and will have more frequent inspections. Facilities that perform hot/cold holding of potentially hazardous foods or do food processing at retail are moderate risk, with a moderate level of inspection. Facilities that primarily handle prepackaged products are low risk and will be inspected infrequently. To counter some of the vagueness of this inspection approach, the new inspection frequency provision also allows FDA to change an establishment's risk level based on repeated deficiencies, a confirmed foodborne illness event, and/or a poor inspection history. Still, many food safety advocates continue to be dismayed by the exemptions granted in the Tester Amendment, and they believe that FSMA's 3-year inspection cycle is inadequate, even though this inspection frequency is an improvement over the 10-year cycle that FDA had been using prior to the passage of FSMA.

In November 2010, toward the end of the 111th Congress, the Senate voted in favor of its version of the bill (S. 510) without realizing that it had a procedural error. However, the House Parliamentarian saw "the revenue raising fee" in the Senate's version and determined that the House could not take up the bill. Under Article 1, Section 7 of the Constitution, bills that raise revenue must originate in the House of Representatives, not the Senate. When this unfortunate issue came to light, many thought that the chance for passage was dead, but FSMA's supporters and champions did not give up. The first revival attempt was offered by the House when it attached S. 510 to the pending "Omnibus Spending Bill." At first, this appeared to be working, but then the Senate announced that it would not be voting on the Omnibus during the remainder of the 111th Congress. The second attempt came only days before the end of the session. To get the food safety bill back to the House without the fee, Senate Majority Leader Harry Reid (D NV) had one last idea—so during a Sunday session on a dreary December afternoon, Reid removed the text from Senate Bill 510. He then stripped the original text from an old House bill (HR 2751, "Cash for Clunkers") and replaced it with a version of S.510 that did not have any fees. Next, he called for a voice vote and the bill was passed by unanimous consent. A few days later, the House passed H.R. 2751 by a vote of 215 to 144. On January 4, 2011, President Obama signed it into law. While there was no public fanfare to honor the moment, the proponents of FSMA celebrated (*see* photos) knowing that the new law would have a huge impact on the future of food safety, both within and outside the USA.

Photos of Food Safety Advocates at FSMA Celebration, January 2011.

Some of MOFS Advocates. (L to R): Polly and Ken Costello, Elizabeth Armstrong with her two daughters Isabella and Ashley; Megan and Barbara Kowalcyk; Patricia Buck

Senator Tom Harkin (D IA) and Carol Tucker Foreman, CFA's distinguished fellow at the Food Policy Institute, discuss food safety

Some of FSMA's Congressional Champions: Senator Harkin (D IA); Senator Klobuchar (D MN); Rep. Pallone (D NJ); Rep. DeLauro (D CT) and Senator Durbin (D IL)

In general, most people believe that the enactment of legislation is the end-point of the policy-making process, but as FSMA moved forward toward implementation, it became clear that enactment of the law, while critically important, is only the first step. After the President signed FSMA, and as FDA's proposed regulations were being moved forward, the supporters found themselves in a battle with the White House Office of Management and Budget (OMB) over the multiple delays that OMB imposed on the proposed rules. Fortunately, within FSMA, there were specific time deadlines for finalizing various proposals, which the Center for Food Safety and the Center for Environmental Health cited in their 2012 lawsuit against FDA (Center for Food Safety 2013). The court agreed and set new deadlines for FSMA's seven priority areas and instructed OMB to not delay FDA in fulfilling its goals (U.S. District Court, Northern District of California 2012). Since then, FDA has finalized its seven major rules (FDA 2016). In addition, the CDC has established the FSMA Surveillance Working Group and six Centers of Excellence for Food Safety (CDC 2015) to meet FSMA's research and surveillance goals.

FSMA's implementation, however, is still a work in process. Letters from consumer groups (Pew Charitable Trusts 2015a, b), as well as support from Congressional leaders and food industry stakeholders, have helped secure appropriations for implementing the new law (Andrews 2015). As of this writing, the major obstacles facing FSMA are some of the changes outlined in the FSMA's Supplemental rules (Zuraw 2014) and the new efforts to condense America's regulatory burden while also reducing spending in food safety agencies. Other challenges have been identified by HHS' Office of Inspector General 2017 Review of FDA, in which the OIG criticized FDA for not taking action in a timely manner, for relying on facilities to voluntarily correct the violations, and for not following up to ensure that facilities had corrected significant inspection violations (Health and Human Services 2017). In other words, there is still a lot that needs to be done.

16.5 Consumer Advocacy and Rule-Making: Adding a Mandatory Label to Beef Products that Have Been Mechanically Tenderized (MT)

Mechanically tenderized (MT) beef steaks and roasts, which are non-intact products, carry a higher risk for pathogenic contamination than a whole, intact cut of beef products (Zuraw 2016). Given this, it is important for the general public and food service cooks to know which products have been treated prior to sale, so that they can handle and prepare them safely. It took 7 years for consumer advocates to move the MT beef labeling rule from a Citizen's Petition to a proposed rule to a finalized rule. In a letter and background memorandum to Secretary of Agriculture Tom Vilsack, the Safe Food Coalition (SFC) petitioned USDA/FSIS asking for a regulation to require meat processors to label MT beef (SFC 2009a, b). The petition focused on three points:

- FSIS was not testing MT beef source materials, including bench trim, and final products.
- FSIS was not requiring a label for MT beef products.
- FSIS' recommendations for beef cooking did not provide important information about cooking MT beef products.

The coalition justified its concerns by citing the 2008 USDA Survey of Beef Operations, which showed that about 18% of all manufactured beef was being mechanically tenderized (Alvares et al. 2008). As a result of this survey, coalition members feared that consumers—who were not aware of the risk posed by mechanically tenderized, non-intact beef steaks and roasts—would be more likely to suffer a foodborne disease. They wanted a label on treated beef products to identify them, and further, they wanted the label to provide validated cooking instructions so that consumers could safely prepare MT beef steaks and roasts.

The 1993 Jack-in-the-Box outbreak clearly demonstrated that ground beef contaminated with *E. coli* O157:H7 could cause widespread and severe illness. As a result, in 1994, *E. coli* O157:H7 was declared an adulterant in ground beef, but it was not declared an adulterant in beef steaks and roasts. The rationale for this distinction was twofold: (1) steaks and roasts were "intact," i.e., not pierced or cut in any way, which made them sterile inside, and (2) killing pathogens on the surface of steaks and roasts was easy to accomplish using any cooking method. Therefore, the industry maintained that once surface pathogens on steaks and roasts were killed by a high cooking temperature, the steak or roast was safe to eat, even if the internal temperature had not reached the "pathogen kill-step" temperature. Further, argued the industry, in 1994 there was very little beef being treated with mechanical tenderization, so the risk of eating a MT beef steak or roast was extremely low—therefore, there was no need to declare *E. coli* O157:H7 as an adulterant in beef steaks and/or roasts.

After the publication of the 2008 beef survey and the additional research on the translocation of surface bacteria to the interior muscle during mechanical tenderization, the MT beef issue became a larger public health concern. According to several studies conducted by USDA/Agricultural Research Service (ARS), MT beef steaks or roasts can become internally contaminated when surface bacteria on carcasses or parts are transferred to the interior (Luchansky et al. 2011, 2012), with translocation frequently occurring during the grinding or mechanical tenderization processes (USDA/FSIS 2016b). Once transfer occurs, it then becomes necessary to kill the pathogens living inside ground beef or MT beef steaks and roasts to make the product safe to eat. Complicating the MT beef issue are two other unrelated but important details: (1) once a steak or roast "rests" after the mechanical tenderization is complete, the product returns to its previous state and visual inspection cannot detect that a treatment has been applied and (2) many consumers prefer to eat beef steaks and roasts undercooked.

To resolve the consumers' inability to identify treated product and to help them prepare MT treated product safely, a label was proposed. Consumer groups now had research showing that pathogenic translocation from surface to the interior posed a public health risk, and they had a survey showing an increase in product availability. In addition, the consumer groups knew that many consumers prefer beef cooked to medium-rare or rare—both of which can pose a risk if consumed—and they knew that CDC had identified several outbreaks as being linked to MT beef (Federal Register 2015). Taken together, the consumer advocates argued that MT beef could very likely cause a foodborne illness, and given USDA's labeling regulations, it was not unreasonable to ask for a label designed to help consumers make informed decisions about their purchases and to provide consumers with important safe handling practices. In order to better protect consumers, SFC members maintained that the following three provisions had to be included in any MT beef labeling rule:

- Mechanically tenderized beef steaks and roasts must be identified.
- Label must provide validated cooking instructions.
- Label must draw attention to the need to use a food thermometer to test for internal doneness.

Petitioning a government agency to consider a regulatory change can be a daunting task. The petitioning group must clearly state the problem and support its proposed change with documented research. In response to a Citizen's Petition, the governmental agency is supposed to provide timely feedback, but in reality, the response time varies considerably. On June 10, 2013, 3 years after the Safe Food Coalition petitioned USDA/FSIS for the MT beef label and 2 years after the Conference for Food Protection (2010) did the same, FSIS submitted a proposed rule to the Federal Register. Included in its proposal was a FSIS cost-benefit analysis that estimated a net benefit of $1,137,000 if the label were adopted, as well as other unquantified benefits of increased consumer information and market efficiency (USDA/FSIS 2013). Interested stakeholders had 60 days to respond. Not surprisingly, the meat industry did not support this proposed rule so they asked for (and received) an extension of the public comment period and then requested and received a second extension to December 24, 2013. After all of the comments were submitted, FSIS took almost another year evaluating them before sending the final proposed rule to the Office of Management and Budget (2015). OMB began its final review in late November 2014, but by this time, it was clear that OMB's required review would not be completed before the end of the year, meaning that the MT beef label would have to wait until the end of the next regulatory cycle—2 years later—to be considered.

The use of "delaying tactics" is a fairly common practice in the legislative and regulatory world. The industry did not want the MT beef label, but if they had to have one, then it should be later instead of sooner. Fortunately, Agriculture Secretary Vilsack, as well as leaders at FSIS and OMB, came to appreciate the consumer advocates' call for immediate action, so a solution was found. Earlier in 2014, the "Added Solutions Rule" had been finalized, and several government reviewers had commented about the cost benefits of implementing these two new labeling rules together. In the end, the MT beef labeling rule was attached to the Added Solutions Rule and was finalized in 2015 with the MT label's implementation to begin in May 2016 (Wheeler and Fabina 2016). Consumer advocacy efforts led by the SFC, the Conference for Food Protection, and the Congressional supporters and media reporters played a huge role in securing this new consumer-friendly rule.

On the other hand—as is true of many reform efforts—the new MT beef label still faces some hurdles. According to the FDA's latest food safety survey (Lando 2010), consumers are not familiar with mechanically tenderized beef steaks and roasts and are fearful of buying or eating them. In fact, this was one of the arguments that the beef industry used in its opposition to the label—consumers simply will not buy the product if it is labeled, meaning that the MT beef label will not be needed because no one will be buying the product. Except, of course, this argument does not apply to the restaurant market, which actually shows a preference for buying mechanically tenderized beef. As mentioned above, CDC has investigated six foodborne illness outbreaks involving needle or blade tenderized beef products (Federal Register 2015). In the 2010 outbreak, many of the 21 people sickened ate a MT beef product at a restaurant (CDC 2010), with several of these victims suffering major life-altering illnesses (McGraw 2012). Since then, CDC has studied outbreak features and has reported that for single food preparation settings, restaurants are the most common setting (Heiman et al. 2015). As a result, consumer advocates are concerned about the role that MT beef steaks and roasts, commonly sold to restaurants, may play in causing foodborne illness.

Unlike steaks and roasts sold at grocery stores, restaurant food is sold in bulk and has the MT beef label on the large box of steaks or roasts, not on the individual packages. As a result, restaurant cooks do not have the label information available as they prepare the steak or roast—in fact, many chefs are not even aware that such a label exists. Nor is the label available to restaurant clients—there is no requirement for retail menus to supply this information—so unless the client specifically asks about the type of steak or roast being prepared, there is no reason to insist on cooking the item to "well done." In other words, while MT beef labeling could empower consumers with more information about the food they plan to eat, the label for restaurant beef is hampered by proximity issues and by the fact that restaurant menus do not inform consumers about food items that may carry a higher risk.

Detailed labeling of food products and increased food safety education are important tools in the prevention of foodborne illness. Far too many people are being sickened with bacterial and viral and parasitic foodborne pathogens. In addition, for those who suffer allergen reactions, labeling information is a crucial part of their health care. Moving forward, consumer groups need to motivate consumers to demand more complete information about their food, including improved labeling information and menu clarifications for high-risk foods.

16.6   Some Consumer Advocacy Achievements

While the above sections have demonstrated how consumer advocates have collaborated to effect change in America's food safety systems, there are also multiple examples of food safety advocates working independently for specific reforms. The list in this area would be extremely lengthy, and without a complete literature review, it is impossible to note all of the contributions. Below is a short list of some of the notable achievements of nongovernmental organizations (NGOs) that have a focus on food safety.

- Center for Food Safety:

    - Collaborated with others to publish a report, Chain reactions I, II, and III: How top restaurants rate on reducing use of antibiotics in the meat their meat supply (Center for Food Safety 2016).
    - Filed a lawsuit—with the Center for Environmental Health—against FDA for not meeting the mandatory deadlines (to promulgate final regulations) under FSMA (Center for Food Safety 2013).

- Center for Foodborne Illness Research & Prevention:

    - Published a white paper (Roberts et al. 2009) on the long-term health outcomes related to foodborne diseases; held an international work shop on LTHOs in 2014 (see Chap. 9).
    - Continues to work to improve data sharing between federal food agencies, in particular for surveillance on animal antibiotic use (Center for Foodborne Illness Research and Prevention 2016).

- Center for Science in the Public Interest:

    - Established Outbreak Alert! Database in 2004 to track foodborne illness outbreaks.

- Petitioned USDA to declare four strains of antibiotic-resistant (AR) *Salmonella* as adulterants in ground meat (Center for Science in the Public Interest 2014).

• Consumer Federation of America:

- The Food Policy Institute hosts an annual conference on timely and important food safety issues.
- Coordinates the interactions between USDA-FSIS and the Safe Food Coalition (SFC).

• Consumers Union and Consumers Reports:

- Called for the development of stronger standards for poultry (Consumers Union 2010, 2013).
- Raised awareness about the challenges associated with mechanically tenderized beef products (Consumers Union 2015).

• Food & Water Watch:

- Supports improvement in America's factory farming policies. *See* "Poultry Litter Incineration: A False Solution to Factory Farm Pollution" (Food &Water Watch 2015a).
- Reported on multiple gaps in America's seafood supply, in particular for imported seafood (Food &Water Watch 2015b).

• Food Associations: various NGOs that focus on food and food technologies to improve food safety. Some notable leaders are the American Frozen Food Institute, Food Marketing Institute, Grocery Manufacturers of America, International Food Information Council, Institute of Food Technologists, Produce Marketing Association, NSF International, and United Fresh Produce Association.

• Food Industry: companies associated with dairy, eggs, grains, fish, meat, nuts, poultry, and produce have developed food safety programs and support various food safety initiatives. Two examples would be ConAgra's Food Safety Council and Walmart's decision to require all of its suppliers (both domestic and foreign) to meet specific food standards in order to sell in their retail outlets.

• Health-oriented associations: various NGOs that focus on the prevention of infectious diseases. Some notable leaders in this arena are the American Academy of Pediatrics, American Public Health Association, Association of Food and Drug Officials, Infectious Diseases Society of America, International Food Protection Association, March of Dimes, and National Association of City and County Health Officials.

• Keep Antibiotics Working:

- Collaborates with the Pew Charitable Trusts and the Infectious Diseases Society of America to host a Working Group on Antibiotic Resistance. This Working Group has helped to improve FDA's guidance about the judicious use of medically important antimicrobial drugs in food-producing animals (KAW 2017).

- Pew Charitable Trusts:

  - Supported FDA's Food Safety Modernization Act's public health goals (Pew's Charitable Trusts 2016); Pew's Safe Food Project also supports funding to implement the new law (Pew Charitable Trusts 2015a, b).
  - Coordinates efforts between FDA and the Make Our Food Safe coalition members.

- STOP Foodborne Illness:

  - Petitioned USDA to declare six non-O157 STECs as adulterants in ground beef. In 2011, USDA declared six additional *E. coli* strains as adulterants in ground beef (USDA/FSIS 2012).
  - Continues to assist victims of foodborne illness by providing outreach and arranging for victims to testify to Congressional Members about their experiences.

- US Public Interest Research Group (US PIRG):

  - Petitioned food service industry leaders to not purchase meat/poultry that have been raised on antibiotics (CalPIRG 2016).

- US Stakeholder Forum on Antimicrobial Resistance—A platform formed by Infectious Diseases Society of America:

  - Supports the National Action Plan for Combatting Antibiotic Resistant Bacteria, 2015 (Obama, President 2014).
  - Provides opportunities for AR stakeholders to investigate AR challenges and solutions, as well as request appropriations for increases in AR research (S-FAR 2017).

16.7 Future Food Safety Advocacy Challenges

Tomorrow's challenges with regard to food and food safety are very large, and the impacts could be significant. We are facing a world where food production and foodborne illness surveillance methods are changing rapidly. We are experiencing growth in the world population, and we have major foodborne pathogens developing antibiotic resistance. Unfortunately, large segments of our national and global communities live/work in substandard conditions with little knowledge about food safety. Meanwhile, there are many others—who live/work in optimum conditions with access to information—that do not follow the recommended safe food practices. If we don't start attending to these challenges, our national and global communities could experience deep societal impacts.

It is vital for nongovernmental food safety stakeholders, both at the national and global levels, to promote stronger food safety policies. These advocacy efforts are essential since governments—whether state, federal or foreign—cannot objectively

Table 16.3 Food safety advocacy for the future

Food safety issue	Potential actions
WHO, CDC, and the European Union have identified antibiotic resistance as a major public health threat in the twenty-first century.	Support legislation in US Congress that promotes more research on antibiotic resistance and its potential overuse in treatments of humans and animals. We also need to provide incentives to drug companies to resume research on new antibiotics and vaccines, which was provided in the 21st Century Cures Act, passed in 2016 (Dall 2016).
Surveillance of foodborne illness is essential for the future control of foodborne diseases, especially those caused by AR bacterial strains.	Financial resources are needed to maintain and enhance US surveillance programs. In addition, it is imperative to have resources to assist state public health departments and laboratories as new technologies change the surveillance landscape. An independent data center for food needs to be developed to integrate existing data sources (IOM 2010).
Currently, there is limited ability to monitor antimicrobial use in humans and animals. The World Health Organization has a plan called Global Antimicrobial Resistance Surveillance System—GLASS, and the G20 Leaders' Declaration included combatting antibiotic resistance as one of the world's challenges (G20 Germany 2017).	The USA needs to develop a mechanism to track the use of animal antibiotics so that standards can be set for animal antibiotic use. As a first step, the USA must implement an on-farm data collection system to monitor the purposes, types, and doses of animal antibiotics being used (Center for Foodborne Illness Research and Prevention 2016).
Salmonella, the US leading foodborne illness killer, is not an adulterant for USDA-inspected food, thereby weakening our ability to control *Salmonella* in meat and poultry products.	Declare the most serious AR *Salmonella* strains as adulterants. USDA has declared several strains of *E. coli* and *Listeria monocytogenes* as adulterants, so they have the authority to do the same for AR *Salmonella*.
The US General Accountability Office's report (GAO 2011) on Seafood HACCP says the program needs to be strengthened. In 2014, US imports of fish and shellfish exceeded $20 billion in value, based on USDA-ERS data (USDA/ERS 2016c). Critics say that FDA's Seafood HACCP has failed to meet its obligations (U.S. House of Representatives 2016).	Retain USDA's residue testing oversight of catfish at US border points of entry. Moving forward, we need to review FDA's Seafood HACCP program and find resources to improve and implement a revised program.
The FDA Food Safety Modernization Act (FSMA) is a huge step forward for food safety. However, some provisions still need to be clarified or implemented.	FDA should develop an interim rule for raw manure applications (Consumer Federation of America 2016). Improperly composted raw manure, used as fertilizer in fields, has been recognized as a biohazard for produce (CAST 2009). FDA also needs to implement FSMA's provisions with regard to public notification of reportable foodborne illness outbreaks and recalls.

(continued)

Table 16.3 (continued)

Food safety issue	Potential actions
Epidemiology drives prevention, but there are relatively few research studies investigating the association between foodborne illness and major long-term health outcomes (LTHO), like arthritis, diabetes, kidney dysfunction, or neurological disorders. Consequently, even though LTHOs are expensive to treat, they are often not included in burden of disease estimates (see Chap. 9).	Support research on foodborne diseases, including the long-term health outcomes (LTHOs) (Roberts et al. 2009). Once we have more information about the LTHOs, the burden of disease estimates will be more accurate, thereby allowing us to develop and implement better preventive food safety policies.
There are safety and health issues in food production and food retail outlets. Dangerous equipment or fast processing lines (Norton et al. 2015), as well as a lack of worker health benefits, have been associated with foodborne crisis events. For example, a 2015 study has shown that sick food service workers cause almost half of all restaurant-related outbreaks (Restaurant Opportunities Centers United (ROCU) 2010).	Support improvement in all areas of food safety management, including worker fatigue, worker injury, worker wages, and the lack of worker sick leave policies. In addition, we need wider access to restaurant inspection reports and food production data.
New technologies are changing the way that we grow, harvest, process, inspect, test, and trace-back food.	New technologies can drive change, but sometimes the innovation exceeds the public's understanding. We must be willing to objectively assess these new technologies and then develop strategies to fill the public's knowledge gaps.
Food safety education needs to include information about the risks posed by certain foods or the risk associated with unsafe food sources. Consumers need this information in addition to the messages about the four core safe food handling practices. Without risk information, consumers will be less likely to adopt the use of the safe food practices into their daily routines.	Support risk communication in food safety messaging. Schools, agricultural cooperative extension, food safety training programs, and public health facilities (such as hospitals) need food safety messages that identify risk and provide information on ways to mitigate the risk.
There is a shortage of professionals entering public health- and food-related sciences. The USA, like many other nations, is finding it difficult to fill the vacancies in these fields. The expectation is that this situation will worsen.	Support the creation of financial incentives for undergraduate and graduate students to enter public health- or food-related fields. Support already developed programs to generate interest in these fields for America's middle and high school students (FDA 2014; Grocery Manufactures Association 2016).

evaluate their own efforts. Therefore, independent food safety advocates must continue to perform the tedious work of reviewing and evaluating food-related documents or complicated data reports. However, there is a looming problem: most of the food safety advocacy groups struggle with sustainable funding, which means that this type of independent, critical analysis is dwindling. Table 16.3 provides some insights on some of the work that still needs to be done.

Obviously, as listed in Table 16.3, there are many areas that need the attention of food safety advocates. To address these future food challenges, we need a worldwide commitment to food safety led by experts in agricultural management (especially veterinarians); food safety/food science professionals; infectious disease researchers; and leaders from nutrition, technology, and the social-economic sciences. Below is a short list of potential priorities:

- Economic analysts need to create new incentives to address the rising costs associated with science-based food safety programs and processes.
- Educators must include risk information into their food safety messages. To be most effective, these new risk communications should be targeted to specific challenges and/or audiences.
- Governments need to develop dedicated resources for improved food safety training at farms, food facilities, and throughout the inspection workforce.
- Innovators of food safety and food production technologies must develop integrated systems and programs aimed at reducing the development and/or spread of foodborne pathogens—especially those that are antibiotic resistant.
- Researchers, both governmental and private, need to dedicate their efforts to food safety, foodborne diseases (including the long-term health outcomes), and other agricultural concerns, such as the preservation of healthy soil and clean water.

References

Achenbach J, Sun L. Trump budget seeks huge cuts to science and medical research, disease prevention. Washington Post. 2017. https://www.washingtonpost.com/news/to-your-health/wp/2017/05/22/trump-budget-seeks-huge-cuts-to-disease-prevention-and-medical-research-departments/?utm_term=.c192b73efff2. Accessed 15 July 2017.

Alvares C, Green K, Lim C. Results of checklist and reassessment of control for *Escherichia coli* O157:H7 in beef operations. USDA Food Safety and Inspection Service (FSIS) and the Data Analysis and Integration Group (DAIG) of the Office of Food Defense and Emergency Response. 2008. http://www.fsis.usda.gov/shared/PDF/Ecoli_Reassement_&_Checklist.pdf?redirecthttp=true. Accessed 29 Aug 2016.

Andrews J. Senators, food industry groups call for more FSMA funding. Food Safety News. 2015. http://www.foodsafetynews.com/2015/05/seven-u-s-senators-major-food-industry-groups-call-for-more-fsma-funding/#.WDMxKvkrLs1. Accessed 21 Nov 2016.

Beck R. 2015 Consumer trust study: impact on food safety. Webinar, Part IV: transparency—food safety. Center for Food Integrity. 2016. http://www.foodintegrity.org/research/consumer-trust-research/research-webinars/. Accessed 4 Nov 2016.

Blum J. Obama says FDA needs review after peanut illnesses. Bloomberg News. 2009. http://www.deseretnews.com/article/705282167/Obama-says-FDA-needs-review-after-peanut-illnesses.html?pg=all. Accessed 21 Nov 2016.

Bottemiller H. Looking back: the story behind banning *E. coli* O157:H7. Food Safety News. 2011. http://www.foodsafetynews.com/2011/09/looking-back-in-time-the-story-behind-banning-ecoli-o157h7/#.WQ4Q8_krLs0. Accessed 6 May 2017.

CAST—Council for Agricultural Science and Technology. Food safety and fresh produce: an update. CAST Commentary QTA20091. 2009. https://www.cast-science.org/download.cfm?PublicationID=2946&File=f030788c37850f31a5287c755f70231b4013. Accessed 6 May 2017.

CDC. Multistate outbreak of *Salmonella typhimurium* infections linked to peanut butter. (Final Update). 2008–2009. https://www.cdc.gov/salmonella/2009/peanut-butter-2008-2009.html. Accessed 16 Feb 2017.

CDC. Multistate outbreak of *E. coli* O157:H7 infections associated with beef from national steak and poultry. (final update). 2010. https://www.cdc.gov/ecoli/2010/national-steak-poultry-1-6-10.html. Accessed 13 April 2017.

CDC. Foodborne illness surveillance systems. 2011. https://www.cdc.gov/foodborneburden/pdfs/factsheet_g_surveillance.pdf. Accessed 5 May 2017.

CDC. Impact of outbreaks traced to contaminated food. Infograph. 2013. https://www.cdc.gov/foodsafety/pdfs/impact-of-contaminated-foods-508c.pdf. Accessed 7 May 2017.

CDC. Integrated food safety centers of excellence. 2015. https://www.cdc.gov/foodsafety/centers/. Accessed 2 Aug 2016.

CDC. Norovirus transmission. 2016. http://www.cdc.gov/norovirus/about/transmission.html. Accessed 12 Dec 2016.

Center for Effective Government. Cutting costs and courting contamination: what food safety budget cuts mean for public safety. 2012. http://www.foreffectivegov.org/node/12044. Accessed 15 July 2017.

Center for Food Safety. Center for Food Safety v. Margaret Hamburg. No. C 124529 PJH. 2013. http://www.centerforfoodsafety.org/files/57-sj-decision_78315.pdf. Accessed 4 Nov 2016.

Center for Food Safety. Chain reaction II: how top restaurants rate on reducing use of antibiotics in their meat supply. 2016. http://www.centerforfoodsafety.org/files/chainreaction2-report-2016-final_13986.pdf. Accessed 24 May 2017.

Center for Foodborne Illness Research & Prevention. Tracking antibiotic use in food animals. 2016. http://www.foodborneillness.org/cfi-library/2016_CFI_Fact.Sheet_Track.animal.antibiotic.use_FINAL.pdf. Accessed 5 May 2017.

Center for Science in the Public Interest. Petition for an interpretive rule declaring antibiotic-resistant *Salmonella* Heidelberg, *Salmonella* Hadar, *Salmonella* Newport, and *Salmonella typhimurium* in meat and poultry to be adulterants. 2014. https://www.fsis.usda.gov/wps/wcm/connect/25551672-a5db-4a11-9253-05f92f6b49a4/Petition-CSPI-ABRSalmonella-2014.pdf?MOD=AJPERES. Accessed 5 May 2017.

Centers for Disease Control and Prevention. Figure 15. Relative rates of culture-confirmed infections with *Campylobacter*, STEC* O157, *Listeria*, *Salmonella*, and *Vibrio*, and overall measure of change, compared with 1996–1998 rates, by year. FoodNet 1996–2015. https://www.cdc.gov/foodnet/reports/data/incidence-trends.html#figure15. Accessed 12 Dec 2016.

Clapp S. Fixing the food system: changing how we produce and consume food. Praeger, An imprint of ABC-CLIO. LLC; 2017.

Conference for Food Protection. Petition to USDA/FSIS to label mechanically tenderized beef products. 2010. http://www.fsis.usda.gov/wps/wcm/connect/7da02e44-712f-4779-aa10-fb1760493261/Petition_CFP_071710.pdf?MOD=AJPERES. Accessed 30 Nov 2016.

Congressional Budget Office. An introduction to the congressional budget office. 2016. https://www.cbo.gov/sites/default/files/cbofiles/attachments/2016-IntroToCBO.pdf. Accessed 29 Aug 2016.

Consumer Federation of America, Center for Foodborne Illness Research & Prevention, STOP Foodborne Illness, Consumers Union, and Center for Science in the Public Interest. Comments on Docket FDA-2016-N-0321: Risk assessment of foodborne illness associated with pathogens from produce grown in fields amended with untreated biological soil amendments of animal origin. 2016. http://foodborneillness.org/cfi-library/SFC_Comment.Untreated.soil.amendments_071916.pdf. Accessed 10 May 2017.

Consumers Union. How safe is that chicken? 2010. http://www.consumerreports.org/cro/2012/05/how-safe-is-that-chicken/index.htm. Accessed 5 May 2017.

Consumers Union. Talking Turkey: our new tests show reasons for concern. 2013. http://www.consumerreports.org/cro/magazine/2013/06/consumer-reports-investigation-talking-turkey/index.htm. Accessed 5 May 2017.

Consumers Union. USDA releases long-awaited proposed rule on labeling mechanically tenderized meat. 2015. http://consumersunion.org/news/usda-releases-long-awaited-proposed-rule-on-labeling-mechanically-tenderized-meat/. Accessed 5 May 2017.

Dall C. How 21st century Cures Act could boost new antibiotics. Center for Infectious Disease Research and Policy. 2016. http://www.cidrap.umn.edu/news-perspective/2016/12/how-21st-century-cures-act-could-boost-new-antibiotics. Accessed 17 Feb 2017.

Donald T. 20 years in food safety: a look backward and beyond. Food Quality & Safety. 2010. http://www.foodqualityandsafety.com/article/20-years-in-food-safety-a-look-back-and-beyond/. Accessed 17 Apr 2017.

FDA. Melamine pet food recall. 2007. https://www.fda.gov/animalveterinary/safetyhealth/recalls-withdrawals/ucm129575.htm. Accessed 5 May 2017.

FDA. Science and our food supply. 2014. http://www.fda.gov/downloads/food/foodscienceresearch/toolsmaterials/ucm430366.pdf . Accessed 12 Dec 2016.

Federal Register. Pathogen reduction; hazard analysis and critical control point (HACCP) systems; final rule. Part II. 9 CFR Part 304, et al. 1996. https://www.fsis.usda.gov/OPPDE/rdad/FRPubs/93-016F.pdf. Accessed 8 May 2017.

Federal Register. Descriptive designation for needle- or blade-tenderized (mechanically tenderized) beef products. 9 CFR 317. 2015. https://www.federalregister.gov/documents/2015/05/18/2015-11916/descriptive-designation-for-needle%2D%2Dor-blade-tenderized-mechanically-tenderized-beef-products. Accessed 5 May 2017.

Flynn D. Jury verdicts: guilty, guilty and guilty in PCA criminal trial. Food Safety News. 2014. http://www.foodsafetynews.com/2014/09/guilty-guilty-and-guilty-in-pca-criminal-trial-in-albany-ga/#.WBuReeArLs0. Accessed 2 Nov 2016.

Food &Water Watch. Poultry litter incineration: a false solution to factory farm pollution. 2015a. https://www.foodandwaterwatch.org/sites/default/files/fs_1510_md-poultry-incineration-web.pdf. Accessed 5 May 2017.

Food &Water Watch. Trans Pacific Partnership (TTP) delivers a rising tide of imported fish. 2015b. www.foodandwaterwatch.org/sites/default/files/tpp_imported_fish_fs_dec_2015.pdf. Accessed 5 May 2012.

Food and Drug Administration (FDA). Fact sheets & presentations. Food Safety Modernization Act. 2016. https://www.fda.gov/Food/GuidanceRegulation/FSMA/ucm247546.htm. Accessed 5 May 2017.

Fortin N. Food regulation: law, science, policy, and practice. 2nd ed. Hoboken, NJ: John Wiley & Sons; 2016.

G20 Germany. G20 Leaders´ Declaration: shaping an interconnected world—building resilience. 2017. http://www.g20.org/gipfeldokumente/G20-leaders-declaration.pdf. Accessed 15 July 2017.

Goetz G. Long history of violations at peanut plant linked to Salmonella outbreak. Food Safety News. 2012. http://www.foodsafetynews.com/2012/11/long-history-of-health-violations-at-peanut-co-linked-to-salmonella-outbreak/#.WQz2j_nyvs0. Accessed 5 May 2017.

Government Accountability Office. Seafood Safety: FDA needs to improve oversight of imported seafood and better leverage limited resources. 2011. http://www.gao.gov/products/GAO-11-286. Accessed 5 May 2017.

Government Accountability Office. Food safety: a national strategy is needed to address fragmentation in federal oversight. 2017. http://www.gao.gov/assets/690/682095.pdf. Accessed 5 May 2017.

Grocery Manufactures Association: A science and education foundation. Hands on classrooms. 2013–2016. http://handsonclassrooms.org/. Accessed 20 Feb 2017.

Hawkins D. Peanut corp. Salmonella settlements approved. Law 360. 2010. https://www.law360.com/articles/191631/peanut-corp-salmonella-settlements-approved. Accessed 10 May 2017.

Health and Human Services. Office of Inspector General. Challenges remain in FDA's inspections of domestic food facilities. 2017. https://oig.hhs.gov/oei/reports/oei-02-14-00420.asp. Accessed 10 Oct 2017.

Heiman K, Mody R, Johnson S, Griffin P, Gould L. *Escherichia coli* O157 outbreaks in the United States, 2003–2012. Emerg Infect Dis. 2015;21(8):1293–301. https://www.ncbi.nlm.nih.gov/pmc/articles/PMC4517704/. Accessed 10 May 2017

Henao O, Jones T, Vugia D, Griffin P. Foodborne diseases active surveillance network—2 decades of achievements, 1996–2015. Emerg Infect Dis. 2015;21(9):1529–36. https://wwwnc.cdc.gov/eid/article/21/9/15-0581_article. Accessed 5 May 2017

Institute of Medicine (IOM) and National Research Council. In: Chapter: Creating an integrated information infrastructure for a risk-based food safety system. In: Enhancing food safety: The role of the Food and Drug Administration. The National Academies Press; 2010, pp. 147–180. http://www.nap.edu/read/12892/chapter/10. Accessed 12 Dec 2016.

International Food Information Council Foundation (IFICF). 2016 Food and health survey. 2016. http://www.foodinsight.org/sites/default/files/2016-Food-and-Health-Survey-Report_FINAL1.pdf . Accessed 10 May 2010.

Keep Antibiotics Working (KAW) et al. Comments to FDA on Docket No. FDA-2016-D-2635. The judicious use of medically important antimicrobial drugs in food-producing animals; Establishing appropriate durations of therapeutic administration. 2017. http://foodborneillness.org/cfi-library/2017_Joint.Comments_FDA_Animal.antibiotic.use&duration.pdf. Accessed 10 May 2017.

King L. Institute of Medicine. Improving food safety through a one health approach: workshop summary. National Academies Press; 2012. http://www.ncbi.nlm.nih.gov/books/NBK114498/. Accessed 12 Dec 2016.

Koehler D. Testimony before U.S. House of Representatives' Subcommittee on regulations and healthcare. Hearing: Impact of food recalls on small businesses. Washington, DC; 2009. https://www.gpo.gov/fdsys/pkg/CHRG-111hhrg47797/html/CHRG-111hhrg47797.htm. Accessed 4 Nov 2016.

Lando A. 2010 FDA Food Safety Survey. 2010. https://www.fda.gov/downloads/Food/FoodScienceResearch/ConsumerBehaviorResearch/UCM407008.pdf. Accessed 5 May 2017.

Layton L. Suspect peanuts sent to schools. Washington Post. 2009. http://www.washingtonpost.com/wp-dyn/content/article/2009/02/05/AR2009020500743.html. Accessed 5 May 2017.

Leighton P. Mass Salmonella poisoning by the Peanut Corporation of America: state-corporate crime involving food safety. Dordrecht: Springer; 2015. http://www.paulsjusticepage.com/library/PeanutCorp-MassSalmonellaPoisoning.pdf. Accessed 22 Aug 2016.

Luchansky JB, Porto-Fett A, Shoyer B, et al. Inactivation of Shiga toxin–producing O157:H7 and non-O157:H7 Shiga toxin-producing *Escherichia coli* in brine-injected, gas-grilled steaks. J Food Protect. 2011;74(7):1054–64.

Luchansky JB, Phebus RK, Thippareddi H, Call JE. Translocation of surface-inoculated *Escherichia coli* O157:H7 into beef subprimals following blade tenderization. J Food Protect. 2012;71(11):2190–7.

McGraw M. Iowa woman nearly died from eating a tenderized beef steak. Kansas City Star. 2012. http://www.kansascity.com/news/nation-world/article79516902.html. Accessed 14 April 2017.

National Association of County and City Health Officials (NACCHO). National profile of local health departments. 2011. http://archived.naccho.org/topics/infrastructure/profile/resources/2010report/upload/2010_profile_main_report-web.pdf. Accessed 16 Nov 2016.

Nestle N. Safe food: the politics of food safety. Berkeley, CA: University of California Press; 2010.

Norton D, Brown L, Frick R, et al. Managerial practices regarding workers working while ill. J Food Protect. 2015;78(1):187–95. https://doi.org/10.4315/0362-028X.JFP-14-134.

Obama, President. Executive Order 13676. Combating antibiotic resistant bacteria (CARB). 2014. https://www.gpo.gov/fdsys/pkg/FR-2014-09-23/pdf/2014-22805.pdf. Accessed 3 Aug 2016.

Office of Management and Budget; Office of Information and Regulatory Affairs. Descriptive designation for needle- or blade-tenderized (mechanically tenderized) beef products. USDA/FSIS RIN: 0583-AD45. 2015. https://www.reginfo.gov/public/do/eAgendaViewRule?pubId=201504&RIN=0583-AD45. Accessed 20 Feb 2017.

Organic Consumers Association. The roots of FSIS workplace violence. 2000. https://www.organicconsumers.org/old_articles/irrad/InspectorView.php. Accessed 8 May 2017.

Pew Charitable Trusts. About the make our food safe coalition. 2009a. http://www.pewtrusts.org/en/research-and-analysis/analysis/2009/09/02/about-make-our-food-safe. Accessed 20 Feb 2017.

Pew Charitable Trusts. 2009 Survey: Pew-commissioned poll finds large majority of Americans want stronger food safety rules. 2009b. http://www.pewtrusts.org/en/about/news-room/press-releases/0001/01/01/pewcommissioned-poll-finds-large-majority-of-americans-want-stronger-food-safety-rules. Accessed 2 Aug 2016.

Pew Charitable Trusts. Letter to U.S. House Appropriations Committee Chairmen Rogers and Cochran, Ranking Member Lowey and Vice Chairwoman Mikulski. 2015a. http://www.pewtrusts.org/~/media/assets/2015/02/fda_coalition_funding_letter_2015.pdf Accessed 21 Nov 2016.

Pew Charitable Trusts. Pew and partners call for food safety funding. 2015b. http://www.pewtrusts.org/en/research-and-analysis/speeches-and-testimony/2015/10/pew-and-partners-call-for-food-safety-funding. Accessed 5 May 2017.

Pew Charitable Trusts. Safe Food Project. Priorities. 2016. http://www.pewtrusts.org/en/projects/food-safety/priorities. Accessed 21 Aug 2016.

Produce for Better Health Foundation. State of the Plate, 2015 Study on America's Consumption of Fruit and Vegetables. 2015. http://www.pbhfoundation.org/pdfs/about/res/pbh_res/State_of_the_Plate_2015_WEB_Bookmarked.pdf. Accessed 7 May 2017.

Produce News. Peanut butter recall likely to push FDA food-safety reforms. The Produce News. 2009. http://www.producenews.com/a-lead-story/9-news-section/story-cat/3511-3308. Accessed 2 Aug 2016.

Restaurant Opportunities Centers United (ROCU). Serving while sick. 2010. http://rocunited.org/wp-content/uploads/2013/04/reports_serving-while-sick_full.pdf. Accessed 12 Dec 2016

Roberts T. Lack of information is the root of U.S. foodborne illness risk. Choices. 2nd Quarter. 2013;28(2). http://www.foodborneillness.org/cfi-library/Roberts_Lack.of.Info_Choices_2013.pdf. Accessed 8 May 2017.

Roberts T, Kowalcyk B, Buck P, et al. The long-term health outcomes of selected foodborne pathogens. White Paper. Center for Foodborne Illness Research & Prevention. 2009. http://www.foodborneillness.org/cfi-library/CFI_LTHO_PSP_report_Nov2009_050812.pdf. Accessed 12 Dec 2016.

Safe Food Coalition (SFC). Background information for letter to Secretary Vilsack about mechanically tenderized (MT) beef products. 2009a. http://foodborneillness.org/cfi-library/2009_SFC_backgrounder_mechanical.tenderized.meat.pdf. Accessed 10 May 2017.

Safe Food Coalition (SFC). Citizen petition to Secretary Vilsack, Secretary, USDA, to initiate labeling of mechanically tenderized beef products, along with a program of educational outreach to consumers and retailers about this product. 2009b. http://foodborneillness.org/cfi-library/2009_SFC_ltr_to_Vilsack_Meat_Tenderization.pdf. Accessed 10 May 2017.

Scallan E, Hoekstra RM, Angulo FJ, Tauxe RV, Widdowson MA, Roy SL, et al. Foodborne illness acquired in the United States—major pathogens. Emerging Infect Dis 2011;17(1):7–15. https://wwwnc.cdc.gov/eid/article/17/1/P1-1101_article. Accessed 6 May 2017.

Scharff R. Economic burden from health losses due to foodborne illness in the United States. J Food Protect. 2012;75(1):123–31. http://citeseerx.ist.psu.edu/viewdoc/download?doi=10.1.1.662.2667&rep=rep1&type=pdf. Accessed 21 Nov 2016.

S-FAR—U.S. Stakeholder Forum on Antibiotic Resistance. S-FAR Hill briefing. 2017. http://s-far.org/home-1/. Accessed 8 Mar 2017.

Supreme Beef Processors Inc. v. United States Department of Agriculture. 2001. 275 F.3d 432. 5th Circuit. Summary: No. 0011008. In: FindLaw. http://caselaw.findlaw.com/us-5th-circuit/1429779.html. Accessed 21 Nov 2016.

Surendran A. An unseen killer's toll: unsafe meat ended or tragically changed their lives. Philadelphia Inquirer. 2003. http://articles.philly.com/2003-05-19/news/25459439_1_coli-barbara-kowalcyk-human-illness. Accessed 15 Aug 2016.

Tucker-Foreman C. Modern meat. Frontline/PBS Interview. 2002. http://www.pbs.org/wgbh/pages/frontline/shows/meat/interviews/foreman.html. Accessed 7 May 2017.

U.S. Department of Justice. Former peanut company president receives largest criminal sentence in food safety case; two others also sentenced for their roles in Salmonella-tainted peanut product outbreak. 2015. https://www.justice.gov/opa/pr/former-peanut-company-president-receives-largest-criminal-sentence-food-safety-case-two. Accessed 6 May 2017.

U.S. District Court, Northern District of California. Center for Food Safety and Center for Environmental Health v. Margaret Hamburg and Jeffrey Zients. CV 12 4529. 2012. http://www.centerforfoodsafety.org/files/2012-08-29-fsma-complaint-filed_78450.pdf. Accessed 4 Nov 2016.

U.S. House of Representatives. Letter on S.J. Res 28, a Congressional Review Act resolution. 2016. http://crawford.house.gov/uploadedfiles/catfish_letter.pdf. Accessed 12 Dec 2016.

U.S. PIRG, California (CalPIRG). Letter to YUM asking them to join other restaurant industry leaders in addressing the overuse of antibiotics in livestock production. 2016. http://calpirg.org/sites/pirg/files/resources/In-N-Out%20Coalition%20Letter%20FINAL.pdf. Accessed 9 Mar 2017.

USDA/Economic Research Service (ERS). Food-away-from home. 2016a. https://www.ers.usda.gov/topics/food-choices-health/food-consumption-demand/food-away-from-home.aspx. Accessed 7 May 2017.

USDA/Economic Research Service (ERS). Import share of consumption. 2016b. http://www.ers.usda.gov/topics/international-markets-trade/us-agricultural-trade/import-share-of-consumption.aspx. Accessed 3 Nov 2016.

USDA/Economic Research Service (ERS). U.S. food import value by food group. 2016c. http://www.ers.usda.gov/data-products/us-food-imports.aspx#25420. Accessed 12 Dec 2016.

USDA/Food Safety Inspection Service (FSIS). Control of Listeria monocytogenes in ready-to-eat meat and poultry products. [Docket No. 97-013F]. 2003. https://www.fsis.usda.gov/wps/wcm/connect/226dd629-d171-4d01-aee9-6dba6e4216ea/04-032N.pdf?MOD=AJPERES. Accessed 20 Feb 2017.

USDA/Food Safety Inspection Service (FSIS). Verifying sanitary dressing and process control procedures by off-line inspection program personnel in slaughter operations of cattle of any age. Directive 6410.1 Rev. 1. 2011a. https://www.fsis.usda.gov/wps/wcm/connect/4d4f2ca7-af74-4879-b385-4c163c0b361c/6410.1.pdf?MOD=AJPERES. Accessed 20 Feb 2017.

USDA/Food Safety Inspection Service (FSIS). Not applying the mark of inspection pending certain test results. [Docket No. FSIS-2005-0044]. 2011b. https://www.fsis.usda.gov/OPPDE/rdad/FRPubs/2005-0044.pdf. Accessed 20 Feb 2017.

USDA/Food Safety Inspection Service (FSIS). Shiga toxin-producing Escherichia coli in certain raw beef products. [Docket No. FSIS-2010-0023]. 2012. https://www.fsis.usda.gov/OPPDE/rdad/FRPubs/2010-0023FRN.pdf. Accessed 20 Feb 2017.

USDA/Food Safety Inspection Service (FSIS). Descriptive designation for needle- or blade-tenderized (mechanically tenderized) beef products proposed rule. [Docket No. FSIS-2008-0017]. 2013. https://www.fsis.usda.gov/wps/wcm/connect/741a8b94-85b0-4800-963e-ad0517838a02/2008-0017.pdf?MOD=AJPERES . Accessed 20 Feb 2017.

USDA/Food Safety Inspection Service (FSIS). New performance standards for Salmonella and Campylobacter in not-ready-to-eat comminuted chicken and Turkey products and raw chicken parts and changes to related agency verification procedures: response to comments and announcement of implementation schedule. [Docket No. FSIS-2014-0023]. 2014a. https://www.fsis.usda.gov/wps/wcm/connect/03910f0e-48f8-4111-ba52-490b07c25c24/2014-0023.htm?MOD=AJPERES. Accessed 12 Dec 2017

USDA/Food Safety Inspection Service (FSIS). Descriptive Designation for raw meat and poultry products containing added solutions. [Docket No. FSIS-2010-0012]. 2014b. https://www.fsis.usda.gov/wps/wcm/connect/942b0716-42a9-4d0e-8e5f-3ba4d2dfc70d/2010-0012.pdf?MOD=AJPERES. Accessed 20 Feb 2017.

USDA/Food Safety Inspection Service (FSIS). Best practices guidance for controlling Listeria monocytogenes in retail delicatessens. [Docket No. FSIS-2013-0038]. 2015a. https://www.fsis.usda.gov/wps/wcm/connect/c6fa2945-4ab5-46ba-8bcb-318c8a85a9f9/2013-0038-2015.htm?MOD=AJPERES. Accessed 20 Feb 2017.

USDA/Food Safety Inspection Service (FSIS). Redesign of FSIS sampling methodologies to improve detection of *E. coli* O157:H7 in raw beef products. Directive 10,010.1 Rev.4. 2015b. https://www.fsis.usda.gov/wps/wcm/connect/c100dd64-e2e7-408a-8b27-ebb378959071/10010.1Rev3.pdf?MOD=AJPERES. Accessed 12 Dec 2016.

USDA/Food Safety Inspection Service (FSIS). Records to be kept by official establishments and retail stores that grind raw beef products. [Docket No. FSIS-2009-0011]. 2015c. https://www.federalregister.gov/documents/2015/12/21/2015-31795/records-to-be-kept-by-official-establishments-and-retail-stores-that-grind-raw-beef-products. Accessed 6 May 2017.

USDA/Food Safety Inspection Service (FSIS). USDA finalizes new food safety measures to reduce Salmonella and Campylobacter in poultry. Press release. 2016a. https://www.usda.gov/wps/portal/usda/usdamediafb?contentid=2016/02/0032.xml&printable=true&contentidonly=true. Accessed 20 Feb 2017.

USDA/Food Safety Inspection Service (FSIS). Mechanically tenderized infograph. 2016b. https://www.fsis.usda.gov/wps/wcm/connect/b7e9dcdd-0d6f-4597-b2d0-b6833f88f336/MTB-Infographic.pdf?MOD=AJPERES. Accessed 6 May 2017.

Wegener H, Hald T, Wong L, et al. *Salmonella* control programs in Denmark. Emerg Infect Dis. 2003;9(7):774–8. https://www.ncbi.nlm.nih.gov/pmc/articles/PMC3023435/citedby/. Accessed 8 May 2017

Wheeler M, Fabina J. Descriptive designations for mechanically tenderized beef and added solutions. USDA/FSIS labeling and program delivery staff. Power point. 2016. http://www.fsis.usda.gov/wps/wcm/connect/9a96a9f0-54b3-4f2a-a95b-a690fc787733/Descriptive-Designations-MTB-Added-Solutions.pdf?MOD=AJPERES. Accessed 30 Nov 2016.

World Health Organization. Global action plan on antimicrobial resistance. 2015. http://apps.who.int/iris/bitstream/10665/193736/1/9789241509763_eng.pdf?ua=1. Accessed 15 Aug 2016.

Zuraw L. Revised FSMA provisions need more tweaks. Food Safety News. 2014. http://www.foodsafetynews.com/2014/11/revised-fsma-provisions-need-more-tweaks/#.V47vXLiAOko. Accessed 20 July 2016.

Zuraw L. What is mechanical tenderizing, and why is it hazardous to your steak? Kaiser Health News. 2016. http://www.pbs.org/newshour/rundown/what-is-mechanical-tenderizing-and-why-is-it-hazardous-to-your-steak/. Accessed 30 Nov 2016.

Chapter 17
A Critical Appraisal of the Impact of Legal Action on the Creation of Incentives for Improvements in Food Safety in the United States

Denis Stearns

Abbreviations

ABC	American Broadcasting Company
CDC	Centers for Disease Control and Prevention
E. coli	*Escherichia coli*
FDA	Food and Drug Administration/US
FOIA	Freedom of Information Act
LTFB	Lean, finely textured beef
PBS	Public Broadcasting System
PFGE	Pulsed-field electrophoresis
US	United States
USDA	United States Department of Agriculture

17.1 Introduction

Since Congress passed the first two major acts of food-related legislation in 1906, there has been no shortage of laws in the United States that govern the safe manufacture, distribution, and sale of food. There has also been no shortage of illness, hospitalization, and death attributable to adulterated and unsafe food. As I have argued before: "Confronted with the problem of adulterated food in the marketplace, the response is always legal, premised upon a kind of regulatory imperative that assumes the effectiveness of inspection and testing as the enforcement mechanism" (Stearns 2014a, b). For that reason, major changes to food laws were not enacted except in response to public outrage and fear, and the most extensive and

D. Stearns (✉)
Stearns Law, PLLC, Port Townsend, WA, USA
e-mail: Denis.Stearns@gmail.com

© Springer International Publishing AG, part of Springer Nature 2018
T. Roberts (ed.), *Food Safety Economics*, Food Microbiology and Food Safety,
https://doi.org/10.1007/978-3-319-92138-9_17

seemingly strict laws passed when the prospect of increased enforcement was needed to restore public confidence. To say that protecting sales has always been as important as protecting the public would be understatement.

For as long as there has been the exchange of food among people, whether by barter or sale, there have been concerns about the quality and safety of the food exchanged. Some of the earliest laws on record involve food, including laws that empowered supervisors to patrol markets to prohibit the sale of adulterated goods. The 1202 Assize of Bread required that a loaf be sold for a fair price and accurate weight, with violators subject to being "drawn upon a hurdle... through the greatest of streets, where the most people are assembled, and through streets which are most dirty, the false loaf hanging from his neck" (Hart 1952). One can easily imagine such a law being highly effective, because the market to be patrolled was small, and the likelihood of being caught was large. In addition, both the discovered violation and resulting punishment were highly visible, serving as a quick and notable lesson to others tempted by the prospects of higher profits at the expense of unsuspecting consumers. While this does not have the same degree of deterrent influence as the likelihood of being apprehended, it can contribute to such food laws being effective incentives for the honest sale of food of a higher quality than would have been available in the absence of such laws.

Despite how far back in time issues of food and law reach, the modern history can be said to have begun in the early months of 1993—when the Jack in the Box *E. coli* O157:H7 outbreak became news. Beginning in the middle of November 1992, and through to the end of February 1993, there were more than 500 lab-confirmed *E. coli* O157:H7 infections and at least 4 related deaths, making it one of the largest reported outbreaks in the history of the United States (Stearns 2005). Reports of the outbreak and its victims dominated the news for well over a year, spurring changes everywhere, from how consumers viewed the safety of ground beef and fast food to how federal government regulated the manufacture and inspection of meat products (Nestle 2003). The media's focus on issues of food and food safety has only increased over time, widely reporting every outbreak that occurs in the news.[1] However, this focus misrepresents as much as it represents, failing to tell the story of how outbreaks with identified sources are but the tip of the unsafe food iceberg.

A study that the CDC issued in 2013 estimated that over nine million persons each year suffer a foodborne illness due to a major pathogen, a category that includes *E. coli* O157:H7 (Painter et al. 2013). In addition, this study confirmed the reason that news of outbreaks misrepresent the true incidence of food-caused illness is because "linking an illness to a particular food is rarely possible except during an outbreak." For this reason, among others, the extent to which there has been progress in improving food safety remains the subject of considerable dispute.[2] What is

---

[1] During my 15 years as a partner at Marler Clark, every time there was a foodborne illness outbreak, the firm would quickly receive calls from media outlets seeking comments and interviews. Nearly from the beginning, we had a full-time person whose job was to handle media contacts and to arrange for interviews with our attorneys and our clients.

[2] See Denis Stearns, *On (Cr)edibility: Why Food in the United States May Never Be Safe*, STANFORD L. & POL. REV. 245, 249 and n. 12 (2010) (Stearns 2010) (explaining how the profitability of food depends in part on the ability to avoid investment in improved safety while causing significant amounts of foodborne illness that is never traced to its source).

not in dispute is that the increased attention, both from consumers and public health officials, has made the safety of food and increasingly high-profile subject of discussion, research, and litigation. In fact, the increased attention is both a cause *and* by-product of such litigation. Thus, just as we can consider whether increased regulation has increased food safety, so too can we consider whether an increase in food litigation—which is also a kind of legal or regulatory response—increased food safety or has the mechanisms and means to have done so.

This chapter is not intended to definitively answer the question of the law's incentivizing effect on food safety. It is safe to assume that there is *some* effect, even if we can only venture educated guesses about the extent of such effect. But regardless of the assumptions made, it is still possible to examine the many ways that litigation—and the legal system—makes more likely and unlikely, through the creation of incentives and disincentives, efforts to improve food safety. Although such improvement efforts might fail (thus the difficulty of determining the *extent* of the effect), the fact that efforts are made show, at least, that the incentives exist and are working. In short, if you are trying to avoid litigation, and best the way to do so is to improve the safety of your food products, litigation is acting as an incentive for safety. It is certainly not the only one, but it is just as certainly an added one, the subtraction of which will tend to make food less safe overall. All that said, how litigation works, and doesn't work, is key to understanding how litigation can act as an incentive.

Before looking at the details of how litigation can create incentives for food safety, or fail to do so, one must understand how litigation currently works. Therefore, the first section will set forth the basic concepts and rules of litigation in the food safety context, basically how an attorney can use the legal system to impose liability on the maker or seller of unsafe food. Once this foundation is built, the second section will discuss how the law has evolved to address the problem of food safety and how that evolution has both bettered and hampered the likelihood that litigation rewards investment in food safety while punishing the producers and retailers of food that choose not to make such investments. The final section will look at the question of whether a significant increase in the amount of litigation would translate into a significant increase in the safety of food in the United States.

17.2   Food-Related Litigation as a Form of Safety Regulation

Litigation is not commonly understood as part of the "regulatory" structure governing the quality and safety of the food supply. Although anyone planning on the manufacture of food products for public sale will understand there exists a host of legal requirements—licensure, regulations, and inspections—that must be met before food can be legally made and sold. One of the defining characteristics of much regulation is that it controls whether one can go into business producing food products.[3] In this regard, regulatory compliance is a positive incentive in that a company wanting to produce food to sell to the public must comply with regulations as a prerequisite to being in business. The presumed effectiveness of regulatory compliance as a safety measure is based on the presumption that a business that has demonstrated compliance will be safer than one that has not been required to make such a demonstration.

In contrast to this positive incentive, ongoing compliance is a prerequisite of continuing to do business and, as such, acts as a negative incentive—that is, one complies to avoid being shut down or fined. Although ongoing efforts to comply with regulations are presumed to result in the food products being safer, this presumption is based on the safety-enhancing characteristics of the regulations. In other words, even though a producer may understand that the goal of the regulations is the production of safe food, the goal of the producer is regulatory compliance, with safe food being a by-product of such compliance efforts. Of course, many producers may both want to achieve compliance and a safe product, but the efforts to achieve these two different goals are not necessarily the same.

To understand the incentive effect of litigation on producer behavior and safe food, it is key is to understand that the efforts to comply with regulations—that is, *the effort to avoid any finding of noncompliance—are not the same thing as making all reasonable efforts to make food as safe as possible.* When the goal is regulatory compliance in and of itself, the question is whether such compliance is sufficient to assure the production of safe food. The answer to this question is plainly no because nearly all foodborne illness outbreaks that have been attributed to a given food product are the result of producer activity that was not, *at the time of production*, deemed to have been in noncompliance with regulations. Certainly, noncompliance is often found after the fact of the outbreak; however, the incentive for compliance was demonstrably insufficient, both for the producer and regulators. Indeed, a foodborne outbreak is, by itself, evidence that the fact of regulation consistently falls short in assuring that the food produced and sold in the United States is, in fact, safe to eat.

17.3   Understanding the Imposition of Liability and Recovery of Damages

Just as one serious consequence of selling unsafe food is a person falling ill, another consequence is the victim filing a lawsuit against the producer. The reason for filing a lawsuit is to seek compensation for damages, which is why such damages are

---

[3] The production or sale of food in a regulated capacity is one way of distinguishing commercial food activity from noncommercial food activity, such as making food for a bake sale or a neighborhood picnic. Being "in the business" of making or selling food is legally significant to litigation, because the rules (about to be discussed) that govern the imposition of liability for the sale of injury-causing food are usually dependent on your ability to show that the entity being sued was a business. In other words, the liability rules applied to food (and other products) do not usually apply to noncommercial activity.

often called compensatory. Such damages are of two kinds: economic and noneconomic. Economic damages—sometimes helpfully referred to as "out-of-pocket" damages—include things like medical bills and lost wages, i.e., things that would not have needed to be paid except for the fact of the foodborne illness. Noneconomic damages include things like pain and suffering, emotional distress, and loss of enjoyment of life. Perhaps not surprisingly, economic damages are easier to calculate and prove than noneconomic.

Although a person filing a lawsuit is seeking to recover damages, another way of putting it is that the person is seeking to hold whomever caused the injury *liable* for damages. Viewed from the perspective of the one being sued (the defendant), the filed lawsuit is a *liability risk*; it is putting the defendant at risk of being held liable for the alleged damages. The lawsuit does not itself create the liability risk (or, ultimately, the liability). Decision-making and conduct that gives rise to production and sale of unsafe food that causes injury creates liability. As will be discussed in greater detail in a future section, the failure to invest sufficient money and insufficient care in the production of safe food are cost-saving methods for the producer, but they impose a cost on those injured, a cost described in economic terms as an externality (Jensen 2003). The term "externality" refers to the fact that consumers did not buy the food with intention of also buying an illness. Seeking the recovery of damages by a lawsuit is simply the mechanism used to obtain a form of reimbursement, transferring the cost back onto those responsible for causing the injuries, who created liability by injuring consumers.

To understand liability risks, one must understand the various theories of liability available to hold a company liable for unsafe food, including how each theory is alone sufficient to impose liability for the sale of unsafe food. It is common for a lawyer to include all available theories of liability[4] when filing a lawsuit. The measure of damages does not vary depending on the theory upon which the plaintiff succeeds, and just because a company can be held liable for damages on the basis of multiple theories, that does not mean that the plaintiff can recover greater damages for proving more than one theory. What must be proven in order to succeed does vary depending on which theory is used to impose liability.

17.3.1   Negligence and the Requirement of Reasonable Care

In my experience, most who consider the prospect of product-related litigation as a risk tend to think of liability based on proof of *negligence*. The commonplace concept of negligence is to act carelessly—that is, without using sufficient care. The legal concept of negligence is not so different at its core; however, turning the everyday concept of negligence into a legal rule that can be applied in a nonarbitrary way

---

[4]A "theory of liability" is also commonly referred to as a "cause of action" or "legal claim." The document that commences the litigation, and that sets for the legal claims, is called the complaint.

to a diverse set of facts has historically been a complicated task. Indeed, US courts have struggled with devising the rules of negligence for well over a century. At the heart of this struggle is the question of whether being able to show that we "did our best" acts as a sufficient defense to a claim of negligence. To the surprise of many, it does not.

Accepting that there are a great many variations depending on the specific facts involved, the rules of negligence are generally as follows:

1. Negligence involves a willful, but not intentional, act and, thus, differs from an allegation that one intended to cause harm (a slap to the face) or caused harm by a non-willful act (a spasmodic leg kick).
2. One alleging negligence (plaintiff) has the burden of proving that the person sued (defendant) was "at fault" for the act or omission that caused injury.
3. The requisite fault that must be proven is determined by asking if the defendant acted with "ordinary care and prudence" given the existing circumstances, and it is this idea of fault that becomes the "reasonable person standard."

The reasonable person standard is not without controversy, even though, as a rule of law, it is well-established. This standard is also where the legal concept of negligence differs from the commonplace one. The legal concept of negligence is based on an objective standard, a standard not based on what a particular person was *capable* of doing under the circumstances, or what a person in fact knew at the time. Instead, the defendant's conduct is judged against an external standard, based on what a reasonable person, under the same circumstances, *should have* known and *should have* done. *Thus, it is no defense to a claim of negligence to answer that one was unaware of a risk if the plaintiff can show that a reasonable person would have been aware of the risk and acted to prevent the injury.* This is the question in food cases: did the producer exercise *reasonable care*—that is, act as a reasonable producer would have acted under the same circumstances had the producer known all that was capable of being known? In other words, the conduct at issue is judged retrospectively, and some criticize this as being unfairly strict and unfair in punishing conduct that seemed reasonable at the time.

However, this criticism is largely without merit when viewing the question of negligence from the perspective of incentive creation. For example, in the Jack in the Box *E. coli* outbreak cases, one of the issues was whether there was negligence in cooking the hamburger patties. Unlike the Washington state regulation that had recently been changed to 155 degrees, the FDA Model Food Code at the time recommended that ground beef be cooked to a minimum internal temperature of 140 °F, a recommendation that restaurants followed by using electronic timers that had been programmed based on extensive cooking tests done at the corporate office. Moreover, at the time, Jack in the Box was operating no differently than other restaurants. No one, except restaurant inspectors, used thermometers to check internal temperatures. Therefore, a defense attorney could credibly argue that Jack in the Box had acted consistent with prevailing *industry standards*, also sometimes called "custom."[5] Just as a plaintiff can point to a failure to adhere to industry standards to prove a lack of reasonable care, so too can a defendant point to compliance to argue that it acted with reasonable care. However, an industry standard, even if uniformly followed, can still be deemed to be unreasonable and so not be a defense. As the famous Judge Learned Hand wrote, "Courts must in the end say what is required; there are precautions so imperative that even their universal disregard will not excuse their omission."[6] The fact of a custom's existence, and compliance or deviation from it, is but one of many things for the jury to consider in determining if there was negligence.

Jack in the Box case provides an excellent real-life example of proof of a negligence claim because it shows a restaurant company making decisions that were all arguably reasonable at the time made. Jack in the Box operated its restaurants in a way that was consistent with industry standards, and it followed the FDA Model Food Code in how it cooked its ground beef patties. Jack in the Box also bought ground beef patties from a USDA-inspected supplier that was subject to the same safety laws as every other meat supplier in the United States. Nonetheless, there is no question that nearly all (if not all) juries, if asked to apply the reasonable person standard, would reach a verdict in favor of the plaintiff. Both industry standards and prevailing regulations provided ample justification for the reasonableness of the company's conduct, but in retrospect that conduct was clearly insufficient, as proven by the sheer size of the resulting outbreak and the deaths and injuries caused.

Ultimately, in a lawsuit the proof of negligence provides an example for how members of a given industry should act in the future. In that regard, the lawsuit—in shaping the narrative of how a company had failed—points the way to avoiding such failure in the future. It is also a cautionary tale for a company who might risk making the same mistakes again, giving an attorney an opportunity to not only allege negligence but gross negligence that might justify the award of punitive damages on top of compensatory damages. The award of punitive damages is quite rare, though, and thus unlikely to provide much additional incentive beyond the threat of facing allegations of negligence. Moreover, the kind of conduct that would justify an award of punitive damages is usually sufficiently egregious that the persons involved would be unlikely, at least as judged on behavior, to have been susceptible to more typical incentives. For example, it is not clear what kind of incentive would have prompted the managers at Peanut Corporation of America to decide against knowingly shipping *Salmonella*-contaminated peanuts, especially with the prospect of criminal punishment as an additional risk.

---

[5] Generally, to be admissible, evidence of "custom" must show that a practice is well-established and broadly followed, making it likely that member of the industry should know of the custom.

[6] *T.J. Hooper*, 60 F.2d 737, 740 (2d Cir. 1932) (Hooper 1932). In the *T.J. Hooper* case, a tugboat operator was sued for negligence as a result of it having lost two barges of coal in a storm. The boats had not been outfitted with radios and, thus, could not receive storm warnings. The defendant argued that it had fully complied with existing custom because most tugboat operators at the time did not outfit boats with radios. Rejecting this argument, the court held that not adopting available new technology could be proof of negligence.

The demonstrable power of negligence claims to create more concrete and compelling stories about what *not* to do is one reason that negligence as a theory of liability likely creates stronger incentives (at least prospectively) than lawsuits that do not attempt to establish liability by proof of negligence. As will be shown, the proof of negligence is often difficult to impossible; thus, the availability of a liability theory that does not require such proof is a boon to consumers and also a reason that lawsuits are, in general, a more significant creator of incentives than would otherwise exist in the absence of an easier-to-prove legal claim like strict product liability.

17.3.2   Strict Liability Claims: Liability Without Proof of Negligence

Even though, in a select few cases (notably those involving restaurants) proving negligence is not too difficult with the benefit of hindsight and the help of a thorough outbreak investigation and report, negligent acts are more often subtle or unseen with mass-manufactured foods like ground beef, peanut butter, or cookie dough. Sometimes the use of a contaminated ingredient will be identified, or specific insanitary conditions in a processing plant will be found. More often though, the fact of the food's contamination is by itself the best evidence that "something" went wrong in the manufacture of the food product, even though that something remains unknown or not subject to more than an educated guess or speculation (neither of which is allowed in a trial). If the cause of the contamination cannot be shown on a more-probable-than-not basis, by some form of evidence or expert testimony, then a jury will be left without an explanation regarding how the contamination occurred. Fortunately, such an explanation is not required to establish liability for a product-related injury when the legal claim is premised on strict liability.

In general terms, strict liability is not complicated. To state a legally viable claim of strict liability, three things (or elements) must be alleged: (1) the company manufactured and sold a product, (2) the product was defective when sold, and (3) the defect caused a plaintiff to be injured. Proving all three of these things is not always easy, in the case of an outbreak linked to an identified food product, such proof is rarely available.

Different Types of Defects and Why They Matter for Incentive Creation. At the heart of a strict liability claim is the concept of "defect," a concept that encompasses the fact of a product being unsafe to an unreasonable degree but also that the product is unsafe beyond that which would be expected by a reasonable consumer. Broadly speaking, there are three kinds of defect: manufacturing defects, design defects, and defects of warning, instruction, or marketing. With food cases, the defect is nearly always a manufacturing one, which is sometimes also referred to in the context of a product not being reasonably safe as constructed. Although the rules are stated variously, the crux remains the same—because of some act, omission, or a failure of some kind, the finished product failed to comply with its producer's specifications. For example, a producer does not intend that its ice cream product contain *Listeria*, and if one was to look at the specification for the product, you would not see *Listeria* as a listed ingredient. For that reason, the product is considered a deviation from what is intended—a mistake. Another important difference between manufacturing and design defects is that the legal claim for a manufacturing defect requires that a personal injury of some kind occurs. Thus, even if a given batch or lot of a food product is found to have been contaminated with a pathogen, and a product recall occurs, only those who were infected and thus injured will have a product liability claim. Of course, purchasers would be entitled to a refund of the purchase price, if they still had a receipt of leftover product, but such a remedy is not really beyond what any consumer is likely to obtain for any returned product.

For every other kind of product except food, the law of product liability imposes a risk of liability that is significantly larger than that faced by the producers of food. Not only do defects in nonfood product affect an entire product line, but there is potential liability to all purchasers. *With food, the defects might affect an identifiable batch or lot but never an entire product line.* And even though an entire batch or lot might be subject to recall (a significant cost), only a portion of the purchasers will be injured such that a legal claim will arise. If you add in that most injury-causing defects in food are never attributed to a specific food as the cause of the injury, then it is not surprising that the economic incentives created by food-related litigation are relatively weak as well as difficult to predict and quantify.

Manufacturers Versus Retailers in Strict Liability Cases. As originally adopted by the American Law Institute in 1965, the Restatement of Torts (Second) §402A applied strict liability to everyone in the chain of distribution. Consequently, it did not matter if you were nothing more than the shipping company transporting boxes of canned peas or the grocer who put the peas on the shelf for sale; if a can of peas ended up making someone sick, the plaintiff was free to sue the grocer, who then would need to file its own lawsuit—what is called an "indemnity action"—against the other companies allegedly at fault, seeking reimbursement for what had been paid in damages to the plaintiffs. Although the emphasis was on expediting the recovery of damages for a plaintiff, an additional and significant rationale for imposition of liability against all product sellers was that retailers were in a position to exert economic influence against suppliers to improve safety. While an individual consumer would have no such influence, the retailer could, in a manner of speaking, represent the injured consumer in demanding safer products.[7]

Imposing liability to all those involved in the sale of the product is referred to as chain-of-distribution liability. It was assumed that, because the retailer could be

---

[7]These rationales for applying strict liability to retailers were first announced in the much-studied, much-debated case of *Vandermark v. Ford Motor Co.*, 391 P.2d 168 (Cal.1964). In an oft-quoted passage from the case, the court explained that, "Retailers like manufacturers are engaged in the business of distributing goods to the public. They are an integral part of the overall producing and marketing enterprise that should bear the cost of injuries resulting from defective products."

sued and held strictly liable, retailers would be more vigilant in only buying from manufacturers who used the greatest care and invested sufficiently in safety. And because retailers were presumed to have significant economic clout in terms of choosing from whom to purchase goods, the economic clout was similarly presumed to create significant economic incentives for safety.

The protection of retailers from strict liability was not uniformly adopted, however. And even in those states where this protection existed, there were usually exceptions. For example, in Washington state, a retailer can still be held strictly liable, as if it was a manufacturer, if the brand name of the retailer is placed on the product, or if the manufacturer is insolvent, or beyond the jurisdiction of the court. Moreover, there remained numerous ways to argue that a retailer was, in fact, a manufacturer. For example, the meat departments in most grocery stores do a final grind on ground beef purchased from a manufacturer—an activity that many courts have found to be sufficient to make the retailer a manufacturer for purposes of strict liability. Finally, there are a number of situations that can give rise to allegations of retailer negligence too, such as failure to use reasonable care in the selection of a supplier by, for example, being unaware of numerous failed inspections or problems at a manufacturing facility that would have been discovered by having the facility inspected. Also, if customers had been injured by a product previously, the prior injury can give rise to an allegation that a retailer was negligent for continuing to sell the product, despite its history of causing injuries.

In addition to creating a presumed economic incentive for the retailers to influence manufacturers, this approach to liability also was more likely to hold a distant manufacturer accountable, once a plaintiff might have difficulty locating or suing in some far-flung locale. Unlike the consumers, those in the chain of distribution are able to identify each other, transferring liability up the chain, ultimately to a manufacturer that might otherwise have avoided being held accountable. *But as retailers more and more were subject to product liability lawsuits, pressure built to offer some protection against liability, which then prompted changes to the law.* In many states, it matters whether a company is a manufacturer versus a retailer because strict liability sometimes does not apply to nonmanufacturing retailers, those who are said to "act as mere conduits in the chain of distribution."[8]

Today, foodservice operations are manufacturers with regard to most products sold. Where the issue can become contentious, however, is when there is more than one potential manufacturer involved. For example, in a case involving an *E. coli* O157:H7 outbreak linked to tacos served to elementary school students, the district argued that it was not a manufacturer because another company was indisputably the manufacturer of the contaminated ground beef used to make the tacos. Both the trial court and the appellate court rejected this argument, deciding that that the school district's actions fit squarely within the definition of manufacturer because it had a design for cooking the meat, a recipe, and the district's cooking process fell

[8] *Almquist v. Finley School District*, 57 P.3d 1191, 1197 (Wash. Ct. App. 2002) (explaining that strict liability should apply to only those who exercise "actual control" over the product versus those who pass the product along unchanged like a distributor or grocer) (Almquist v. Finley School District 2002)

neatly within the dictionary definitions for produce, make, fabricate, and construct.[9]

In other cases, the question of whether a restaurant is a manufacturer for a given product may not be so clear as the cooking of taco meat and its assembly into tacos. For example, a restaurant could have a self-serve salad bar where all of the items on it are purchased prewashed and pre-cut and, thus, ready to use, with the only step taken being to open the packages and pour the contents into containers on the salad bar. In a *Salmonella* outbreak case linked to a buffet restaurant in Georgia, a trial court decided that strict liability did not apply because the restaurant was not a manufacturer, only a seller. Such a decision is likely an anomaly, however. Still, the case shows that the issue is not always clear-cut, and some plaintiffs can be forced to prove negligence to recover damages. It is for that reason, among others, that complaints always include claims for strict liability and negligence, and sometimes breach of warranty too.[10]

Proving the Product Was Defective. One peculiarity of unsafe food cases is that there is rarely direct evidence of the product and its contamination. There is a simple reason for that: food must be eaten if it is going to make someone sick. As a result, persons who claim to be injured by unsafe food must prove the fact of contamination by way of circumstantial evidence.

What must be proved is that the food was in fact unsafe. This is where the fact of the injury comes in to be used as proof of the unsafe (or defective) condition of the product. This manner of proof is often referred to as the "malfunction doctrine," and it usually applies in food cases.[11] Only the likelihood of other possible causes is enough for a dispute when a plaintiff has a confirmed case of foodborne illness that investigators have linked to an outbreak. On the other hand, non-outbreak cases are notoriously difficult to prove based on the use of circumstantial evidence alone. Therefore, even with the malfunction doctrine as a tool, its successful use is primarily restricted to persons injured as part of an outbreak. Consequently, to the extent that the legal viability of a food-related claim depends on it being part of an outbreak that has been investigated and a source identified, food-related litigation will always depend to a great degree on the effectiveness of surveillance and outbreak investigation.

---

[9] *See Almquist*, 114 Wn. App. at 405 ("to make means to bring a material thing into being by forming, shaping, or altering material; to fabricate means to form into a whole by uniting parts; and to construct means to form, make, or create by combining parts or elements").

[10] This chapter has not delved deeply into breach of warranty as a claim much used for food cases; mostly it is a claim that is rarely needed, except in states, like Michigan, that rely on a form of implied warranty claim instead of strict liability. Historically, the doctrines and rules that would evolve into strict product liability were developed in warranty cases, particularly those involving unsafe food products (Stearns 2015).

[11] Although direct evidence of food contamination is not common, it is not unheard of either. For example, in cases involving a sushi restaurant in Arkansas, more than one Mexican food restaurant, and a national fast-casual restaurant (like an Applebee's), I have had clients who have taken home leftovers that ended up testing positive for a pathogen. Similarly, outbreak investigations sometimes find ingredients in restaurants that test positive, allowing a link to the persons who got sick eating at the restaurant.

Proving Causation-in-Fact. Whether asserting strict liability or negligence, a plaintiff must prove that the injury complained of would not have occurred "but for" eating the unsafe food. The "but for" causation analysis is often more confusing in theory than in practice, making an example from a challenging real-life case helpful here. An 8-year-old girl and her cousin of the same age share a hamburger, each eating half. Both fell ill within 2–3 days. The cousin has a confirmed *E. coli* O157:H7 infection that, fortunately, is not severe and recovers quickly. The 8-year-old girl suffers symptoms like her cousin, going on to develop hemolytic-uremic syndrome (HUS), a deadly complication almost always associated with *E. coli* O157:H7 infections in children. Although both of the children consumed other food items known to be at risk for contamination with *E. coli* O157:H7, the only food that the two children had in common was the hamburger. Therefore, based on these facts, a convincing argument could be (and was) made that, *but for* consuming the hamburger and *but for* its consumption, neither of the children would have been infected with *E. coli* O157:H7 and injured.

A plaintiff need not prove causation beyond a reasonable doubt—the standard applicable to criminal trials. Neither must a plaintiff prove causation on a statistically significant basis—that is, to a 95% confidence level. Instead, causation only needs to be proven on a "more probable than not basis," which is sometimes also referred to as the "preponderance of the evidence" or "balance of the evidence" standard. Put simply, this standard requires any amount more likely than not. Thought of as a scale, even the slightest of tips in favor of something being the cause is enough to meet the burden of proof under this standard. It is only when some other causes are more probable, or if one or more causes are equally probable, that the standard is not met.[12]

Defenses to Liability. When it comes to food cases, the best defense is to sell only safe food. For the first 15 years that the Marler Clark law firm operated, it represented thousands of victims of foodborne illness and filed hundreds of lawsuits. During that time, only three cases went to trial, and only one all the way to a jury verdict.[13] That means nearly all of the other cases settled. That is not to say that everyone who contacted the firm became a client—most did not. The main reason that someone did not become a client was because there was a lack of evidence of the cause of the illness. However, assuming that a person was ultimately successful in finding an attorney to file a lawsuit, despite having been turned down by a law firm like Marler Clark, the primary (and likely successful) defense would be the

---

[12] When there are multiple causes that have contributed to causing the same injury, some courts use what is called the "substantial factor" test, which asks whether one or more defendants "contributed in a material way" to causing the injury. The jury is then asked to determine the extent of each defendant's contribution, and the verdict is allocated according to percentages. In practice, the "but for" causation test is not really any different, because it has never been the rule that there can only be one "but for" cause.

[13] The case that went all the way to verdict, and then on appeal, involved the lawsuits arising from the *E. coli* O157:H7 outbreak at the elementary school in eastern Washington. After a nearly month-long trial on the issue of liability (the trial on damages was to follow), the jury was asked to allocate fault as between the school district and the supplier of the contaminated ground beef. The jury allocated 100% of the fault to the school district for its improper cooking.

defending attorney's ability to convince the jury that the "more probable than not" standard had not been met or that a cause other than the alleged food was a more probable cause of illness.

For those cases that are unlikely to be defeated on causation grounds, thus leading to no recovery for the plaintiff, a defendant may nonetheless attempt to reduce the amount of recovery by showing that there are other entities at fault for causing the damages. The rule of *joint and several liability* once allowed a plaintiff to obtain a full recovery from any entities regardless of the relative amount of fault that might be attributable to the entity. Today several (but not joint) liabilities are the more common rule, allowing defendants to argue that a greater percentage of fault is attributable to another entity, including the plaintiff. Thus, for example, if it can be shown that plaintiff failed to cook ground beef to a recommended temperature, a defendant can seek to have the recovery reduced by the percentage of fault that the jury attributes to the plaintiff. Even though such "blame the plaintiff" defenses can sometimes seem like a good idea to a zealous defense attorney seeking something to argue on behalf of a client, the potential for the defense to anger a jury is a real risk, and the likely reason that few defendants follow through beyond using the defense as a negotiating tactic.

17.3.3 Negligence: The Case of the Contaminated Ingredients

One of the more common liability scenarios is where a restaurant receives an ingredient that is contaminated or otherwise unsafe. When such an ingredient is added unchanged, with no cooking or other intervention, there are few grounds for alleging negligence. For example, imagine a restaurant like Jimmy Johns that uses alfalfa sprouts to make subway sandwiches. If those sprouts are contaminated with *Salmonella*, there is little chance for the restaurant to prevent susceptible customers from falling ill. Would it therefore be a complete defense to liability to point out that (1) the restaurant did not do anything to the sprouts except put them on a sandwich, and (2) there was no way to know that the sprouts were contaminated, given that *Salmonella* cannot be detected by sight or smell. The short answer is as follows: No, the restaurant could still be liable.

First, the question of negligence is not so cut-and-dried to allow a restaurant to escape liability because it neither caused, nor actually knew of, the contamination. With something like sprouts, the risk of contamination has been known for some time. For example, a study published in 2001 noted, "at least 12 reported sprout-related disease outbreaks involving a total of more than 1500 cases have been reported since 1995"[14] (Proctor et al. 2001). Outbreaks linked to contaminated sprouts happened with such regularity that it no longer took a creative attorney to

---

[14] Mary E. Proctor, et al., *Multistate Outbreak of Salmonella Serovar Muenchen Infections Associated with Alfalfa Sprouts Grown from Seeds Pretreated with Calcium Hypochlorite,* 39 Journal of Clinical Microbiology, Vol. 10, 3461, 3461 (Oct. 2001) (Proctor et al. 2001). These outbreak figures are from the United States only and, thus, omit one of the largest outbreaks of all time—also linked to sprouts. This historic outbreak involved *E. coli* O157:H7 and occurred in Japan in 1996, sickening 9441 people, mostly school children.

argue that a restaurant using sprouts was negligent just in choosing to do so. In other words, certain foods can acquire enough of an outbreak record to be deemed inherently risky.[15] For example, in the lawsuits arising from a 1996 unpasteurized apple juice outbreak, a coffee shop retailer was added as a defendant on the grounds that it had sold the juice, pouring it into a cup before serving it. In arguing that it should be dismissed from the lawsuit, the retailer argued that it had not made the juice, did not know that it was contaminated, and could not as a matter of law be held negligent. Opposing the dismissal, the plaintiffs argued (in a brief that I wrote) that there had been numerous outbreaks linked to unpasteurized apple juice, making its sale potentially negligent. The court agreed.[16]

In addition to using an ingredient that is or should be known to be risky, the second way negligence can arise from the unknowing use of contaminated ingredients is through the careless selection of a supplier. Just like certain ingredients can have track records that should put a user on notice of the risk being taken by the ingredient's use, suppliers have track records too—especially if the industry involved is subject to inspection. For example, in an outbreak linked to a supplier of peanuts and peanut products, one food company had inspected the supplier's plant and its testing records and then decided not to purchase from the supplier. Another equally large food company relied upon a private third party inspector that the supplier hired itself to prepare a report. After passing this inspection, the other food company began buying peanuts to use in its own products. A multistate *Salmonella* outbreak linked to peanuts was announced by the CDC, leading to the supplier's bankruptcy filing and, ultimately, criminal charges. Yet, despite the clear culpability of the peanut supplier, the company that had decided to buy peanuts on the basis of a third party inspection was still subject to lawsuits and significant defense costs and liability risks, based on both negligence and strict liability theories.

17.4 The Mixed Message of Strict Liability: An Easier Standard But a Blunted Impact

The advent of a food-focused litigation practice like that of the pioneering law firm, Marler Clark, created incentives for food safety in only the most indirect of ways. For example, even if one assumes that the risk of being sued promotes increased concerns by food producers that translate into improved safety efforts, it is more

---

[15] In February 2012, Jimmy John's permanently stopped selling sprouts on its sandwiches, this after the fifth sprout-related outbreak of illnesses linked to sandwiches sold at the restaurant chain. Eight months later, Kroger, the nation's largest grocery store chain, also announced that it would stop selling sprouts. *See* Elizabeth Weise, *Kroger stores stop selling sprouts as too dangerous*, USA Today, Oct. 20, 2012.

[16] It should be emphasized that such a ruling is not a finding that the retailer was negligent; it was a legal conclusion that the plaintiffs had alleged sufficient facts to have earned the *opportunity* to prove negligence at trial. In this particular case, a settlement was eventually reached with the maker of the juice.

reasonable to assume that the risk of negative media coverage and heightened regulatory attention that inevitably attends a lawsuit is a bigger cause of concern. Marler Clark has honed a number of legal strategies in over two decades of practice, but it has also honed just as many media strategies as well. Drafting a press release thus becomes as important as drafting a complaint in commencing a lawsuit. Because if there is one thing that has changed as much as the litigation landscape since the 1993 Jack in the Box *E. coli* outbreak, it is the media landscape. In fact, the hunger of the media for food-related stories has increased over the years, with overarching narratives that a succession of lawsuits and attorney interviews has done much to shape. Bill Marler, in particular, has managed to make himself a kind spokesperson for the public health, advocating for the importance of food safety without being a nonstop critic of the food industry and its failings. By praising, in general terms, industry efforts, Marler carves out a media space that allows him to hold up certain food companies as the exception not the rule—notwithstanding the fact that the exceptions sometime seem to exceed the rule. In this way, the companies that cause outbreaks are depicted as outliers, allowing Marler Clark to come across as both a public advocate and a friend of industry.

Ultimately, there is no question that litigation sends a message to the food industry that has had an impact, albeit immeasurable, on food safety. But if litigation sends the message, it is the media that amplifies it. It is that amplification that also puts pressure on the agencies to take stronger enforcement actions, both with regard to companies linked to a given outbreak but also as to companies in the industry segment linked to a given outbreak. When the incidence of *E. coli* O157:H7 began to again increase in the 2000s, the enforcement efforts that followed had a notable effect. Although outbreaks of *E. coli* O157:H7 infections linked to ground beef are not unheard of, they are relatively uncommon now. The same cannot be said about this pathogen and fresh produce, though. Neither can the same be said about *Listeria*, which in the last several years has remained stubbornly on the rise.

The bottom line is this: litigation drives media attention, which in turn drives heightened regulatory enforcement and sometimes also new rulemaking. This latter point is important given that the "rules of the game"—which is to say regulations that might affect the ability of a lawyer to argue that a given food product is defective—are what can tilt the playing field in favor of one side in the litigation over the other. For example, USDA for some time appeared to take a policy position that treated *E. coli* O157:H7 as an adulterant only when found in ground beef, and not any other kind of meat products, especially so-called "intact" cuts of meat. In the wake of a 2000 *E. coli* O157:H7 outbreak linked to cross-contamination between intact cuts of meat and fruit served on a buffet, the meat supplier defendant sought and obtained dismissal of all claims against it on the ground that the intact cuts of meat that it had sold "legally" were not adulterated under federal law, even if in fact contaminated with *E. coli* O157:H7. Thus, according to the argument, because only federal law had the power to define what constitutes adulteration, not state law, the meat in question could not give rise to liability under state product liability law (Stearns 2005). Backed by industry, the issue made its way all the way to the US Supreme Court, which declined to review the Wisconsin Court of Appeals decision that reversed the state court decision dismissing all claims. But the legal battle took years, thus providing an example how the relative success of legal claims in creating incentives is always in a state of flux, affected by multiple factors. It is for that reason, among others, that Marler Clark has remained unique in terms of investing the resources into all aspects of the legal fight, not just focusing on individual cases on "one-shot" affairs, like most personal injury practices do with injury claims (Galanter 1974).

17.5 Reversing Externalities: Litigation of Foodborne Illness Cases as the "Cleanup Crew"

There have been a few papers that purport to address the issue of foodborne illness as an incentive for increased food safety, but most are either impaired by a lack of data or otherwise rest on premises or offers as conclusions, propositions that are more axiomatic than insightful. Two examples make these points well.

In the first report of its kind, the USDA's Economic Research Service published a report in 2001 that examined—and attempted to measure—the impact of food-related litigation in creating economic incentives for firms to produce safer food (Buzby et al. 2001). Although the report effectively identified a number of important issues, including the dampening effect on incentive creation of confidential settlements and liability insurance, the conclusions reached were largely equivocal and justifiably so given the report's concession that "[r]eliable estimates of the annual number of foodborne illness claims and lawsuits are unavailable" (Buzby et al. 2001). To compensate for the unavailability of key date, the report focused on what was available and seemingly reliable: "jury verdict data on foodborne illness lawsuits [in the United States] for 1988–1997" (Buzby et al. 2001).[17] However, when it comes to lawsuits involving foodborne illness, a case would not proceed to jury trial unless there was a serious question whether the plaintiff was going to be able to prove the facts necessary to establish liability. It is no accident that the report concludes: "Legal incentives probably work better in outbreaks and less well for sporadic cases. Mass outbreaks have greater potential to damage firms, both in terms of financial damages and of damaging a firm's reputation" (Buzby et al. 2001). Here in this key finding is a truth that can be stated in unequivocal terms today. Legal incentives do not *probably* work better in outbreaks; they *only* work in outbreaks. And legal incentives do not work *less well* in sporadic cases; they do not work *at all* in sporadic cases.

The timeframe for the data relied upon in this report was from 1988 to 1997. The Marler Clark law firm was created in 1998, a creation that was founded in large part on the experience and expertise that I and William Marler had gained beginning in 1993 litigating the Jack in the Box outbreak cases and then the 1996 Odwalla outbreak cases. The creation was also founded on the recognition that the investigative techniques that had made it possible to trace the source of these two outbreaks

---

[17] *Id.* at 13.

would inevitably improve, leading to the identification of sources of future outbreaks. It also seemed increasingly clear that foodborne illness would continue to be a subject of great public concern, and that concern would drive increased regulatory enforcement of the food industry and with that increased microbiological testing. All of these factors made it more likely that there would be a need for more foodborne litigation in the future, and therefore it made sense to create a law firm that would develop that practice area as a new specialty.

The subsequent (and correctly predicted) rise in foodborne illness litigation paralleled the rise in prominence of Marler Clark and its specialized practice. In addition to becoming the dominant firm in practice area of its own creation, from the start the firm positioned itself as a kind of public interest law firm, litigating foodborne illness cases but also advocating for a host of food safety issues. For example, shortly after a 1991 outbreak of hepatitis A infections linked to an infected restaurant employee, Marler Clark began to advocate for mandatory vaccinations of all foodservice workers. In August 2002, William Marler wrote an op-ed for the Denver Post, responding to an outbreak of *E. coli* O157:H7 infections linked to contaminated ground beef, and its title was "Put me out of business – please." This phrase would become a trademark, of sorts, indicating the firm's admixture of litigation directed at companies that had caused foodborne illness outbreaks, and straight-spoken advocacy intended to depict foodborne illness as the actual enemy, with the government's regulatory shortcomings just as responsible as food companies who cut corners or otherwise failed to live up to the public's trust in a safe food supply.

Ultimately, this two-prong approach never accepted that litigation alone was much of an effective incentive. Instead, litigation's true function was as a kind of "cleanup crew," dealing with the aftermath of a foodborne illness outbreak much like first responders called to a house fire. Just as building codes are intended to make house fires less likely, regulations governing food product are intended to make foodborne illness less likely. It follows then just as a housebuilder may decide to use cheaper materials, or employ unqualified workers, as a means of increasing profits, so too might a food producer put off needed plant maintenance or purchase ingredients of lower quality. In both instances, companies are externalizing the risk while internalizing the profits. The utility of a lawsuit, then, is really about reversing that externality, and reimbursing losses sustained, while imposing the costs of corner-cutting back on those who cut them. If the fact of a lawsuit being filed creates an incentive for other companies to cut fewer corners, that is an additional benefit of litigation, but certainly not the primary one.

17.6   The Correct Measure of Damages: What Does It Cost to Settle a Claim?

The relationship between the decision of a lawyer and a client to settle a claim versus going to trial is relatively simple. Since September 1998, when Marler Clark was established, of the over 100,000 claims that have achieved compensation for the client, a significant majority were filed as lawsuits. But of those lawsuits filed, only *one* set of claims, from a single outbreak, went to trial and resulted in a verdict. That means that of the "over $600 million" (Marler Clark 2016) recovered for clients in the history of the firm to date, that money solely was the by-product of settlements, not trial. It is only questionable or overvalued claims that go to trial. Claims that both sides view as having a solid evidentiary basis and a potential for a jury to award significant, albeit difficult-to-precisely estimate, damages result in settlements. As such, it is worth looking at how the settlement value of a foodborne illness claim is established.

First, because settlements are typically confidential, a law firm that settles lots of a given kind of cases has an informational advantage over a law firm that does not settle the same kind of cases or does not do so in sufficient quantity to aid in the proper evaluation of settlement value. It is this factor that has over time created a competitive advantage for a firm like Marler Clark. It not only has a relatively unrivaled level of experience derived from focusing on these cases since 1998, but the firm and its attorneys are in a better position to obtain maximum settlement value for a given claim based on knowledge of what comparable claims have settled for in the past. Indeed, a significant source of referrals over the years for the firm is the result of a law firm seeking the assistance of Marler Clark in valuing a given claim for purposes of settlement. An outside firm is much more inclined to refer a case for handling by Marler Clark, sharing the fee, when a referral is likely to result in a higher overall settlement.

Second, the model created and followed by Marler Clark is consistent with what Galanter identified as a "repeat-player." In forming the law firm, Marler Clark intended to create a niche practice that did not previously exist, and because the intent was to dominate that niche, it acted as a repeat-player always acts, with a longer view and with no incentive to prioritize quick "one-shot" settlements. Aware that the settlement of claims was creating a value range for future settlements, the firm had strong incentives to resist valuation models commonly used for injury claims. For example, such valuation models often used the amount of economic damages as the basis for a determination of noneconomic damages—pain and suffering, with a multiplier of two or three commonly used. Thus, if I was injured in a car accident, an insurance company might offer me twice my medical bills as compensation for my pain and suffering, creating settlement value of $30,000 for a claim where there was $10,000 in medical bills. An insurer is a classic example of a repeat-player; the insurer has a long-term incentive to protect this model of valuation to prevent the inflation of damage payments.

In a typical *Salmonella* outbreak, the injuries involved can vary a lot, from death to 3 days of diarrhea and related symptoms. A non-insignificant number of persons in the outbreak might not seek medical care and, thus, might not obtain culture confirmation of an infection. Others might have a single visit to a physician, while others might end up going to an emergency room. Looked at through the perspective of symptoms alone, the severity of injury is not necessarily reflected, at least not accurately, by the amount of medical bills incurred. There might be a person who would have been admitted to the hospital, based on severity of symptoms, if the person had presented to an emergency room. Another person might have two visits to the emergency room with, arguably, symptoms that did not fully justify this course of action. As a result, you might have two claims, one with medical bills exceeding $10,000 and one with no medical bills at all. Add to this the fact that the person with no medical bills also lacks "proof" of the *Salmonella* infection. From a legal (and negotiating) perspective, the challenge becomes how to obtain a fair settlement on behalf both of these claimants. Although the strategies can vary, the important takeaway point here is that, in the absence of good lawyering, a food producer (or its insurer) might avoid paying any cost for a majority of claims arising from an outbreak by trying to take the position that it would only pay for claims of those people who had culture confirmation of infection or incurred medical bills. A firm like Marler Clark actively resists such attempts to minimize the costs of a given outbreak by asserting claims on behalf of all varieties of claimants and in presenting claims based on symptoms not medical bills.

17.7 Confidentiality Agreements and the Suppression of Information Related to Food Safety

Given the media attention focused on outbreaks and the litigation that so often follows, it is not as if the public record lacks information about the facts of an outbreak, the extent of injury or death, and the results of investigations into the cause of the outbreak. The source of this kind of information is often official, which is to say, reports that the CDC, federal agencies, or state health department issues during or after an investigation. Requests for public records made under the federal Freedom of Information Act (FOIA), or state public record laws, are another important vehicle for getting the detailed facts of an outbreak and its investigation known. News reporters also commonly quote from key pleadings in a lawsuit, such as the complaint that commenced the action and motions filed with the court on some contested factual or legal issue. Indeed, when a lawsuit is filed, and a press release issued, copies of the complaint are always made available for reporters to review and quote.

Some confidentiality arises during the course of litigation. A plaintiff is entitled to obtain a significant amount of information and internal documents about the sued company as part of what is called the "discovery process." As a condition of disclosure, a defendant demands (and usually obtains) that the court issue a protective order that strictly limits the dissemination of the information that is deemed confidential or proprietary. Attorneys participating in the litigation are duty-bound to adhere to the requirements of confidentiality, although the degree and means of adherence are case-specific and subject to interpretation. For example, a document that was marked "confidential" would never be disclosed to someone not authorized to see the document, but information contained in the document might still be set forth in a legal pleading and are quoted in open court. Although courts were once much more willing to have documents filed under seal (and thus not part of the court-file's public record), court rules do not restrict this practice. Thus, as litigation continues, the ability of the company sued to keep things secret tends to lessen, while the opportunity for embarrassing details to be disclosed tends to increase. There is, as a result, an increasing incentive to settle the lawsuit to avoid further bad publicity.

When a lawsuit is resolved, almost always by settlement, there is inevitably an attempt to limit the amount of information available about the outbreak and its causes, an attempt that is motivated by company sued and its desire to erase the record. Strict requirements of confidentiality are often part of the bargain, with a settling company willing to pay more to settle a claim to obtain the plaintiffs' agreement to the strictest of confidentiality terms. Agreeing to not disclose the settlement amount is usually a standard term. A settling company might also demand that the facts of the outbreak not be discussed publicly and that the plaintiffs will not disparage the company or its products. There can be requests that a press release posted on the law firm website be removed. Of course, all of the information already in the public record will so remain. And, in general, plaintiffs resist any effort to restrict free-speech rights beyond maintaining confidentiality about the settlement terms. *Nonetheless, the lack of available information about settlement amounts plainly undermines the incentive effect of food-related litigation.*

There is, however, little to be done about the problem, given that settlements are the by-product of an agreement between private parties. A law could be passed that prohibits confidential settlements, but the likelihood of such a law is miniscule.

17.8   Strategic Litigation Against Public Participation

Since 1990, 13 states, from Georgia to Idaho, have adopted food defamation laws to limit public discussion of food safety (PBS 1998). The most famous case was the Texas beef industry suing Oprah Winfrey over her discussion of the practice of using dead cattle in animal feed with a guest from The Humane Society who raised the possibility that this practice could spread mad cow disease, if an infected animal were processed. Lower-profile cases have also been filed, including by an Ohio company, Agri General, that was caught repacking eggs into a carton showing a later expiration data (PBS 1998). In 2012, ABC News ran an 11-segment investigation on a low-cost meat product, named "lean, finely textured beef" (LFTB) in USDA's regulations approving the product (high-fat beef trim and cartilage rendered at low temperatures to remove the fat and then treated with ammonia to kill pathogens). The meat company, Beef Products, Inc., was sued when the term "pink slime" was used on the TV show to describe the product made from beef "scraps, once used only in dog food and cooking oil" (Sullivan 2014). In each of these three cases, there was economic injury to the companies, but there were also risks to human health being discussed.[18] Where do we draw line between scientific inquiry

---

[18] Pathogens of concern: Mad cow disease in cattle, prion not destroyed by heat during rendering into feed; eggs, *Salmonella* can proliferate in the yolk over the extended time due to repackaging; LFTB, high-risk ingredients with a likely heavy load of pathogens, including *E. coli* O157:H7.

and public policy discussion vs. defamation of products and the protection of corporate profits?

In addition to this type of lawsuit, the section above discussing the *silence required* about the amount payed in settlements (the price of recovering damages for the illness endured), this appears to be a case of "hush money." The injured person suffering the foodborne illness gets paid extra to keep his/her mouth shut. Thus, in both situations, public discussion of foodborne illness is limited to the benefit of the food company. Limits on public discussion do two things, reduce class-action suits on repeat violators of food safety regulations and reduce public information on what is happening in the food industry to control or amplify pathogens.

17.9 The Rise of Indemnity Agreements as an Impediment to Improved Food Safety

A relatively recent contributor to the economic disincentives for food safety is the near ubiquity of indemnity agreements that retailers require of food suppliers. The concept of indemnification is not a new one, having been a feature of contract law for many hundreds of years. The concept is a fairly simple one, based as it is an aspect of the law called equity. Most laypeople have some innate sense of principles of equity, created as they were out of common sense notions of fairness. For example, if I went on to the Farmer Jane's land and lead off one of her steers to sell to the local butcher, who slaughters the steer and sells the meat, I have thus committed an act for which I am strictly liable, which is to say, without any defense except an attempt to prove that I did not commit the act. Farmer Jane can bring an action against me in court, and—obviously unable to return the steer now—the court will order that I pay to Farmer Jane the amount of money I received from the butcher, an amount by which I was said to be unjustly enriched. Farmer Jane is thus said to have been made whole, with the value of that which she was deprived now reimbursed to her.

The legal concept of indemnity works in much the same way. If by some wrongful act I can be said to have caused someone to have incurred a loss through no fault of their own, then I can be required to reimburse that loss. For example, imagine that this time I am a butcher, and Farmer Jane sells me a steer that I slaughter and butcher, and from which I then sell the meat. But imagine too that I am careless in how I slaughter the steer, and I fail to properly refrigerate the meat, and the contaminated and rancid meat kills someone, who then sues me and Farmer Jane for damages. Assuming that Farmer Jane can show that she is blameless in causing the person's death, she would be entitled to be indemnified by the butcher—that is, the butcher would be required to pay the damages on Farmer Jane behalf, or reimburse the portion of the damages that she ended up paying, including the costs of defending the lawsuit.

Although the concept of indemnification is relatively simple, the reality of how such indemnification can occur has often been complicated and unpredictable.

There is nothing in indemnification that prevents one from being sued in the first place, or from incurring costs and paying damages, only then to be required to file a lawsuit of one's own to be reimbursed.

Given that retailers—especially those operating nationwide, and thus subject to a great variation in legal protection from state to state—remained subject to significant product-related litigation, another more effective means of protection was sought. Once found, that means of protection turned out to be a familiar one—the law of indemnity. But instead of relying on the common law form of indemnity on a case-by-case basis, a mechanism that had always been more than a little imperfect, retailers ultimately used their economic clout in a different way. Instead of increased vigilance and oversight with regard to safety, pressuring manufacturers to invest what it took to ensure maximal product safety, *retailers used their economic clout to require suppliers to sign indemnity agreements and to purchase insurance on their behalves.* Attorneys were hired to draft detailed agreements with language that requires the supplier to "indemnify and hold harmless" the buyer from "all liabilities and any losses, including lost profits, recall costs, loss of reputation or injury to good will, attorney fees and costs, and any other economic loss or damage that is caused by, or in any way attributable to, the contamination or injury-causing quality of the product, seller's negligence, breach of warranty, or any other liability creating act or omission."

Such language had the perverse effect of giving the retailers the protection from liability they had long sought, while relieving retailers of the responsibility to be gatekeepers on behalf of consumers, ones that consumers might assume were watching out for their best interests and only buying products of the highest, near-guaranteed safety and quality. Instead, retailers had effectively looked out for their own financial interests, suddenly free to be less concerned about product safety. For example, if an outbreak of hepatitis A infections occurred because bags of frozen berries proved to be contaminated, even bags bearing the retailer's brand name, the retailer could simply tender (hand over) the defense of the lawsuits to the supplier and its insurance company, safe in the knowledge that it was protected from all economic costs related to the outbreak and resulting litigation. Of course, the retailer was not protected from the reputational costs; however, in the last decade, retailers have proved to be much less concerned about these noneconomic costs when it seems to be otherwise confident about its secure position in the marketplace and its relative protection from liability. Thus, for example, if you look at a company like Costco, which has established a solid reputation for its oversight of suppliers, one can see how it would perceive the risk of recalls and product-related litigation as more of an economic risk to be managed and much less a risk to its own reputation.

The inability of indemnity agreements to protect against reputational damages is one reason the blunting effect of such agreement is seen less in the restaurant industry and with other branded products. The recent series of outbreaks linked to Chipotle Mexican Grill restaurants is one of the examples of how outbreaks, and the amplifying effect of lawsuits, can damage the market value of a company and erode market share. After a series of 5 outbreaks linked to their restaurants in the fall of 2015, the stock price of Chipotle dropped from about $700 to $400 per share from

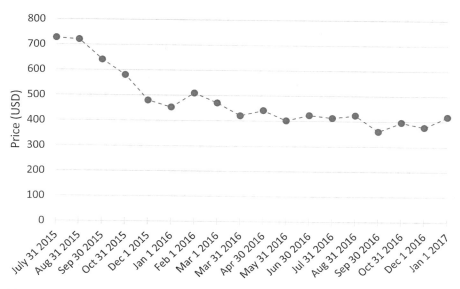

Fig. 17.1 Chipotle's stock price fell after outbreaks in 2015. Data from: Yahoo Finance, Summary for Chipotle Mexican Grill, Inc. Co, http://finance.yahoo.com/quote/CMG?p=CMG. Accessed 20 Jan 2017 (Yahoo Finance 2017)

July 2015 to July 2016 (Fig. 17.1) (Flynn 2016). On December 1, 2017, the stock price had fallen further to $309. The economic damages to Jack in the Box far exceeded the estimated $100 million paid in settlement of injury claims, an amount largely paid by insurance companies in any case. These reputational economic damages are most likely to drive investment in food safety, especially after an outbreak has occurred. And because restaurants and manufacturers of branded products are more likely to be identified as a source of an outbreak, as compared to the seller of commodity products that are incorporated into products prior to sale, the liability risk analysis for these kinds of producers is likely to be much more susceptible to influence by litigation.

17.10   The Rise of Criminal Prosecution as a Means of Focusing on Culpability

Not long after the formation of Marler Clark, I wrote an opinion piece for a publication (now defunct) called the Food Protection Report. The piece was titled, *The Courage to Criticize*, and in it that explained that civil litigation is not primarily about assigning blame in the moral sense, noting:

> People who are unaware of the doctrine of strict liability tend to view lawsuits in moralistic terms, as an exercise in finger pointing, and as a means of assigning *blame*. In fact, there is

not always a "bad guy" to blame, and in most cases it simply doesn't matter, because a lawsuits' primary purpose is compensation for the victim, not retribution. As a society, we have concluded that manufacturers should be held legally (*i.e.*, financially) responsible, regardless of blameworthiness, if and when a person is injured by a defective product.

Having made this explanatory point, I then went on to question the relative lack of criticism from industry and public health official when outbreaks occur, especially outbreaks where the injuries occurred because of notable negligence or even malfeasance. I thus argued that "it is time for health department officials, and responsible people in the food industry, to avoid the reflexive defensiveness that so often accompanies the announcement of another foodborne illness outbreak." Over 15 years later, having seen outbreak after outbreaks occur linked to everything from ground beef to cantaloupes, I modified the argument and its indictment of complicity into a kind of conspiracy of silence that is designed to maintain consumer confidence in the overall safety of the food supply, writing:

> Without sufficient consumer confidence in the safety of [food], both sales and the government's credibility suffer. As a result, the interests of industry and government align in protecting the credibility of the regulatory system as a whole, even where that alignment of interests is at the expense of public health. (Stearns 2015)

One way that this alignment has been recently buttressed is by the rise in the criminal prosecution of executives and owners of food companies found to have been at fault for the sale or manufacture of contaminated food products. Although this rise might at first glance be seen as a response to my plea for a less accommodating and more critical approach to those found to be responsible for causing foodborne illnesses, these criminal prosecutions actually support the "safest food supply in the world" narrative that is used to distract from "the ubiquity of unsafe food in the United States," undercutting the "visibility and transparency [that] are prerequisites to increased food safety" (Stearns 2015). Admittedly, a criminal prosecution can pull back the curtain and reveal a reality of blameworthy practices in putting profit above public safety; however, that revelation is intended, to depict the blameworthy individuals as criminals, i.e. outliers or exceptions, individuals who can be held up as proof that the food system is, in general, still as safe as safe can be. Just as a community that had been plagued by an arsonist can let go of fear once news of the arsonist's arrest is made public, so too can the consuming public stop fearing *Salmonella* in eggs once two egg producers are sent to jail for their crimes in selling contaminated eggs and in running a notably unsafe operation.

None of this is to suggest, however, that criminal prosecutions serve no positive purpose. Executives in the food industry may pay more attention to the details of operations if there exists a chance—albeit slight—that something otherwise overlooked could become the basis for future criminal prosecution. This increased attention is the by-product of what I have been calling the amplifier effect of litigation—civil and criminal, an effect which is an evidence that the regulation of food safety is insufficient, on its own, to create the incentives needed for the elimination of unsafe food. Of course, none of this should be surprising. Where a desired outcome is being driven by a negative incentive, a cost that is sought to be avoided, the decisional calculus will always depend on an assessment of the likelihood of paying the cost. If food safety could be achieved by a tax levied against the activity of food production, making it a cost certain to be paid, then plainly the food supply could be made certainly safe. However, the tax to be paid as a result of producing unsafe food is primarily paid by those made sick, and it is only the small minority of persons made sick as part of an investigated outbreak that identifies the source infection who are in a position to be reimbursed for the tax payments. A successful civil action accomplishes this reimbursement on behalf of the person made sick, while a criminal action accomplishes the reimbursement on behalf of the general public. Since both reimbursements add to the costs paid by the responsible food company, and the responsible company executives, there must be a net increase in the incentives created. By amplifying the consequences of failing to invest enough in food safety, or in failing to exercise sufficient care in food production, some success is achieved.

17.11 Conclusion: Civil Litigation and Food Safety—An Imperfect Deterrent

It has long been a truism in debates about criminal prosecution that its deterrent effect depends on the perceived likelihood of a criminal being apprehended, convicted, and punished. If I am planning to commit a murder, and I am relatively certain that I will never be apprehended, then the existence of life imprisonment is not likely to affect my decision-making. Similarly, if I am a food producer weighing an investment in food safety that will assure that my products do not cause death, the economics of my decision depends on the likelihood that my product will be identified as the cause of the death. Thus, just as when we talk about imposing punishment, when we talk about a civil lawsuit against a food producer who is alleged to have caused a death, what must be understood is that the target of litigation (just as the person on trial) has already been identified and accused. Consequently, the primary purpose of legal action is to hold the accused accountable in some sense, which is arguably sufficient justification on its own. The question of deterrence—whether the effort to hold one accountable causes a secondary benefit—is a question about the systemic benefits of attempting to identify and hold accountable those who cause death by criminal or negligent acts, whether by handgun or contaminated food products.

One reason that foodborne illness litigation is a relatively recent phenomenon is because effective epidemiology and active disease surveillance is a relatively recent phenomenon. When the Jack in the Box *E. coli* O157:H7 outbreak occurred in 1992–1993, the use of pulsed-field electrophoresis (PFGE) testing was just beginning to be used on a wider basis. And it was only in the wake of the Jack in the Box outbreak that PulseNET was developed, along with other tools that began to make it more likely that multistate outbreaks would be noticed and identified, even if involving only a few confirmed cases.

The true impact of civil litigation on food safety has always been about its amplifying effect, an effect that is not inherent in litigation, but that was a hallmark of the business model that Marler Clark created (largely through the vision of William Marler). Just as he had seen how media coverage in the Jack in the Box litigation could create incentives toward settlement in the food companies being sued, so too did Marler recognize the power of media to motivate possible cases to come forward, contacting health departments and attorneys. When discussing economic incentives and foodborne illness litigation, one topic that has been ignored to date has been the importance of economic incentives to those injured by foodborne illness. A significant amount of gastrointestinal illness was for very long understood to be part of the price of living. As media coverage of foodborne illness outbreak continued and evolved after Jack in the Box, one message that was being transmitted to the public was that diarrhea and vomiting was something that should be reported to the health department and was worth a trip to the doctor's office. And these same symptoms might also merit a phone call to an attorney.

Although the rise of a law firm like Marler Clark was not a sufficient cause of the new perspective on foodborne illness in the United States, and with that an increased expectation of food safety efforts on the regulatory and public health fronts, it is difficult to imagine how the narrative of food safety as it expanded and was sustained over the next 25 years would have done so without the "drumbeat" created by the repeated and successive filing of lawsuits, along with press releases and interviews, a multiplying number of websites and blogs, and insistent lobbying of politicians and regulators "on behalf of" public health. It is important to keep in mind too that the creation of a prominent public profile through the use of media and otherwise is also an important aspect of obtaining clients. One cannot file a lawsuit without an injured person for whom a legal claim can be asserted; given that there are ethical restrictions on contacting injured persons directly, potential clients must be reached through the media or online. Of course, as the law firm became more prominent and its successes in litigation more well known, the ability to attract future clients improved, thus reinforcing the firm's reputation. Without this reputational effect, the ability to create incentives would be much weaker.

Acknowledgments My sincere thanks to Robert Scharff and Abigail Kolenbrander, each of whom reviewed a draft of this chapter and provided many helpful suggestions.

References

Almquist v. Finley School District, 57 P.3d 1191, 1197 (Wash. Ct. App. 2002).

Buzby J, Frenzen P, Rasco B. Product liability and microbial foodborne illness. In: USDA: agricultural economic report no. (AER-799). Washington, DC: U.S. Dept. of Agriculture, Economic Research Service; 2001. p. 1–45.

Flynn D. Foodborne illness settlements might signal Chipotle's recovery. Food Safety News. 2016.

Galanter M. Why the "haves" come out ahead: speculations on the limits of legal change. Law Soc Rev. 1974;9(1):95–160.

Chapter 18
International Food Safety: Economic Incentives, Progress, and Future Challenges

Tanya Roberts

Abbreviations

CDC	Centers for Disease Control and Prevention, USA
FDA	Food and Drug Administration, USA
FSMA	Food Safety Modernization Act, USA
HACCP	Hazard Analysis at Critical Control Points
IFSAC	Interagency Food Safety Analytics Collaboration
MPDG	Meat and Poultry Dialogue Group
OIG	Office of Inspector General, USDA
USDA	US Department of Agriculture

18.1 Introduction

The authors have covered a variety of topics addressing food safety in the supply chain and the economic issues involved with improving it. This chapter highlights key issues which need to be addressed and the potential for economic incentives to provide safer food. It also looks toward the future, identifying some of the challenges to producing safer food worldwide. While the emphasis within many of the chapters is primarily based upon the US perspective, the issues and economic incentives are appropriate to dealing with improving food safety worldwide. In some instances, programs to improve food safety implemented in countries either broadly or focused on specific pathogens are cited as examples of viable approaches to be considered in other countries.

This book starts with an introduction to the economic frameworks and tools commonly used to examine and support public and private sector strategies to mitigate food safety problems. It also summarizes the state of the art of private and public initiatives to improve the level of food safety in food supply chains and

T. Roberts (✉)
Economic Research Service, USDA (retired), Center for Foodborne Illness
Research and Prevention, Vashon, WA, USA
e-mail: tanyaroberts@centurytel.net

© Springer International Publishing AG, part of Springer Nature 2018
T. Roberts (ed.), *Food Safety Economics*, Food Microbiology and Food Safety,
https://doi.org/10.1007/978-3-319-92138-9_18

reduce outbreaks of foodborne diseases. This final chapter highlights key insights from the book: (1) Food safety is a major public health problem worldwide with large economic costs; (2) lack of information about pathogens in the marketplace is the economic problem; (3) the private and public sectors need to collaborate and share information to identify the foods and companies causing foodborne illness; and (4) pathogen control is not prohibitively expensive, and governments have the responsibility of protecting the public and of providing economic incentives for industry to minimize the amount of unsafe food in the markets.

Looking into the future, there are challenges to maintain or improve the current levels of food safety. These challenges, discussed in the second section of this chapter, include the increasing level of antibiotic resistance in foodborne pathogens worldwide, the impact of climate change and environmental degradation on conditions for pathogen transmission, and the increasing global connectivity of food chains and urbanization and the impact of market power on food safety. While there is ample reason to be concerned, it is also clear that we are now globally more aware and capable to deal with these problems. There is already technology to significantly reduce pathogen contamination; also it is often uninformed or careless human behavior that significantly increases risks of contamination. This is again where economists can provide an important contribution to the solution of food safety conundrums. Understanding the consumer, business, and governmental agency motivations and actions will help define the set of incentives to stimulate more preventive and responsible attitudes and behaviors in food production, transport, handling, and consumption.

18.1.1 Food Safety Is a Major Public Health Problem Worldwide with Large Economic Costs

The global burden of foodborne disease is considerable and of the same order of magnitude as the major infectious diseases such as HIV/AIDS, malaria, and tuberculosis (Chap. 7). The costs of foodborne illness in the United States are estimated to be between $61 and 90 billion annually (Chap. 8). This estimate is conservative as it does not include all acute and long-term health outcomes caused by foodborne pathogens (Roberts 2013; Chap. 8). From an economic perspective, unwanted foodborne illness is an unwanted "externality" to the purchase of food in the marketplace. In the future, costs of human illness estimates will likely increase if foodborne pathogens gain more antibiotic resistance (Chap. 15).

18.1.2 Lack of Information About Pathogens Is the Economic Problem in the Marketplace

Information is the key to economic incentives for increased food safety. If there were perfect information and consumers could easily and readily distinguish between safe and unsafe, there would be no need for food safety regulations.

Companies producing unsafe food would have no demand and leave the market. However, when regulators with sophisticated pathogen detection technology and scientific expertise, such as the US Centers for Disease Control and Prevention (CDC), can identify a food, pathogen, and company name in only 1 out of 1,000 cases of foodborne illness, there are reasons to question the ability of markets to address attribution of illness to its cause (Chap. 2). Given that consumers generally lack information to make rational decisions in purchasing safe food, it is critical to increase economic incentives for companies to provide safer food.

Companies may have information on pathogen levels of contamination in their food supply chain or can develop it by implementing a Hazard Analysis and Critical Control Points (HACCP) system (Chap. 4). Companies do not need to know whether each food item they produce contains a hazard or not. Instead, they must know that they have a functioning HACCP system in place which should ensure that hazard levels do not exceed acceptable levels in their food items, on average, that are produced and sold. Note that acceptable levels of safety may differ for industry, consumers, and government. However, for government and consumers knowing that such systems are in place could provide sufficient information to maintain trust in the food supply. Despite federal/state regulations with inspections, recalls, and litigation including fees and criminal penalties, occasionally foodborne illness outbreaks occur. Infrequently, criminal charges are brought which help shape the food safety system. Nonetheless, most companies want to do the right thing for their customers and provide safe food to protect the integrity of their brand or earned reputation (Chap. 10).

Given that companies are not identified with contaminated food they sell in 999/1,000 cases of foodborne illness, this is a low level for economic incentives in the United States. Companies selling contaminated food and not caught are getting an unfair competitive advantage in the market place. If not challenged, this could lead to a race to the bottom where an entire industry becomes unsafe (Chap. 2). This highlights the need for better data to link illnesses with the companies that cases these illnesses. We must provide stronger economic incentives to produce safer food.

18.1.3   The Private and Public Sectors Need to Collaborate and Share Information to Identify the Foods and Companies Causing Foodborne Illness

The US CDC is using the latest pathogen testing technologies to increase the testing of outbreak pathogens (Chap. 2). Whole-genome sequencing (WGS) provides more information that is useful to identify strains from ill persons and link them to a common source. The trend toward sales of multiple-ingredient food items in a meal, however, can make identification more difficult to determine which ingredient contains the causative pathogen. A project at the University of California, Davis is creating a database on pathogen test results. This database links foods and pathogens from farm to fork to help quickly address any foodborne illness incident and develop improved food safety strategies (Chap. 15). The database will provide insight into molecular methods of infection and drug resistance of pathogens worldwide.

A useful addition to such a database would be routine pathogen information gathered by government in their regulatory programs and by the private sector in their HACCP plans and other monitoring efforts. The Meat and Poultry Dialogue Group (MPDG) recommended that "Voluntary data sharing between the public and private sectors should be incentivized" (Meat and Poultry Dialogue Group 2017, p. 3). The group also recommends various actions that the US Congress should take, including the development of new legislation to provide USDA with oversight authorities on the farm. The MPDG is a type of public-private partnership discussed in Chap. 14.

A recent report by the US Interagency Food Safety Analytics Collaboration (IFSAC) reveals the usefulness of pathogen database and analysis efforts. IFSAC found that outbreaks due to *Salmonella*, *Campylobacter*, *E. coli* O157, and *Listeria monocytogenes* for 2013 were not just due to animal products (IFSAC 2017). Rather the primary cause was seeded vegetables (such as tomatoes) for *Salmonella*, vegetable row crops (such as leafy greens) for *E. coli* O157:H7, and fruits for *Listeria monocytogenes* as the most common cause of outbreaks (Table 18.1). Traditionally, the first three pathogens have mainly been associated with food animals, but over the past 20 years, it has been documented that *Salmonella* outbreak sources are very diverse and frequently include many nonprotein food items like produce and grains. A study by Luber found cross-contamination to be of greater importance than

Table 18.1 US foodborne source attribution estimates for *Salmonella*, *Campylobacter*, *E. coli* O157, and *Listeria*, 2013

Pathogen/Rank	1st	2nd	3rd	4th	5th	6th	7th	8th
Salmonella	16.6% seeded veg	11.5% eggs	10.4% chicken	9.8% other produce	9.3% pork	9.1% beef	8.9% sprouts	6.9% turkey
Campylobacter	29.2% chicken	15.4% other seafood	13.8% seeded veg	12.9% veg row crops	6.5% other meat/ poultry	6.5% pork	4.9% turkey	4.3% other produce
E. coli O157:H7	42.1% veg row crops	37.9% beef	8.0% dairy	5.5% fruits	2.6% seeded veg	1.6% game	0.6% other produce	0.6% other meat/ poultry
Listeria m.	50.3% fruits	35.9% dairy	4.9% sprouts	4.8% turkey	3.4% veg row crops	0.4% pork	0.3% chicken	–

Note: Seeded veg includes tomatoes. Veg row crops include leafy vegetables
Source: Data from IFSAC (2017)

undercooking of poultry meat in determining the risk of salmonellosis and campylobacteriosis (Luber 2009). If cross-contamination in the kitchen or storage is a greater source of risk, then a wide variety of foods might be transmitters. Studies have shown that food mingling can occur at multiple locations, on a cutting board, utensil, or plate, by misplacement of a product in the refrigerator or by not separating raw products from other food items when placing them in a grocery cart.

Cross-contamination can also occur in production and transporting of fruits and vegetables. In fields of produce and fruit orchards, pathogen sources include dust from nearby feedlots or from roads with trucks transporting food animals, wild animals contaminating soil, manure applications to fields, workers hygiene habits and access to toilets, and contaminated water (Berry et al. 2015; Jahne et al. 2016; Kumar et al. 2017; Yanamala et al. 2011). One commenter on the Food Safety News website mentioned that supermarkets often rehydrate produce in a shared water tank, the only separation being a different tank for organic produce. This comingling of produce in the water tank could be another source of pathogen cross-contamination.

18.1.4   Pathogen Control Is Not Prohibitively Expensive, and Governments Have the Responsibility of Protecting the Public and of Providing Economic Incentives for Industry to Minimize the Amount of Unsafe Food in the Markets

Taxpayers and consumers expect companies and governments to provide safer food (Chap. 10). For example, a survey of taxpayers indicated that they want a much larger portion of the USDA budget to be allocated to food safety (Chap. 1). Yu and colleagues surveyed consumers asking their willingness to pay for a 50% lower risk of illness and found that consumers were willing to pay an additional $1 per container of bagged salad (Yu et al. 2018). Yet, what consumers say in surveys, particularly in willingness to pay studies conducted in a laboratory environment, versus what they do in purchasing is known to differ, often substantially. Industry costs of additional testing are only 3 cents per bagged salad, according to Earthbound Farms (The Packer 2011). And Costco has required produce suppliers to test for six *E. coli* strains since 2011, to verify that their supplier's food safety programs are working (The Packer 2011).

Taxpayers and consumers expect companies and governments to provide safe food, to some extent, whether the country is Sweden or New Zealand (Chaps. 11 and 12). Support for the Food Safety and Modernization Act (FSMA) by both US consumers and retailers was important in passing the act in 2011 (Chap. 16). Benefit/cost analysis of HACCP and of FSMA showed a net gain to society due to these US regulations (Chaps. 4 and 14). Sweden and New Zealand also showed a net gain to society from their *Salmonella* regulations and *Campylobacter* regulations, respectively (Chaps. 11 and 12).

While foodborne illnesses are very costly, pathogen control is not expensive. Sweden produces *Salmonella*-free broilers at only 1 cent/pound additional retail cost (Chap. 12). Texas American's testing of each batch of hamburgers for *E. coli* O157 cost 1 cent/pound or less and gained them a contract with Jack in the Box, instead of selling in the spot market (Chap. 10). Costs of withdrawing antibiotics from animal feed are not statistically different from zero according to an ERS analysis, assuming improved animal husbandry practices and supplementing feed with probiotics or natural plant ingredients having antimicrobial properties (Chap. 15). Perdue has long provided the Panera Bread restaurant chain with poultry products produced without using antibiotics. More recently Perdue has adopted a policy and a label of "no antibiotics ever" in its name-brand products. In both instances, they found no significant cost impact after adjusting feeding and animal husbandry practices (Chap. 15).

New incentives have been added for companies to improve pathogen control in fruits and vegetables sold in the United States with FSMA (Chaps. 14 and 16), for example, moving the Food and Drug Administration (FDA) from a reactive to proactive approach, shortening the regulatory inspection cycle from 10 to 3 years, granting FDA authority to set standards for imported food, and giving FDA mandatory recall authority. However, works remain to be done: FDA has not been timely with its implementation of FSMA's provisions (public notification of foodborne outbreaks and food recalls is one example); FDA has not set a water standard yet and to date has failed to set an interim rule on raw manure applications to growing produce.

Strict regulations like Sweden have for *Salmonella* in the supply chain could be applied in other countries, and these strict controls could be applied to other major foodborne pathogens as well. The European Union has strict regulation for two serotypes of *Salmonella* in chickens. Major pathogens could be declared adulterants in meat and poultry under USDA authority. Since 1994, declaring two pathogens as adulterants has helped to reduce *E. coli* O157 in ground beef products and *Listeria monocytogenes* in ready-to-eat meat and poultry.

Last, USDA's Office of the Inspector General (OIG) found that USDA makes minimal effort to assure that importing companies are meeting US standards (OIG/USDA 2017). In addition, processed chicken meat is now allowed into the United States without labeling the country of origin, despite known hazards in production in China (Dewey 2017). Inspection of US imports could be strengthened.

18.2   Looking Toward the Future

Several trends in food production could strain the international food safety market. The current trend toward industrialized food production in the world will increase, since the costs of production and distribution are cheaper at larger scale (Chap. 4). The trend toward convenience foods will increase, including highly processed foods and frozen foods. The more highly processed foods become, the more difficulty consumers have in determining the exact ingredients, some of which may be higher risk. Another trend is toward greater consumption of raw fruit and vegetables where pathogens are not killed by heat. However, the increasing availability of technologies to allow more rapid and accurate identification of potential pathogens as well as vulnerability in the supply chain may more than offset these concerns.

The globalization of food companies will continue to increase (Food Engineering 2017). New mergers increasingly occur with beef, pork, and chicken companies all

Table 18.2 Top 10 global meat processing corporations, annual food sales 2017, billion US dollars

Company name and rank	Food sales (billion $)	Country HDQ	Products and comments
1. JBS	38.1	Brazil	Beef slaughter, boxed beef, beef further processing, ground beef, pork slaughter, fresh pork, poultry slaughter, poultry further processing, lamb, prepared foods, case-ready, private labeling, export, natural/organic
2. Tyson Foods	36.9	USA	Beef slaughter, boxed beef, beef further processing, ground beef, pork slaughter, fresh pork, fresh sausage, cured sausage, ham, deli meat, bacon, poultry slaughter, poultry further processing, prepared foods, portion control, case-ready, private labeling, export, natural/organic
3. Cargill	30.0	USA	Revenue sources: 36% North America, 27% Asia Pacific, 13% Latin America, 24% other. Beef slaughter, boxed beef, beef further processing, ground beef, fresh sausage, deli meat, poultry slaughter, poultry further processing, portion control, case-ready, private labeling, export
4. Smithfield Foods	20.3	China	Pork slaughter, fresh pork, ham, deli meat, bacon, prepared foods, case-ready, private labeling, export, natural/organic
5. NH Foods	9.9	Japan	Fresh meats, processed foods, marine products, dairy products, natural flavorings, and health foods 90 facilities in 18 countries
6. Hormel Foods	9.5	USA	Beef further processing, pork slaughter, fresh pork, fresh sausage, cured sausage, ham, deli meat, bacon, poultry slaughter, poultry further processing, prepared foods, portion control, case-ready, private labeling, export, natural/organic
7. Danish Crown	8.9	Denmark	Pork, beef, and processed foods sold in 100 countries; processing plant locations include China
8. BRF Brasil Foods	8.4	Brazil	BRF started in 2009; products reach more than 150 countries
9. OSI	6.1	USA	65 facilities in 17 countries
10. Marfrig Global Foods	4.8	Brazil	Beef, lamb, poultry, processed food 45 facilities in 13 countries; food products sold in 100 countries

Sources: Sharma and Schlesinger (2017), The National Provisioner (2017), and company websites

under one organization in some cases, with production and supply chains located in multiple countries (Table 18.2). Economies of scale in marketing and production are one reason for these mergers. Spreading financial risks over different products in different countries is another reason. The larger the companies become, the more market power they have as well as possible influence over country regulations and policies. There is a parallel development toward more local production and distribution such as farmers' markets, though this does not guarantee safer food.

Long-run climate change will reduce the amount of food grown per acre, on average, and high CO_2 levels will cause certain foods to lose some of their nutritional value (Miller 2017). If climate change melts portions of the Himalayan glaciers, this would influence water availability and grain production in Asia (Chakraborty and Newton 2011). In the United States, water shortages and degradation are becoming more common, due largely to reduced water melt from the glaciers, water use by animal production, fertilizer runoff from grain fields, and run-off from feedlots that contaminates water supplies.

World population pressure often leads to increased contamination of land and water with pathogens and chemicals. In addition, the projected population growth along with growing wealth means that food production must increase by 50% within the next 50 years and simultaneously meet greater demand for high-quality products, including more processed foods (Chakraborty and Newton 2011).

Antibiotic resistant diseases, including those associated with animal production, will increase as more foods enter international commerce and as people increase travel, with the possibility of bringing diseases home from abroad. This could be countered by increasing control over antibiotics in animal feed as has happened in parts of Europe and the US. Controls need to be strengthened in both animal use of antibiotics and human medicine in most countries. Of particular concern are the populous countries, India and China, that are the major manufacturers of the world's antibiotics. All these trends will require continuing advances in technology and strategies to assure that food safety stays at the forefront of attention by all those in the supply chain from producers through to the consumer. Producers, supply chain companies, government regulators, and consumers themselves will all need to be educated, results monitored, and improvements developed to shore up any weaknesses identified in delivering safe, reliable, and high-quality food in the future.

Acknowledgment I greatly appreciate the input from coauthors Walter Armbruster, Diogo M. Souza Monteiro, Arie Havelaar, and Patricia Buck. Any mistakes, of course, are mine.

References

Berry ED, Wells JE, Bono JL, Woodbury BL, Kalchayanand N, Norman KN, Suslow TV, López-Velasco G, Millner PD. Effect of proximity to a cattle feedlot on *Escherichia coli* O157:H7 contamination of leafy greens and evaluation of the potential for airborne transmission. Appl Environ Microbiol. 2015;81:1101–10. https://doi.org/10.1128/AEM.02998-14.

Chakraborty S, Newton AC. Climate change, plant diseases and food security: an overview. Plant Pathol. 2011;60:2–14. https://doi.org/10.1111/j.1365-3059.2010.02411.x.

Dewey C. The dark side of Trump's much-hyped China trade deal: it could literally make you sick, The Washington Post. 2017. https://www.washingtonpost.com.

Food Engineering. 2017 Top 100: The World's Top 100 Food & Beverage Companies. 2017. https://www.foodengineeringmag.com/articles/96916-top-100-the-worlds-top-100-food-beverage-companies

IFSAC (The Interagency Food Safety Analytics Collaboration), Foodborne illness source attribution estimates for 2013 for *Salmonella, Escherichia coli* O157, *Listeria monocytogenes*, and *Campylobacter* using multi-year outbreak surveillance data, United States. 2017. https://www.cdc.gov/foodsafety/pdfs/IFSAC-2013FoodborneillnessSourceEstimates-508.pdf

Jahne MA, Rogers SW, Holsen TM, Grimberg SJ, Ramler IP, Kim S. Bioaerosol deposition to food crops near manure application: quantitative microbial risk assessment. J Environ Qual. 2016;45:666–74. https://doi.org/10.2134/jeq2015.04.0187.

Kumar GD, Williams RC, Qublan HM, Sriranganathan N, Boyer RR, Eifert JD. Airborne soil particulates as vehicles for *Salmonella* contamination of tomatoes. Int J Food Microbiol. 2017;243:90–5. https://doi.org/10.1016/j.ijfoodmicro.2016.12.006.

Luber P. Cross-contamination versus undercooking of poultry meat or eggs – which risks need to be managed first? Int J Food Microbiol. 2009;134(1–2):21–8. https://doi.org/10.1016/j.ijfoodmicro.2009.02.012. Epub 2009 Feb 23

Meat and Poultry Dialogue Group, Recommendations to modernize the meat and poultry oversight system in the United States. 2017. https://docs.merid.org/SITECORE_DOCS/Meat%20and%20Poultry_final.pdf

Miller SG. Climate change is transforming the World's food supply. Live Sci. 2017. https://www.livescience.com/57921-climate-change-is-transforming-global-food-supply.html

Office of Inspector General (OIG)/USDA, Evaluation of Food Safety and Inspection Service's Equivalency Assessments of Exporting Countries, Audit Report 24601-0002-21. 2017. https://www.usda.gov/oig/webdocs/24601-0002-21.pdf

Roberts T. Lack of information is the root of U.S. Foodborne illness risk, Choices, 2nd Quarter 2013;28(2). http://www.choicesmagazine.org/UserFiles/file/cmsarticle_300.pdf

Sharma S, Schlesinger S. The rise of big meat, Institute for Agriculture and Trade Policy, Europe. 2017. https://www.iatp.org/the-rise-of-big-meat

The National Provisioner, The 2017 Top 100 meat & poultry processors. 2017. https://www.provisioneronline.com/2017-top-100-meat-and-poultry-processors

The Packer. Costco mandates produce testing by suppliers. 2011. https://www.thepacker.com

Yanamala S, Miller MF, Loneragan GH, Gragg SE, Brashears MM. Potential for microbial contamination of spinach through feedyard air/dust growing in close proximity to cattle feedyard operations. J Food Safety. 2011;31:525–9. https://doi.org/10.1111/j.1745-4565.2011.00330.x.

Yu H, Neal JA, Sirsat SA. Consumers' food safety risk perceptions and willingness to pay for fresh-cut produce with lower risk of foodborne illness. Food Control. 2018;86:83–9. https://doi.org/10.1016/j.foodcont.2017.11.014.

Glossary of Economic Terms

Adverse Selection Adverse selection is an information problem that exists when a purchaser of product or service cannot observe the quality of the product being purchased. The incentives created by this can lead to a collapse of quality in the market. For example, a principal may hire an agent as a quality assurance specialist, but may not be able to fully observe that person's activities to that end, and lead to the agent's incentive to shirk their duties.

Agent An agent is a person who is employed to do an act on behalf of another called the principal. The principal is generally the purchaser of the goods or services the agent offers.

Asymmetric Information A situation where the seller of a product has more information about the product than the buyer (or vice versa). For example, the seller has more information about the food safety characteristics of the food it sells, and whether it has a HACCP system that includes pathogen testing of its food that will produce safer food. The consumer or downstream processor is unable to know this information, unless there is some form of government rating system that publishes information, such as the restaurant rating system in Chap. 5 or the information published on pathogen levels in food sold by company in Chap. 2.

Benefit/Cost Analysis (BCA) or Cost/Benefit Analysis (CBA) Benefit/cost analysis is a collection of methods and rules for assessing the social costs and benefits of alternative public policies. It promotes efficiency by identifying the set of feasible projects that would yield the largest positive net benefits to society. The willingness of people/society to pay to gain/avoid policy impacts is the guiding principle for measuring benefits. Opportunity cost is the guiding principle for measuring costs.

Cost of Illness (COI) COI method is the most widely used approach among US regulatory economists to estimate the economic burden associated with food-borne disease. COI adds up the expected medical and productivity costs for illnesses and deaths. Often the QALYs or DALYS are monetized and included to account for the human suffering caused by the illnesses/deaths.

Demand Curve The demand curve is a downward-sloping economic graph that shows the relationship between the quantity of product demanded by a market and the price the market is willing to pay. Quantity is always graphed horizontally on the x-axis while price is graphed vertically on the y-axis. The curve demonstrates that as prices for a product decrease, the quantity demanded by consumers increases. In other words, as the product becomes less expensive, more consumers will want or be able to purchase it.

Disability-Adjusted Life Years (DALYs) DALYs measure the health gap from a life lived in perfect health and quantify this health gap as the number of healthy life years lost due to morbidity or mortality. A disease burden of 100 DALYs is a total loss of 100 healthy life years. Diseases, hazards, or risk factors accounting for more DALYs have a higher public health impact.

Economic Incentives An incentive is something that motivates a consumer or company to perform actions. The incentive can be positive or negative. An example of a positive incentive could be increased sales or contracts for a safer food product. Examples of negative incentive could be a new regulation, litigation, or adverse publicity of a foodborne outbreak associated with the company's products. Incentives can also result in undesired outcomes. For example, a large temporary external increase in the price of a commodity would give companies an incentive to deliver more product to market before required testing is completed.

Externalities Externalities are *indirect* effects of consumption or production activity, such as air pollution or foodborne illness. These are essentially costs or benefits of economic activity that are not incorporated into the price of the good or service being sold. For example, an external cost of foodborne illness is the cost to the insurance pool for those who seek insured medical care. An external benefit would be the reduced risk from disease to an unvaccinated person that results from others around them receiving vaccinations.

Free Rider The free rider problem is a situation that can arise when market participants have access to a good or service whether they pay for it or not. Because there is no incentive to contribute, people can "free ride" on the efforts of others. The generation of nonproprietary information is one area where free rider problems are common. This is often used as a reason for public investment in basic research. A public sanitation system is another example.

Information Facts: Perfect information is usually thought of as complete knowledge of facts relevant to an economic transaction. This is not achieved where one or both parties to a transaction do not have all of the facts that would help them make the decision to buy or sell the good or service in question.

Innovation Innovation can be a new idea, device, or method as well as a better application of existing technology to meet new requirements.

Market Power Market power refers to the ability of a firm (or group of firms) to raise and maintain price above the level that would prevail under competition. The exercise of market power leads to reduced output and loss of economic welfare.

Moral Hazard Moral hazard occurs when one shirks contracted levels of effort because those efforts are not directly observable. For example, liability insurance may require producers to engage in or avoid certain practices to reduce food safety risks, but insurers are not able to observe whether those practices are routinely observed. Typically, moral hazard occurs when a negative outcome (e.g., *Salmonella* contamination) is only partially determined by required behaviors. As a result of this uncertainty, the insured is incentivized to exert a level of effort below the contracted level.

Principal Agent Models/Contract Theory A principal agent model is a framework used in the economic theory of contracts to determine how to design a contract between a principal aiming to incentivize an agent to perform a task or produce a product he/she needs or demands. These models generally assume that the principal has incomplete information about the agent ability in and diligence in completing the task; in other words the principal faces adverse selection and moral hazard. The problem of the principal is to define an incentive structure such that (1) only agents with the ability to perform the task accept the contract and (2) the agent exerts maximum effort performing the task.

Public Good The defining characteristics of a public good are that (1) nobody can be excluded from use of the good and (2) consumption of it by one individual does not actually or potentially reduce the amount available to be consumed by another individual. The provision of a lighthouse is a standard example of a public good, since it is difficult to exclude ships from using its services. No one's ship's use detracts from use by other ships. In the absence of government provision, there will be too little production of public goods as a result of the free rider problem.

Quality-Adjusted Life Years (QALYs) QALYs are estimates of health-related quality of life ranging between 0 and 1, where 1 = perfect health and 0 = death. These are often measured using multi-attribute surveys. For example, the commonly used EuroQol EQ-5D evaluates mobility, self-care, usual activities, pain/discomfort, and anxiety/depression.

Supply Curve A supply curve shows the relationship between price and quantity provided by firms. The curve slopes up and to the right, demonstrating how higher prices incentivize greater production.

Trade-Off Curve A trade-off curve is a graph that explains what happens to the performance of something when you change something else. In Chap. 10 a trade-off curve is used to examine interventions for the reduction of pathogen contamination versus the cost of the intervention.

Value of a Statistical Life (VSL) VSL does not place a dollar value on individual lives. Rather, it is a measure based on individuals' willingness to pay for small reduction in mortality risk, extrapolated to reflect a death probability of 1. For example, suppose people in a survey were asked how much they would be willing to pay to reduce their mortality rate by 1 in 100,000 over the next year. Now suppose that the average response to this hypothetical question was $100. The total dollar amount that the group would be willing to pay to save one statistical life in a year would be $VSL \times 1/100,000 = \$100$. Thus $VSL = 100,000 \times \$100 = \$10$ million.

Index

A

Acidified sodium chlorite (ASC), 225
Adulteration, 193
Agency theory
    adverse selection, 34
    businesses and consumers, 31
    competitive markets, 31
    foodborne disease, 30
    food supply chain, 32, 33
    food systems, 32
    government *vs.* industry, 39–40
    Hazard Analysis and Critical Control
        Points, 33
    interdependency, 35
    International Private Standards, 41–46
    legal element, 30
    limitations, 38
    moral hazard, 34
    multiple agents, 40–41
    probability of food safety, 36
    strict liability, 37
    suppliers, 37
    symmetric imperfect information, 32
    types of information, 30, 34
    well-known issues, 35
Agricultural and Resource Management
    Survey (ARMS), 301, 311
Agricultural Research Service, 232
American Meat Institute, 184, 190
American Meat Institute Foundation, 191
Animal and Plant Health Inspection Service
    (APHIS), 316
Animal Drug Availability Act (ADAA), 304
Animal health industry, 295
Animal Health Institute (AHI), 296

Animal Products Act 1999, 214
Antibiotic-resistant (AR), 24, 328
Antibiotic-resistant determinants
    (ARDs), 315
Antibiotic-resistant infections, 300
Antibiotics in animal feed
    ban, growth promotion, 301
    benefit/cost analysis, 301–303
    challenges, 296–298
    consumer demand for labeling, 309–310
    data gaps, 310–312
    FDA, 297, 298
    global responses, livestock
        economic incentives in regulations, 315
        EU ban, 313
        metagenomics study, 314
        microbiologists, 314
        UK Medical Research Council
            (MRC), 313
        UK review, 313
        WHO, 315
    producer education, 316
    transparency, 310
    US animal agriculture, 304
    US policy actions, 305
    US policy response, 304–306
Antimicrobial resistance
    agricultural technology, 295
    animal agriculture, 294
    FDA-approved label, 296
    food-producing animals, 295
    future trade agreements, 298
    global public health, 300
    human infections, 294
    industry stakeholders and responses

© Springer International Publishing AG, part of Springer Nature 2018
T. Roberts (ed.), *Food Safety Economics*, Food Microbiology and Food Safety,
https://doi.org/10.1007/978-3-319-92138-9

Antimicrobial resistance (*cont.*)
    consumer and other interest groups, 309
    feed companies, 308
    integrators and meat processors, 307, 308
    pharmaceutical companies, 307
    restaurant chains, 308
    low-dose antibiotic use in animal feed, 300
    medically important antibiotics, 295
    serious public health problem, 294, 295
    sick animals in agriculture, 295
    side effect, 296
    societal costs, 303–304
    stakeholders, 294
    surveillance and data sharing, 311, 312
    UK Medical Research Council (MRC), 313
    US livestock industry, 311
The Aquatic and Aquaculture Food Safety
    Center (AAFSC), 279
Asian-Pacific Economic Community
    (APEC), 284
Autoimmune disorders, 151

B
Bacillus cereus, 133
Bacterial Pathogen Sampling and Testing
    Program (BPSTP), 180, 196
Base model, 262
Beef carcass steam pasteurization, 200
The Beef Industry Food Safety Council
    (BIFSCo), 9, 191
Beef Products, Inc., 378
Beef steam pasteurization system (BSPS)
    Cargill/Excel's contribution, 189
    cost-plus contract, 184
    efficacy, 187
    Frigoscandia Equipment, 185, 189
    steam temperature, 187
Benefit/cost analysis (BCA), 246–247
    Chinese Chicken Production, 59–64
    framework, 6
    HACCP program, 54–59
    HACCP systems, 52
    meat and poultry industry, 50–51
    private investment, 59–64
    public decision-making, 52–54
Benefit-cost ratio (BCR), 228
Better processing controls (BPCS), 284
Bottom-up approach, 88
Brucella spp., 161
BSPS-Static Chamber unit (BSPS-SC), 188
Burden of disease, 148
Burden of illness
    food attribution data, 258

    food safety efforts, 258
    NNDSS, 252
    pathogen-specific surveillance systems, 253
    surveillance systems, 252, 253, 256

C
Campylobacter Risk Management Strategy
    (*Campylobacter* Strategy), 210
Campylobacter spp., 9, 10, 21, 22, 24, 74, 125,
    133, 148, 150, 157, 210–211, 394
Campylobacteriosis
    ante- and postmortem examination,
    214–215
    consumption of chicken meat, 212–213
    New Zealand regulatory framework,
    213–214
    risk of foodborne, 216
Capacity-building efforts, 283
Categorical attribution, 155
Celiac disease (CeD), 150
Center for Environmental Health, 343
Center for Food Integrity, 327
Center for Food Safety, 343
Center for Science in the Public Interest
    (CSPI), 246
Centers for Disease Control and Prevention
    (CDC), 8, 53, 90, 178, 198, 252,
    268, 324
Chicken meat
    Campylobacter, 216
    on-farm measures, 217
    VLT, 217
Clostridium botulinum, 15
Clostridium perfringens, 126
Code of Practice (COP), 216
Codex Alimentarius, 283
Complaint surveillance systems, 252
Confederation of India Industry Food and
    Agriculture Center of Excellence
    (CII-FACE), 280
Conference for Food Protection, 345
Congressional Budget Office (COB), 337
Consumer advocacy
    antibiotic-resistant strain, 326
    challenges, strategies and tools, 328–330
    federal agencies, 324, 326
    food and water watch, 348
    food associations, 348
    foodborne disease, 325
    foodborne pathogens, 324
    food industry, 348
    food safety reform, 327, 329, 339
    health-oriented associations, 348

Pew Charitable Trusts, 349
STOP Foodborne Illness, 349
US food safety system, 324
Consumer Goods Forum, 283
Cooke's Classical Model, 161
Cost-benefit analysis
    administration and enforcement costs, 226
    administrative burdens, 223
    campylobacteriosis, 210
    compliance costs, 222–226
    cost framework, 219–221
    cost of illness, 221–222
    direct labour costs, 223–224
    equipment costs, 224
    external services costs, 225
    implementation costs, 223
    limits of detection (LOD), 229
    materials costs, 224, 225
    microbiological monitoring costs, 225
    NMD data and company performance,
        218–219
    NMD programme, 228
    overhead costs, 224
    regulatory interventions, 218
    substantive compliance costs, 223
    very low throughput (VLT), 228
Cost of illness (COI) method, 8, 97, 263
Costs to industry, 101
Council to Improve Foodborne Outbreak
        Response (CIFOR), 252
Critical control point (CCP), 190
Crohn's disease (CD), 91, 150
Cross-sectional surveys, 147, 148
Cryptosporidium, 157
Culture-independent diagnostic testing
        (CIDT), 19, 146

D
Delta Professional Consultancy, 280
Disability-adjusted life year (DALY), 8, 92,
        109, 116, 221
Disease burden of chronic sequelae
    autoimmune disorders, 151
    functional gastrointestinal disorders and
        inflammatory bowel disease, 149–151
    HUS, 151, 152
    malnutrition and growth impairment, 154–155
    neurological dysfunction, 152–153
    psychological disorders, 153
    UTIs, 153
DNA fingerprinting, 16
DNA profiling systems, 254–255
Domestic importers, 271

E
E. coli O157:H7
    contamination of beef, 192
    enforcement of regulatory programs, 185
    farm contamination, 191
    federal and state actions, incentives, 192–194
    Texas American's BPSTP sampling and
        testing protocol, 182
Economic impact
    CDC illness model, 128–130
    cost of foodborne Illness, U.S., 130–140
    cost-of-illness estimates, 140–141
    cost-of-illness modeling, 123–125
    economic model, 124
    methods for estimating costs, 126–128
    Scharff approach, 123
    sensitivity analysis, 139
    state-level cost, 138–139
Economic incentives, 237, 240, 247
    beef industry reaction, 190–192
    company response, outbreak risk, 197–198
    consumer, 14
    downstream retailers, 14
    E. coli O157:H7, 192–194
    Federal and State Actions, 192–194
    Jack in the Box outbreak, 178–179
    link illnesses, 15–18
    market outcomes, 14
    modeling economic costs vs. risk
        reduction, 194–197
    moral hazard, 14
    optimal deterrence, 25–26
    pathogen performance standards, 20
    pathogen testing, 18–20, 180
    policy implications, 24–25
    process verification, 194
    public pathogen information, 23–24
    Texas American Foodservice, 180–184
    third-party audits, 14
    traceability system, 14
Economic Research Service (ERS), 311
Economics of information, 14
Environmental enteric dysfunction (EED), 154
Environmental Protection Agency (EPA), 247
Environmental Services and Research Ltd.
        (ESR), 211
Escherichia coli, 9
European Centre for Disease Prevention and
        Control (ECDC), 163
European Food Safety Authority (EFSA),
        85, 163
Evidence-based decision-making process, 281
Exposure assessment focuses, 87
Extraintestinal pathogenic E. coli (ExPEC), 153

F
Farm-to-Consumer Legal Defense Fund, 339
FDA Global Engagement report, 274
FDA Guidance 209, 305
FDA Model Food Code, 364
FDA's International Food Safety Capacity-
    Building Plan, 274
FDA's International Program, 285
Federal Freedom of Information Act
    (FOIA), 377
Federal Meat Inspection Act, 331
Federal Register, 332
Firm-level traceability, 272
Food and Drug Administration (FDA), 85,
    127, 268, 296, 324, 334–344, 394
Food and Health Survey, 327
Food and Water Watch (FWW), 21
Food Animal Concerns Trust (FACT), 309
Foodborne disease
    burden assessment, 89–91
    burden of disease, 85
    chronic sequelae (*see* Disease burden of
        chronic sequelae)
    cost estimation
        cost-of-illness method, 97–99
        costs to industry, 101
        WTP, 100
    costs associated, 94–96
    evidence-informed policy, 144
    food safety, 84
    health impacts, 84, 91–94
    health outcomes, 150, 155–158
    public health and regulatory agencies, 84
    public health surveillance systems,
        145–149
    risk-based food safety, 86–89
    risk-informed framework, 85
    source attribution methods, 158–161
    strengths and limitations, source attribution
        methods, 159–160
    types of transmission, 161–162
Foodborne Disease Active Surveillance
    Network (FoodNet), 254
Foodborne Disease Burden Epidemiology
    Reference Group (FERG), 8, 109
Foodborne Disease Outbreak Surveillance
    System (FDOSS), 254
Foodborne illness, 272, 273, 362
Foodborne illness surveillance systems
    burden of illness, 252
    CDC, 252
    CIFOR, 252
    complaint surveillance systems, 252
    effects of, 252

Food and Drug Administration, 252
    pathogen-specific systems, 252
    public health efforts, 251
    syndromic surveillance, 253
Foodborne Outbreak Online Database
    (FOOD), 326
Foodborne pathogens
    company strategies, 179
Foodborne UTIs (FUTIs), 154
Food Hygiene Information Scheme (FHIS), 69
Food Hygiene Rating Scheme (FHRS), 69
Food labeling, 310
Food market, 272
Food policies, 328
Food Safety Analytics Collaboration
    (IFSAC), 326
Food Safety Cooperation Forum, 284
Food safety economics
    agency/contract theory, 6
    consumers, 4
    economic and noneconomic factors, 5
    economic concepts and frameworks, 5
    evaluation, health costs, 5
    foodborne disease, 6, 7
    foodborne illness, 7–10
    food policy, 4
    political economy, 10
    risk management, 10
    Salmonella-free labels, 4
Food Safety Inspection Service (FSIS), 191,
    278, 297
Food Safety Modernization Act (FSMA), 10,
    85, 269, 274, 281, 334–344, 393
Food Safety Preventive Controls Alliance
    (FSPCA), 282
Food Safety Scotland (FSS), 69
Food Safety System Certification (FSSC), 43
Food supply chain, 273
FSIS Office of Public Health Science
    Administrator, 332
FSMA Surveillance Working Group, 343
Functional dyspepsia (FD), 150
Functional gastrointestinal disorders
    (FGDs), 149

G
GFSI's Global Markets Program, 283
Global burden, FBD
    chemical hazards, 119
    complexity of transmission pathways, 108
    FERG methodological framework, 119
    global estimates and regional comparisons,
        114–117

methodological framework, 109–114
microbiological hazards, 118
public health/economic impact, 108
Global Burden of Disease 2010 study, 113
Global Food Safety Initiative (GFSI), 42, 283
Global Food Safety Partnership (GFSP), 284
Good agricultural practices (GAP), 269, 278
Good aquacultural practices (GAqP), 269, 279
Good hygienic practice (GHP), 216, 218
Good manufacturing practices (GMP), 278
Government Accountability Office, 324
Grocery Manufacturers Association (GMA), 284
Guillain-Barré syndrome (GBS), 91, 125,
        151, 222

H

Hazard Analysis and Critical Control Points
        (HACCP) system, 7, 11, 30, 179,
        184, 187, 190, 193, 215, 277, 391
Hazard characterization focuses, 87
Hazard identification focuses, 87
Hemolytic uremic syndrome (HUS), 57, 91,
        125, 151, 152, 370
Hierarchical Model of Training Outcomes, 285
High-value agricultural (HVA) products, 268

I

Illinois Institute of Technology's Institute for
        Food Safety and Health
        (IIT IFSH), 282
Indemnity agreements, 379–381
Infectious Diseases Society of America, 348
Inflammatory bowel disease, 91, 149–151
Inter-American Institute for Cooperation on
        Agriculture (IICA), 284
Internal Rate of Return (IRR), 228
International Features Standard (IFS), 43
2016 International Food Information Council
        Foundation (IFICF), 327
International food safety
    antibiotic resistance, 390
    economic costs, 390
    economic frameworks and tools, 389
    economic incentives, 389
    foodborne diseases, 389
    food production, 394
    globalization of food companies, 395
    lack of information, free marketplace,
        390–391
    long-run climate change, 396
    pathogen control, 393–394
    private and public sectors, 391–393
    supply chain and economic issues, 389
    supply chain companies, 396
    US foodborne source attribution
        estimation, 392
International food safety capacity building
    aid-driven agencies, 270
    cost-benefit analyses, 269
    FDA-funded international training and
        food import values, 279
    foodborne illness, 268
    food safety interest, 271–272
    FSMA, 281–283
    global agricultural trade, 268
    HVA, 268
    JIFSAN's metrics approach, 286
    JIFSAN's programs, 279
    monitoring and evaluation efforts, 285–289
    PPP, 274–277
    private sector, 269
    private sector involvement, 283–285
    public and private sectors, 269, 270
    public intervention, 272–274
    public-private partnership, 270
    public sector intervention, 271–272
    regulatory tools, 269
    secondary datasets, 287–288
    supply chain management, 275–276
    US land-grant system, 269
International Life Sciences Institute (ILSI),
        280, 289
Irritable bowel syndrome (IBS), 91, 149, 246

J

JIFSAN's Global Collaborative Training
        Center Initiative, 279
Joint Institute for Food Safety and Applied
        Nutrition (JIFSAN), 269, 278, 280

K

Kansas State University (KSU), 185, 186, 189

L

Lean, finely textured beef (LFTB), 378
Legal incentives for food safety
    blunted impact, 372–374
    civil litigation and food safety, 383–384
    cleanup crew, 375
    compensatory, 363
    confidentiality agreements and
        suppression, 377–378
    contaminated ingredients, 371–372

Legal incentives for food safety (*cont.*)
　correct measure of damages, 375–377
　criminal prosecution, 381–383
　E. coli O157:H7 infections, 360
　economic damages, 363
　foodborne illness, 374–375
　food-related legislation, 359
　food safety, 377–378
　form of safety regulation, 361–362
　indemnity agreements, 379–381
　liability risks, 363
　negligence and requirement, 363–366
　noneconomic damages, 363
　quality and safety, 360
　reasonable care, 363–366
　strategic litigation, public participation,
　　378–379
Listeria monocytogenes, 57, 117, 186, 394
Livestock-associated Methicillin-resistant
　　Staphylococcus aureus
　　(LA-MRSA), 23
Long-term health outcomes (LTHOs), 17, 53,
　　57, 149

M
MacDougall Risk Management, 190
Make Our Food Safe (MOFS), 337
Maryland Extension programs, 286
Meat and Poultry Dialogue Group
　　(MPDG), 392
Mechanically separated chicken (MSC), 22
Mechanically tenderized (MT), 344–347
Medically important antibiotics
　antimicrobial resistance, 306
　chicken production, 308
　description, 295
　domestic sales, 295, 311
　evidence-based recommendations, 315
　FDA-approved labels, 311
　FDA ban, 295
　FDA-specified, 301
　food animal production use, 311
　food-producing animals, 315
　future trade agreements, 298, 318
　growth promotion, 306
　illegal use, 311
　judicious use, 306
　OTC, 311
　policy framework, 305
　sub-therapeutic uses, 297
Memorandum of understanding (MOU), 200
Miller Fisher syndrome, 91
Ministry for Primary Industries (MPI), 211

Ministry of Agriculture and Forestry
　　(MAF), 213
Ministry of Health (MoH), 211
Multilocus sequence typing (MLST), 213

N
The National Academy of Science's
　　Institute of Medicine
　　(IOM), 23
National Aeronautics and Space
　　Administration (NASA), 277
National Agricultural Statistics Service
　　(NASS), 311
National Animal Health Monitoring System
　　(NAHMS), 301, 312
National Antimicrobial Resistance Monitoring
　　System—Enteric Bacteria
　　(NARMS), 312
National Cattlemen's Beef Association
　　(NCBA), 190
National Chicken Council (NCC), 241
National Disease Surveillance system
　　(NNDSS), 127
National Electronic Disease Surveillance
　　System (NEDSS), 254
National Inpatient Sample (NIS), 126
National Microbiological Database
　　(NMD), 215
National Notifiable Diseases
　　Surveillance System (NNDSS),
　　252, 253
National Research Council (NRC), 23
National School Lunch Program (NSLP),
　　198, 200
National Sustainable Agriculture
　　Coalition, 339
Net Present Value (NPV), 228
Neurological dysfunction, 152–153
New Zealand Food Safety Authority
　　(NZFSA), 210
New Zealand Treasury, 228
Nongovernmental organizations
　　(NGOs), 347

O
Office of Information and Regulatory Affairs
　　(OIRA), 51
Office of Management and Budget (OMB),
　　51, 343
Omnibus Spending Bill, 341
Outbreak surveillance systems, 254
Over-the-counter (OTC), 311

P

Partnership Training Institute Network (PTIN), 284
Pathogen Reduction (PR), 54
Pathogen Reduction/Hazard Analysis and Critical Control Points Final Rule (PR/HACCP), 330–334
Pathogen-specific surveillance systems, 252, 253
Pathogen testing, 180, 184, 186
Peanut Corporation of America (PCA), 336, 365
Peroxyacetic acid (POAA), 225
Pew Charitable Trusts, 337, 348
PFGE testing, 383
Poisson model, 262
Polymerase chain reaction (PCR), 181
Post-infectious irritable bowel syndrome (PI-IBS), 150
Poultry-caused salmonellosis, 247
Poultry Industry Association New Zealand (PIANZ), 215
Poultry Products Inspection Act, 331
Preventive Controls Rule, 281
Private insurance plan, 233
Private sector involvement, 283–285
Process change model, 259, 260, 263
Produce International Partnership (PIP) training program, 281
Produce Safety Alliance, 282
Produce Safety Rule, 281
Psychological disorders, 153
Public health sector, 96
Public Health Service Act of 1944, 277
Public-private cooperative programs, 277
Public-private partnerships (PPPs), 270, 274–277
Public resources, 273
Public sector involvement, 276
Pulsed-field gel electrophoresis (PFGE), 255, 258, 383
PulseNet
    base model, 262
    economic benefits and costs, 263–264
    process change model, 259, 260, 263
    recall model, 259
    spillover effects model, 262, 263

Q

Quality-adjusted life years (QALYs), 8, 92, 94, 127

R

Raised without antibiotics (RWA), 307
Rapid Alert System for Food and Feed (RASFF), 85
Reactive arthritis (ReA), 149, 151
Recall model, 259
Restaurant hygiene rating
    economic theory, 68–69
    foodborne illness, 73
    market failure, 68, 78
    outline, 68
    pathogens, 67
    quality uncertainty, 68, 78
    UK Evaluation Evidence, 70–74
    UK Experience, 69–70
    US Experience, 74–77
Risk Management Programme (RMP), 214, 218
Rural Development Initiative partnership, 285

S

Safe Food Coalition (SFC), 344, 345
Safe Quality Food (SQF) Code, 43
Safe Tables Our Priority (STOP), 193
Salmonella, 9, 18, 22, 24, 32, 58, 74, 128, 133, 150, 157, 326, 331, 371, 372, 394
Salmonella control
    BCA, 246
    broilers, 232–234
    Denmark, 237
    economic costs, 236–237
    Finland, 237–238
    Norway, 237
    political economy, 245–246
    principles, 234–236
    private insurance, 236
    Sweden's *Salmonella*-Free Production *vs.* US Broiler Production, 242–245
    US Broiler Industry, 238–241
Salmonella enterica, 115
Salmonella typhi, 116
Salmonella typhimurium, 186
Salmonellosis, 232, 244, 246
Secretary of Agriculture Tom Vilsack, 344
Shiga toxin-producing *E. coli* (STEC), 150, 151, 194
Shigella, 9, 133, 150
Source attribution methods
    challenges and future directions, 163–164
    epidemiological approaches, 160, 161
    expert elicitations, 161
    microbiological approaches, 160
    microbiological hazards, 158

Spillover effects model, 263
Sprout Safety Alliance (SSA), 282
Strict liability claims
    chain-of-distribution liability, 367
    defenses to liability, 370
    food product, 366
    malfunction doctrine, 369
    manufacturers *vs.* retailers, 367–369
    mass-manufactured foods, 366
    non-outbreak cases, 369
    proving causation-in-fact, 370
    types of defects, 366
    unsafe food, 369
Summary measures of population health
    (SMPHs), 92
Surveillance systems
    economics of information problems
        burden of foodborne illness, 258
        burden of illness, 256, 258
        consumers' inability, 255
        foodborne illness surveillance
            efforts, 256
        food safety deficiencies, 255
        government efforts, 258
        identify illness sources, 256
        industry incentives, control foodborne
            illness, 256
        industry trade groups, 256
        market and public health agencies, 255
        outbreak data and illnesses, 258
        traceback investigation, 257
    Federal government, US
        DNA profiling systems, 254–255
        illness surveillance systems, 253–254
        outbreak surveillance systems, 254
    Foodborne illness (*see* Foodborne illness
        surveillance systems)
    PulseNet (*see* PulseNet)
Swedish Board of Agriculture (SBA), 232
Syndromic surveillance, 253

T
Tester-Hagan Amendment, 339
Toxoplasma gondii, 117, 119, 130
Trichinella spp., 150, 157
Tyson Foods, Inc., 190

U
Ulcerative colitis (UC), 91, 150
United Nations Industrial Development
    Organization (UNIDO), 284

United Nations Millennium Development
    Goals, 87
United States Department of Agriculture
    (USDA), 4, 232, 330–334
Urinary tract infections (UTIs), 23, 153
US Agency for International Development
    (USAID), 278
US Centers for Disease Control and
    Prevention (CDC), 57, 391
USDA Branded Food Products Database, 289
USDA Cooperative State Research,
    Education, and Extension Service
    (CSREES), 278
USDA-Foreign Agricultural Service (USDA/
    FAS), 278
USDA National Nutrient Database for
    Standard Reference, 289
USDA's Agricultural Research Service (ARS),
    192, 345
USDA's Economic Research Service (ERS),
    20, 54, 191, 301, 374
USDA's Office of Inspector General (OIG), 197
US Department of Agriculture's Food Safety
    Inspection Service (USDA/FSIS),
    23, 182, 270, 310, 324
US Department of Justice (DOJ), 193
US Food and Drug Administration (FDA), 10, 16
US government agencies, 274
US Interagency Food Safety Analytics
    Collaboration (IFSAC), 392
US National School Lunch Program
    (NSLP), 20
US Public Interest Research Group
    (US PIRG), 349

V
Value for a statistical life year (VSLY), 98
Value of a statistical life (VSL), 98, 127, 134
Verification Services (VS), 226
Veterinary feed directive (VFD), 305, 316
Vibrio vulnificus, 133

W
Waters Corporation, 280
Whole genome sequencing (WGS), 19, 163,
    164, 255, 391
Willingness to pay (WTP), 99–100
Willing to accept (WTA), 99
Working Group on Antibiotic Resistance, 348
World Animal Health Information System
    (WAHIS), 301

World Health Organization (WHO), 8, 109, 268
World Organization for Animal Health
    (OIE), 312
WTO Sanitary and Phytosanitary Agreement, 273

Y
Years lived with disability (YLD), 92, 113
Years of life lost (YLL), 92, 113
Yersinia, 9, 133, 150